D1259951

About Island Press

Island Press is the only nonprofit organization in the United States whose principal purpose is the publication of books on environmental issues and natural resource management. We provide solutions-oriented information to professionals, public officials, business and community leaders, and concerned citizens who are shaping responses to environmental problems.

In 2000, Island Press celebrates its sixteenth anniversary as the leading provider of timely and practical books that take a multidisciplinary approach to critical environmental concerns. Our growing list of titles reflects our commitment to bringing the best of an expanding body of literature to the environmental community throughout North America and the world.

Support for Island Press is provided by The Jenifer Altman Foundation, The Bullitt Foundation, The Mary Flagler Cary Charitable Trust, The Nathan Cummings Foundation, The Geraldine R. Dodge Foundation, The Charles Engelhard Foundation, The Ford Foundation, The Vira I. Heinz Endowment, The William and Flora Hewlett Foundation, The W. Alton Jones Foundation, The John D. and Catherine T. MacArthur Foundation, The Andrew W. Mellon Foundation, The Charles Stewart Mott Foundation, The Curtis and Edith Munson Foundation, The National Fish and Wildlife Foundation, The National Science Foundation, The New-Land Foundation, The David and Lucile Packard Foundation, The Pew Charitable Trusts, The Rockefeller Brothers Fund, Rockefeller Financial Services, The Surdna Foundation, The Winslow Foundation, and individual donors.

About the Center for Science, Policy, and Outcomes

The Center for Science, Policy, and Outcomes is a source of interdisciplinary knowledge and ideas aimed at: assessing the nation's research enterprise in the context of social, economic, environmental, and other outcomes; contributing to the design of research programs that meet the nation's most important needs while preserving scientific autonomy and excellence; ensuring that scientific and technological progress yields its benefits equitably; and enhancing the relationship between scientific knowledge and political decision making. The Center is a project of Columbia University, in collaboration with the Georgia Institute of Technology School of Public Policy. For more information, visit the Center's web site at www.cspo.org, or contact us at cspo@cspo.org.

PREDICTION

PREDICTION

Science, Decision Making,

and the Future of Nature

Edited by
Daniel Sarewitz, Roger A. Pielke, Jr., and Radford Byerly, Jr.

ISLAND PRESS

Washington, D.C. • Covelo, California

Library of Congress Cataloging-in-Publication Data

Prediction : science, decision making, and the future of nature / Daniel Sarewitz, Roger A. Pielke, Jr., Radford Byerly, Jr., editors.
        p.   cm.
Includes bibliographical references and index.
ISBN 1-55963-775-7 (cloth : alk. paper) — ISBN 1-55963-776-5 (pbk. : alk. paper)
    Science and state.   2. Decision making.   3. Forecasting.
    I. Sarewitz, Daniel R.   II. Pielke, Roger A., 1968–   III. Byerly, Radford.
        Q125.P928 2000
        363.1′07—dc21   00-008179

Printed on recycled, acid-free paper ∞ ✿

Manufactured in the United States of America
10 9 8 7 6 5 4 3 2 1

This book is dedicated to the memory of George E. Brown, Jr., who devoted his life to the proposition that science should be servant to peace, justice, and human dignity

Contents

List of Figures

List of Tables and Boxes

Acknowledgments

First and above all, we thank D. Jan Stewart, who has held this project together and kept it moving forward since its inception.

We are very grateful to the participants at two workshops who helped us refine the case histories and view the "prediction problem" in a more coherent way: Thomas Anderson, William Ascher, Chris Bernabo, Richard Bernknopf, Donald Bryan, Jack Fellows, John Firor, Hugh Lane, Lee Larson, Karen Litfin, Jerry Mahlman, Shirley Mattingly, Priscilla Nelson, Fielding Norton, Robert Ravenscroft, James Smithson, Thomas Tobin, Dennis Walaker, Robert Wesson, and Lee Wilkins.

The case histories were greatly improved by the critiques of more than thirty anonymous reviewers, to whom we offer our sincere thanks. We are also grateful to Todd Baldwin, at Island Press, for his enthusiasm, responsiveness, and editorial wisdom.

Finally, we thank the following, whose help on a variety of fronts made this book possible: Tanya Beck, Michael Crow, Donald Davidson, Baat Enosh, Rachelle Hollander, Vicki Holzhauer, Jan Hopper, Justin Kitsutaka, Bobbie Klein, Julie Lundquist, Miles Mercer, Jennifer Oxelson, Tiffany Snook, and Ben Taylor.

Major funding for this project was provided by The National Science Foundation's program in Societal Dimensions of Engineering, Science, and Technology. Additional support was provided by: the Geological Society of America, the National Center for Atmospheric Research, the Center for Science, Policy and Outcomes, and BHP Minerals.

Introduction:
Death, Taxes, and
Environmental Policy

Daniel Sarewitz, Roger A. Pielke, Jr., and Radford Byerly, Jr.

"In this world," said Ben Franklin, "nothing is certain but death and taxes." More than two centuries later, his prediction has lost none of its robustness, and we take it as our unlikely starting place. Looming like a dark cloud on the American political horizon is the nation's Social Security program. The problem, in brief, is death and taxes—a predicted deficiency of both. The dynamics are obvious: As people live longer, they spend more time in retirement; as birth rates go down, the number of retirees increases faster than the number of workers. Ultimately, the government's financial obligations to an ever expanding population of retired people outstrip the ability of the working population to finance those obligations. Either taxes must be raised, or retirement benefits must be curtailed. As a political matter, what dilemma could be more horrific?

Predictions allow us to come to grips with the problem. For a single individual, death and taxes are utterly certain, but the date of the former and the magnitude of the latter are considerably less so. At the level of a population, however, regularity and predictability begin to emerge. If the nation is to manage the financial future of Social Security, policy makers must know something about future revenues, life expectancy, fertility rates, unemployment, and immigration. Historical trends provide helpful guidance in anticipating how these factors will behave in coming years. And demographic reality determines an inescapable future: Fertility rates, the age distribution of the population, and current life expectancies dictate that middle-aged baby boomers will before long fuel unprecedented growth of the retired population. Only catastrophe can alter this fate: pandemic, global war, environmental collapse, asteroid impact—in which case, the solvency of Social Security will be the least of our problems.

Efforts to craft a nonpartisan solution to the Social Security dilemma are invariably torn asunder by a political process that often penalizes those who make hard decisions. The vitriolic partisanship that characterizes American politics at the turn of the millennium has rendered neither political party willing to take the lead on a proposal that could make it vulnerable to attack. At the same time, politicians of opposing stripe latch on to the uncertainties in the economic and demographic predictions—perhaps the economy will grow faster, or people will work until they are older—to argue that the problem may not be as serious as it seems. But when it comes to Social Security, our picture of the future is fairly sharp, and all reasonable projections of future demographic and economic trends converge on a similar conclusion: insolvency thirty years or so into the new century. Such insolvency would create an economic and political crisis of enormous proportions. Yet politicians are thus far entirely unable or unwilling to come to grips with this reality and craft a workable solution.

The difficulty of turning reasonably reliable knowledge about the future into positive action on Social Security highlights the central theme of this book: The value of predictions in public policy is not simply a technical question—it is much more than a problem of reducing uncertainties, of getting the numbers "right." Rather, it is a complex mixture of interdependent scientific, political, and social factors. Technically reliable predictions in and of themselves do not translate into successful decisions. That being said, we must further muddy the waters by observing that even the very idea of technical "reliability" can be damnably difficult to evaluate at the time that a decision must be made. To illustrate this point, we offer a parting shot about Social Security before abandoning the issue for good. Once upon a time, Congress actually did manage to grapple successfully with an impending insolvency crisis. It sought to guarantee actuarial solvency well into the future by raising taxes and making other structural changes to the program. Nonpartisan technical analysis, based on solid demographic and economic data, predicted that such changes would ensure solvency for seventy-five years. Secure in the reliability of that prediction, politicians took action and passed the 1983 Social Security amendments. But by 1998, nonpartisan technical analysis showed that the correct number would have been closer to fifty years. The original prediction was off by 33 percent. This enormous error was revealed only by the passage of time. Even predictions made with great confidence and understanding can be wrong.

With these lessons in mind, we now turn to the issue at hand: the application of scientific predictions to problems of the environment. Fundamental to our concerns is the observation that *decision making is a forward-looking process.* From an individual choosing whether to carry an umbrella, to a large international body debating a global environmental treaty, the basic idea is to achieve a goal in the future—keeping one's hair dry or protecting the planet from global warming. And if decision making is the attempt to achieve a desired future, then any such attempt must include, implicitly or explicitly, a vision of what that future will look like.

Dewey (1991) captured the issue well: "The very essence of civilized culture is that we . . . deliberately institute, in advance of the happening of various contingencies and emergencies of life, devices for detecting their approach and registering their nature, for warding off what is unfavorable, or at least for protecting ourselves from its full impact, and for making more secure and extensive what is favorable."

Two general strategies are employed in this effort: adaptation and prevention. If an event seems inevitable and unavoidable, we often have no choice but to adapt. Thus, we carry an umbrella if there is a threat of rain, or we provide federal insurance for houses on floodplains (or perhaps we prohibit construction in such locations), or we enforce seismic building codes in California, where earthquakes are common. The rain, the flood, the earthquake still come, and we are prepared. In other cases we may actually try to change an anticipated future through conscious action, and thus prevent undesirable impacts. Flood-control projects, for example, are aimed at actually preventing dangerous floods, through hydraulic engineering. Similarly, we may try to use technologies and regulations to reduce carbon dioxide emissions so that we can forestall—rather than adapt to—global warming.

In reality, of course, adaptation and prevention are not mutually exclusive. Earthquakes can be neither prevented nor predicted, but clever engineering can ensure that buildings withstand severe ground motions. This is an adaptation to earthquakes that also prevents an earthquake hazard—that is, the propensity of buildings to fall down.

But a fundamental question remains: If decision making is a forward-looking process, what enables us to look forward? What allows us to anticipate the "contingencies and emergencies of life" and make successful decisions to prepare for or forestall them? Experience and judgment, of course. We know from experience that floods occur on floodplains, and we can exercise our judgment in determining how to reduce flood losses through zoning, construction practices, or hydraulic engineering. But experience and judgment can now be augmented by explicitly predictive scientific information. The path and behavior of a major storm are predicted by weather forecasters; the height of the flood crest and its rate of movement downstream are predicted by hydrologists. Based on such scientific information, we can decide where we need to pile sandbags, how high the pile will need to be, and whether evacuation will be necessary—at least in principle.

Science increasingly predicts occurrences that cannot easily be inferred from experience and judgment alone. An early example was Thomas Malthus's predictions that exponential population growth would outstrip agricultural capacity and lead to famine. Thankfully, the prediction has thus far been wrong; indeed, it has helped to stimulate advances in agriculture that ensure its continued erroneousness. More recently, scientific documentation of acidified lakes in the northeastern United States helped sound the alarm over the prospective environmental dangers of acid rain, and the measurement of rising atmospheric carbon dioxide levels fueled predictions of global warming. Meanwhile, telescopes sweep the sky in search of asteroids that could collide, catastrophically, with the earth. The future of nature has become the business of scientists.

Science, with its promise of prediction, seems to be a perfect mate for decision making, with its forward-looking essence. Environmental policy making tests this new marriage. If the marriage were obviously and inevitably a harmonious one, our book would not be necessary. In fact, the relationship between predictive science and environmental decision making is rocky. To begin with, each activity is complex and difficult in its own right. The theoretical and technical difficulties of predicting complex natural systems are immense, and the magnitudes of the uncertainties associated with such predictions may be not only large, but also themselves highly uncertain. At the same time, the process of making environmental decisions, which often brings together a mix of violently conflicting interests and values, has given rise to some of the most intractable political disputes of the last half century. The idea that predictive science can simplify the decision-making process by creating a clearer picture of the future is deeply appealing in principle, but deeply problematic in practice.

Prediction: Science, Decision Making, and the Future of Nature attempts to paint a comprehensive portrait of the troubled relationship between predictive science and environmental decision making. The goals of the book are to provide insights into the promise and limitations of prediction as a tool for decision makers, to explore alternatives to prediction, to present fresh perspectives about the interface between science and environmental decision making, to develop a usable analytical framework—including specific principles—that govern that interface, and to make concrete recommendations that can increase the likelihood of effective environmental decisions.

At the outset, let us clarify a few basic terms. The phrase *decision making* in our title encompasses the broadest scale of action, from individual responses to a weather forecast, to multinational bodies concerned with global environmental threats. However, we also frequently use the term *policy making*, by which we mean organized decision making that is often codified in regulations, laws, formal recommendations, and treaties and other types of agreements. The phrase *the future of nature* captures a diverse set of interactions between society and the environment, including phenomena that are largely unmediated by human intervention (such as asteroid impacts and earthquakes) and those that bear a significant anthropogenic signature (such as acid rain and the behavior of waste from hard rock mines and nuclear reactors). We have limited our concerns to policy problems involving physical and chemical earth-system processes and resources, and excluded biological ones, in part to make our job more manageable, but also because some progress has already been made in understanding the difficulties of using predictive information to manage biological systems (for example, see Lee 1993, and Gunderson, Holling, and Light 1995). We have also, as a matter of practicality, focused our discussion on prediction activities in the United States.

The following three chapters investigate the marriage of prediction and decision making from political, historical, and behavioral perspectives. These chapters first establish that science and prediction are not the same thing; they then go on to explore the attributes of nature, on the one hand, and *human nature*, on the other, that define the promise and the limits of prediction in environmental policy. These foundation-setting discussions are followed by ten narrative case studies that constitute the heart of the book. The cases fall rather naturally into three separate but related groups:

1. "natural" hazards that are perceived by decision makers as largely unavoidable: short-term weather, floods, asteroids, and earthquakes;

2. environmental problems for which predictions are generated to support a course of action that already has strong political momentum: beach erosion, mining impacts, and nuclear waste disposal;

3. multifaceted environmental issues that respond to—and raise—complex, unresolved policy dilemmas: oil and gas reserves, acid rain, and global climate change.

Central to this disparate and almost biblical-sounding assortment of plagues and problems is society's effort to generate reliable scientific predictions that can be used as a basis for making decisions about humanity's relationship with complex natural systems. The ten cases provide the empirical base for the final chapters of the book, which focus on policy implications, analytical frameworks and principles, and recommendations.

We have worked hard to eliminate jargon and superfluous technical information from the book. The reader does not have to be a hydrologist to understand the chapters on floods and nuclear waste disposal, or a climatologist to understand the chapters on weather and climate change. Basic literacy and interest in science, the environment, and public policy are the only prerequisites for full comprehension and, we hope, for enjoyment, edification, and practical value.

Before turning the reader over to our capable authors (whose biographical sketches can be found at the end of the book), a word on the issue of uncertainty is appropriate. Scientists, decision makers, and analysts have often suggested that effective linkage between science and environmental decisions depends upon the achievement of two goals: First, scientific uncertainties must be reduced (that is, predictions need to be more accurate); and second, technical experts must effectively communicate the nature and magnitude of those uncertainties to people who must take action. This intuitively attractive perspective treats uncertainty as something to be overcome, and prediction as a technical product that must be successfully integrated into the decision-making process prior to taking effective action. It also explicitly justifies tens of billions of dollars of publicly funded scientific research into problems as diverse as hydrocarbon reserve estimates, the behavior of nuclear waste in geological repositories, and the future behavior of the earth's climate.

All the same, it is often impossible to assign meaningful uncertainties to predictions of complex natural processes. One good hurricane can obliterate a beach that was predicted to last for a decade, a single debris-clogged bridge can cause a flood to rise far above its predicted crest, and a huge volcano can instantly negate a decades-long global warming trend. Such "surprises" are an expected—yet unpredictable—reality of open natural systems. Moreover, the complex interactions among the multiple components of such systems may render their detailed behavior unpredictable even in principle, as the continued inability to predict weather more than two weeks in advance starkly demonstrates.

But such technical concerns—which have been addressed in many scholarly and popular studies of "complex" phenomena (e.g., Gallagher and Appenzeller 1999)—are only part of the problem, and perhaps the less significant part. The case studies in this book indicate little obvious correlation between the quality of a prediction as judged by scientific standards and the success of decisions as judged by the achievement of

desired societal outcomes. Earthquakes and acid rain provide conceptual bookends to this point. A complete failure to predict the occurrence of specific earthquakes—infinite uncertainty—forced decision
makers to turn their attention from prediction to prevention of earthquake damage, thus stimulating successful policy action. Conversely, a
scientifically successful acid rain research program yielded predictions
that were largely irrelevant to the information needs of policy makers.
In many cases, reducing and communicating scientific uncertainty associated with predictions are neither necessary nor sufficient conditions
for creating a decision environment that is conducive to beneficial
action. Thus, as we will see, a central challenge for decision makers is to
understand how to distinguish problems that are likely to be amenable
to prediction-based solutions from those that demand alternative
approaches. Meeting this challenge requires an understanding of the
broad context—the interrelated scientific, socioeconomic, and political
environments—in which decisions are made.

In other words, the idea of a prediction as a disembodied number
modified by an uncertainty is entirely too abstract to have any meaning
in the real world. This book is not about numbers, but about the social
and political processes in which numbers are inextricably enmeshed. In
the end, we hope to provide usable insight about how desired societal
outcomes can emerge from these processes. Given the magnitude of the
environmental challenges that face society today, and the likelihood that
this magnitude will grow in the future, our decision-making capability
needs all the help it can get. The predictive promise of the earth sciences,
and the forward-looking character of decision making, tempts us to simply turn to scientists and say: "Tell us what will happen in the future, so
we can know what action to take now." If only it were so simple.

References

Dewey, J. 1991. *How We Think*. Amherst, NY: Prometheus Books, 16.
Gallagher, R., and T. Appenzeller. 1999. "Beyond Reductionism." *Science* 284: 79.
Gunderson, L.H., C.S. Holling, and S.S. Light (eds.). 1995. *Barriers and
    Bridges to the Renewal of Ecosystems and Institutions*. New York:
    Columbia University Press.
Lee, K. 1993. *Compass and Gyroscope: Integrating Science and Politics for
    the Environment*. Washington, DC: Island Press.

PART ONE

Prediction as a Problem

M olecular biologists have not, as far as we know, identified a "prediction gene," but the quest to predict seems as deeply instinctive to the human condition as language, self-consciousness, and artistic expression. Unlike these other characterizing traits, however, the instinct to predict has not always been expressed in effective performance. Oracles, prophets, and stock market forecasters have been accorded a status in society that is commensurate with the promise—not the delivery—of tomorrow revealed.

Scientists today seek to turn prediction into a reputable profession. They bring impressive tools to the quest: powerful theoretical understanding of fundamental processes; advanced monitoring technologies that digitize nature in all its rich profusion; supercomputers that crunch gigabyte-sized databases and spit out a vision of the future. Indeed, these days, science without prediction hardly seems like science at all.

Still, even the most sophisticated scientific predictions are plagued with uncertainties. But unlike predictions based on entrails or the stars, these uncertainties can be quantified (although quantifications of uncertainty are often themselves highly uncertain). We may therefore ask: What characteristics of a scientific prediction will allow us to make a decision that is better than the one we would have made without the prediction? (Of course, the answer to this question, too, may be highly uncertain.)

Prediction in Science and Policy

Daniel Sarewitz and Roger A. Pielke, Jr.

Policy makers have called upon scientists to predict the occurrence, magnitude, and impacts of natural and human-induced environmental phenomena ranging from hurricanes and earthquakes to global climate change and the behavior of hazardous waste. In the United States, billions of federal dollars are spent each year on such activities. These expenditures are justified in large part by the belief that scientific predictions are a valuable tool for crafting environmental and related policies. But the increased demand for policy-relevant scientific prediction has not been accompanied by adequate understanding of the appropriate use of prediction in policy making.

In modern society, prediction serves two important goals. First, prediction is a test of scientific understanding, and as such has come to occupy a position of authority and legitimacy. Scientific hypotheses are tested by comparing what is expected to occur with what actually occurs. When expectations coincide with events, it lends support to the power of scientific understanding to explain how things work. "[Being] predictive of unknown facts is essential to the process of empirical testing of hypotheses, the most distinctive feature of the scientific enterprise," observes biologist Francisco Ayala (1996).

Second, prediction is also a potential guide for decision making. We may seek to know the future in the belief that such knowledge will stimulate and enable beneficial action in the present. Such beliefs are supported by a long—if often mythic—history, predating modern science. For instance, armed with knowledge of the coming flood, Noah was able to build the ark and avoid the catastrophic end that befell those without such foresight. Today, as decision makers debate alternative courses of action, such as the need for a new law or the design of a new

program, they are actually making predictions about the expected outcome of this law or program and its future impact on society: "Decision
making is forward looking, formulating alternative courses of action
extending into the future, and selecting among the alternatives by
expectations of how things will turn out" (Lasswell and Kaplan 1950).

Persistent and pervasive calls for scientific prediction as a basis for
environmental policy making—documented throughout this book—
suggest confusion about these two motives for why we predict. The
value of predictions for validating scientific hypotheses does not imply
a commensurate value for dictating public policy. Moreover, as we will
explore in greater detail, all scientific predictions are not the same: those
used to support environmental decision making are different in essence
from those traditionally used to validate hypotheses. Confusion about
why and how we predict can prevent appropriate allocation of intellectual and financial resources for science and environmental policy. It
also sets the stage for a policy problem: Policy makers lack knowledge
that can help them to anticipate—to predict, that is—the circumstances
in which predictive research can contribute to effective decision making. As a consequence, some environmental policies may rely inappropriately on predictions and thus run the risk of failing to achieve their
intended effects. No process exists for assessing whether particular
environmental issues might or might not be amenable to solution aided
by predictions, and no systematic analysis exists to support such a
process. In this chapter, we try to dissect and define the problem of prediction in policy in a way that is useful for decision makers and
researchers, and we begin to develop a framework for understanding
the case histories, policy analysis, and recommendations that follow.

Types of Prediction

The essential context for prediction in traditional science is reductionism: the effort to break down reality into describable component parts
or processes with an ultimate objective of specifying the "laws of
nature." Such laws are fundamentally predictive, because they describe
behavior of phenomena that is *independent of time and place* (e.g.,
Popper 1959). That is, the behavior is always consistent, and thus predictable. In this sense, the word *prediction* as used in the reductionist
natural sciences is simply a synonym for *explanation* or *inference*
(Toulmin 1961). In reductionist science, moreover, prediction pertains
to the invariant behavior of individual parts, not to the processes of
interaction among natural systems that contain those parts (e.g., Wilson

1998). Thus, for example, progress in physics is often measured by success in identifying and describing increasingly fundamental components of matter; in biology, by finding increasingly fundamental building blocks of life.

Natural Systems

As Oreskes discusses in the next chapter, those disciplines of the natural sciences that seek to understand complex systems—integrative earth sciences (including solid-earth, ocean, and atmospheric sciences)[1]— have not traditionally been involved with prediction (although weather prediction has been a notable exception). Rather, such disciplines have been the source of verbal, graphical, and mathematical portrayals of nature that yield insight into earth processes. This insight can allow humans to better understand, anticipate, and respond to the opportunities and constraints of the natural world. For example, historical interpretation of earthquake occurrence, combined with present-day monitoring, has led to successful strategies for mitigating earthquake losses through appropriate engineering, land-use planning, and emergency management. Such strategies do not require the prediction of specific earthquakes to deliver social benefit.

Integrative earth science disciplines have sought to understand nature "as it is," rather than as reduced to its component parts. That is, while traditional physical science isolates phenomena from their context in nature in order to understand the invariant characteristics of the phenomena, the integrative earth sciences study the context itself. In the case of geology, for example, Baker (1996) writes: "Geology does not predict the future. Its intellectual tradition focuses on the contingent phenomena of the past. . . . Contingency holds that individual events matter in the sequence of phenomena. Change one event in the past, and the sequence of subsequent historical events will change as well." This focus on interpretation, contingency, and sequence is distinct *in its essence* from the reductionist goal of identifying and describing invariant phenomena.

Over the past several decades, however, prediction has increasingly become a goal of integrative earth science disciplines. A proliferation of new technologies for the study of the oceans, atmosphere, and solid earth have led, as well, to the proliferation of massive volumes and new types of data about the environment. At the same time, rapidly increasing computer-processing capabilities permit the analysis of larger and more sophisticated data sets. These changes have allowed earth scientists to develop more intricate conceptual and numerical models about earth-system processes ranging from the flow of toxic plumes in groundwater to the global circulation patterns of the atmosphere and oceans.

While such models can be used to test the validity of hypotheses about earth processes, they are also being used to predict the behavior of complex natural phenomena as input to policy decisions.

This type of prediction is fundamentally different from the predictive aspect of traditional, reductionist scientific inquiry. Rather than identifying the invariant behavior of isolated natural phenomena, prediction of complex systems seeks to characterize the contingent relations among a large but finite number of such phenomena. In contrast to prediction in reductionist science, *these types of predictions are highly dependent on time and place.*

Most generally, efforts to predict the behavior of complex systems use two approaches:

1. mathematical characterization of the significant components of a system and the interactions of those components according to governing laws (often called first principles), to yield a quantitative predictive model; and

2. identification of specific environmental conditions that are statistically significant precursors of a particular type of event.

Prediction of ongoing, evolving processes, such as groundwater flow or atmospheric circulation, is predominantly approached through mathematical modeling. Prediction of episodic, temporally discrete events, such as earthquakes and seasonal hurricane activity, often focuses on the identification of precursors that have shown a statistical linkage but are not necessarily causal. Most predictive efforts actually involve both approaches: the development of quantitative models and the search for correlations between past and future events.

In reductionist science, predictive validity is constantly being tested through the application of theory to scientific and engineering problems. In the integrative earth sciences, testing the usefulness or precision of a predictive model usually requires a comparison with observational data. Models can be tested through "retrodiction," that is, determining the ability of the model to reproduce the behavior of past phenomena (e.g., changes in global atmospheric temperature), or through in situ measurements of ongoing behavior (e.g., sampling to determine if the behavior of a toxic groundwater plume is consistent with the model). Oreskes, Shrader-Frechette, and Belitz (1994), among others, have argued that such tests do not amount to a "verification" of the predictive capability of the model, because natural systems are not "closed." That is:

> Even if a model result is consistent with the present and past observational data, there is no guarantee that the model will

perform at an equal level when used to predict the future. First, there may be small errors in input data that do not impact the fit of the model under the time frame for which historical data is available, but which, when extrapolated over much larger time frames, do generate significant deviations. Second, a match between model results and present observations is no guarantee that future conditions will be similar, because natural systems are dynamic and may change in unanticipated ways.

Still, earth scientists commonly argue that advances in theory, data collection, and computer power will deliver increasingly accurate and useful predictions of complex environmental phenomena in the future (Mahlman 1992; Wyss 1997). That such arguments occupy an important role in policy making is well illustrated by the examples of global warming (see chapter 13) and natural disaster preparedness (see chapter 7), as well as by the billions of dollars spent each year to support predictive research in these and related areas.

Social Systems

As predictions have become central to the notion of what is scientific, so have they become fundamental to the social sciences. Social scientists have long sought to emulate their physical scientist counterparts in developing invariant laws of human behavior and interaction (Ross 1991), an emulation that has often been called "physics envy." (Even in the humanities, some have sought to develop "scientific" methodologies, characterized by predictive skill [Fogel and Elton 1983].) Within the social sciences, scholars have for years debated the usefulness of aspiring to replicate the "scientific" success achieved by the physical sciences. For instance, Nobel Prize–winning economist Milton Friedman has suggested that a theory should be judged on its power to predict (Friedman 1953), whereas another Nobel Prize winner, Herbert Simon, suggests that such power is elusive even for some of our most well-accepted social science theories (Simon 1982). Indeed, although much social science research is supported to develop predictions, such predictions may prove unsuccessful for all but the most simple (and therefore obvious) social situations (Ascher 1979).

Economics has been viewed by many as the "imperial" social science, one that "will always remain valid for analyzing and *predicting* the course of human behavior and social organization" (emphasis in original) (Hirshleifer 1985). Part of its stature derives from the resemblance between the quantitative emphasis and methodologies of economics and those of physics. Sociology, on the other hand, was modeled on the biological sciences. I.B. Cohen (1994) has observed that:

Curiously enough, the biological science of the nineteenth century has weathered the years somewhat better than the physics, requiring revisions and expansions but not the same degree of radical restructuring, while the sociology built on the biology has not done as well as the economics which was (in part, at least) linked with the physics. Apparently, the correctness of the emulated science is not intrinsically connected with the permanent value of the social science.

Within the social sciences, most disciplines have in either small or large part sought to model themselves after economics, with other methodological approaches viewed as "alternatives" (Simon 1985; Dahl 1961). In political science, a large literature exists on developing various theories of political activity based on the "rational actor" theory of economic behavior (Petracca 1991). For instance, a classic text in political science is Anthony Downs's *An Economic Theory of Democracy* (1957). More recently, scholars have used economic methods in pursuit of a predictive model of presidential elections (Lewis-Beck and Rice 1992). For some in political science, the development of predictive theories is what makes the discipline "scientific." According to David Brady, a leading political scientist, "Unless we, as a profession, can offer clear theories of how elections, institutions, and policy are connected and deduce predictions from these stories, we shall simply be telling *ad hoc* stories" (Brady 1993). Cohen (1994) argues that it is "not a fruitful question" whether or not the social sciences are "scientific" in the sense of the physical sciences. Nevertheless, he notes:

> A social science like economics—which looks somewhat like physics in being quantitative, in finding expression of its principles in mathematical form, and in using the tools of mathematics—tends to rank higher on a scale of both scientists and non-scientists than a social science like sociology or political science which seems less like an "exact science."

Thus, in social sciences, as in the case of the natural sciences, predictive capabilities are widely viewed as authoritative and legitimating. Here as well the subtext of such research is that predictive science will add to the development of fundamental knowledge of human behavior, which—aside from its intrinsic value—will enhance society's capability to organize and govern itself.

As the scientific community seeks to predict the behavior of complex systems, the boundaries between physical and social sciences are blurring, or at least overlapping. For instance, in the case of global warming, predictions of future climate impacts are, in part, based on predictions of future population growth and energy consumption, both of which fall squarely in the realm of the social sciences. Similarly, understanding the impacts of natural hazards such as floods and earthquakes depends on future trends in economic development and demography, which are functions of broader social and policy processes.

To summarize, prediction has long been central to the process and validation of modern science. Prediction is also necessarily implicit in the process of decision making. In recent years, coincident with the rapid development of data acquisition and storage and processing capabilities, researchers and policy makers alike have looked to science as a source of predictions about the evolution of complex natural and social systems. We have argued that such activities are distinct from traditional, reductionist scientific prediction. We now look more closely at the relationship between decision making and the prediction of complex systems.

Two Birds with One Stone: How Prediction Simultaneously Fills a Policy Role and a Science Role

The predictive capacity of science holds great inherent appeal for policy makers who are grappling with complex and controversial environmental issues, by promising to enhance their ability to determine the need for and outcomes of particular policy actions. However, this appeal is partly rooted in the conflation—and perhaps confusion—of two conceptually and methodologically distinct activities: predictions as a means to advance science, and predictions as a means to advance policy. We emphasize that the traditional rationale for prediction in science was to validate reductionist theory. Only in recent years, with the rise of high-technology integrated earth science, have policy makers and scientists alike been tempted to extend this rationale to include the support of policy decisions. Today, the value of scientific predictions is increasingly viewed not just in terms of scientific understanding, but in terms of policy making, as well.

This newer, political role for prediction is seductive. If predictive science can improve policy outcomes by guiding policy choices, then it can as well reduce the need for divisive debate and contentious decision making based on subjective values and interests. Prediction, that is, can become a substitute for political and moral discourse. By offering to improve policy outcomes, scientific predictions also offer to reduce political risk, and for policy makers worried about public support and reelection, avoiding political risk is very appealing indeed. This appeal has an additional attribute: The very process of scientific research aimed at prediction can be portrayed as a positive step toward solving a policy problem. Politicians may therefore see the support of research programs that promise to deliver a predictive capability in the future as an alternative to taking politically risky action in the present.

Supply and demand for federally funded research on prediction of environmental phenomena are tightly coupled. As environmental problems become more politically complex—and response options become more controversial and costly—decision makers look toward scientists to help reduce uncertainties and dictate "rational" policy paths. Simultaneously, the growing analytical and computational sophistication of the earth sciences leads to an increased confidence in the capacity of these disciplines to predict the behavior of the environment. Furthermore, finite federal research funding dictates that decision makers and scientists naturally converge on areas of research that are expected to be mutually beneficial.

The short-term benefits for both scientists and politicians are clear: scientists receive federal funding to develop predictions; politicians can point to predictive research as "action" with respect to societal problems, while deferring difficult decisions as they await the results of research. Such an arrangement is seen in a number of nationally important policy issues, such as global climate change, nuclear waste disposal, and natural hazard mitigation.

Over the long term, will this arrangement lead to improved policy making, disappointed expectations, or some combination of both? Prospects for success will almost certainly vary depending on the phenomenon being predicted and the policy problem being addressed. An analytical framework that allows policy makers and scientists to evaluate how and when scientific prediction can benefit the policy process would help ensure an effective allocation of financial and intellectual resources. In particular, a useful framework must evaluate the capacity of predictive research to contribute to positive policy outcomes in light of the following six concerns:

1. Phenomena or processes of direct interest to policy makers may not be easily predictable on useful geographic or time scales. For example, early optimism about the predictability of earthquakes (Press 1975) has been eroded by several decades of scientific failure (see chapter 7).

2. Accurate prediction of phenomena may not be necessary to respond effectively to political or socioeconomic problems created by the

phenomena. For example, better mitigation of natural hazards such as hurricanes and floods may be achieved through effective planning that does not depend on better predictive information (Pielke 1997; chapter 5, this volume; chapter 4, this volume). In the case of acid rain, the political solution of using tradable permits to reduce sulfur oxide emissions did not depend on the predictive results emerging from a ten-year, half-billion-dollar federal research program (see chapter 12).

3. Necessary or feasible political action may be deferred in anticipation of predictive information that may not be forthcoming in a useful time frame. For example, societal adaptation to inevitable future climate impacts has been held in abeyance by the expectation that predictions of global climate change will guide policy choices (chapter 13). Similarly, action may be delayed when scientific uncertainties associated with predictions become politically charged, as seen in the case of both global climate change and high-level nuclear waste disposal (chapter 10).

4. Predictive information may be subject to manipulation and misuse, because the limitations and uncertainties associated with predictive models are often not readily apparent to nonexperts, and because the models are often applied in a climate of political controversy and/or high economic stakes (Rushefsky 1984). For example, in such cases as mining on federally owned land and replenishment of sand on public beaches, mathematical models are used to predict costs and environmental impacts. The scientific assumptions that guide the use and interpretation of such models may be influenced by powerful economic and political interests (chapters 8 and 9).

5. Criteria for scientific success in prediction may be different from criteria for policy success. For example, efforts to model global climate change have led to considerable increases in scientific insight over the past decade. During this time, however, global political controversy over appropriate responses to climate change has not eased and has probably increased. Progress in the science has therefore not translated into commensurate progress in the public realm (chapter 13). As well, scientifically reputable predictions that are not developed with the needs of policy makers in mind can in fact backfire and inflame political debate, as seen in the case of oil and gas resource appraisals (chapter 11).

6. Emphasis on predictive sciences moves both financial and intellectual resources away from other types of scientific activity that might

better help to guide decision making, such as monitoring, assessment, and small-scale policy experiments. Resource allocation for science can therefore influence policy options. If decision makers lack data about present environmental trends, or lack insight into the implications of different policy scenarios, they are less likely to use adaptive approaches to environmental problems, and more likely to wait for a predictive "prescription" (Lee 1993; Brunner and Ascher 1992; chapter 14, this volume).

These concerns suggest that the usefulness of scientific prediction for policy making and the resolution of societal problems depends on relationships among several variables, such as the time frame within which predictions are sought (e.g., tomorrow's weather vs. the next century's climate conditions), the intrinsic scientific complexity of the phenomena being predicted, the political and economic context of the problem, the compatibility of scientific and political goals, and the availability of alternative scientific and political approaches to the problem. If policy makers wish to design environmental research policies that are fiscally responsible, scientifically efficient, and socially beneficial, they will need to evaluate environmental phenomena and problems in the context of these and related variables. Such an evaluation process must begin with a clear picture of the prediction process itself. The ten case histories that constitute the heart of this volume are intended to paint this picture in all its richness, diversity, and complexity.

· **Conclusion**

Scientific prediction is commonly portrayed as a necessary precursor to—and a desirable determinant of—action on environmental policy. In such portrayals, scientific prediction is a source of objective information that can cut through political controversy and help define a path for "rational" action. Because policy making is itself a forward-looking process, this view of prediction may seem plausible. In practice, however, there have been few systematic evaluations of the performance of prediction in the policy realm.

Short-term predictions, especially those associated with discrete, extreme weather events such as floods and hurricanes, have often proven useful in supporting emergency management strategies. Attempts to provide longer predictive lead-times for discrete events such as earthquakes have generally been unsuccessful, although they have heightened public awareness. Efforts to predict events or phenomena with complex, diffuse, and regional impacts, such as acid rain, energy supply and consumption, the behavior of radioactive waste in a geological repository, and global climate change, have rarely contributed to the resolution of policy debates and have often contributed to political gridlock. This experience in part reflects the intrinsic scientific challenge of prediction, but it also derives from the complex scientific and policy context within which the predictive research takes place.

The idea that research programs focused on prediction will catalyze political action requires an extrapolation of the concept of scientific prediction itself, from its traditional significance as a test of fundamental and reductionist laws of nature, to a newer role as a technique that seeks to extract policy-relevant predictive certainty from research on complex processes. Given the difficulties of achieving such relevant certainty, the role of scientific prediction in policy making is itself highly uncertain. A better understanding of prediction in science and policy can help define a more realistic and positive role for science in society and a clearer path toward resolution of the many environmental challenges that face humanity.

Notes

1. We include solid-earth, ocean, and atmospheric sciences under the term *integrative earth sciences*.

References

Ascher, W. 1979. *Forecasting: An appraisal for policy-makers and planners.* Baltimore: Johns Hopkins University Press.

Ayala, F. 1996. The candle and the darkness. *Science* 273:442.

Baker, V. 1996. The geological approach to understanding the environment. *GSA Today* 6(3):41.

Brady, D. 1993. Review essay: The causes and consequences of divided government: Toward a new theory of American politics? *American Political Science Review* 87(1):194.

Brunner, R.D. 1991. Global climate change: Defining the policy problem. *Policy Sciences* 24:291–311.

Brunner, R.D. 1996. Policy and global change research: A modest proposal. *Climatic Change* 32:121–147.

Brunner, R.D., and W. Ascher. 1992. Science and social accountability. *Policy Sciences* 25:295–331.

Cohen, I.B. 1994. *Interactions: Some contacts between the natural sciences and the social sciences.* Cambridge, MA: MIT Press, p. 6.

Dahl, R.A. 1961. The behavioral approach in political science: Epitaph for a monument to a successful protest. *American Political Science Review* 55:763–772.

Downs, A. 1957. *An economic theory of democracy.* New York: Harper.

Fogel, R.W., and G.R. Elton. 1983. *Which road to the past? Two views of history.* New Haven, CT: Yale University Press.

Friedman, M. 1953. *Essays in positive economics.* Chicago: University of Chicago Press.

Hirshleifer, J. 1985. The expanding domain of economics. *American Economic Review* 75:54.

Lasswell, H.D., and A. Kaplan. 1950. *Power and society: A framework for political inquiry.* New Haven, CT: Yale University Press.

Lee, K. 1993. *Compass and gyroscope: Integrating science and politics for the environment.* Washington, DC: Island Press.

Lewis-Beck, M.S., and T.W. Rice. 1992. *Forecasting elections.* Washington, DC: Congressional Quarterly Press.

Mahlman, J. D. 1992. Testimony before the Committee on Science, Space, and Technology, U.S. House of Representatives, on *Priorities in Global Climate Change Research.* Washington, DC: U.S. Government Printing Office, pp. 26–36.

Oreskes, N., K. Shrader-Frechette, and K. Belitz. 1994. Verification, validation, and confirmation of numerical models in the earth sciences. *Science* 263:641–646.

Petracca, M. P. 1991. The rational choice approach to politics: A challenge to democratic theory. *Review of Politics 1991* 53:289–320.

Pielke, R.A., Jr. 1997. Reframing the U.S. hurricane problem. *Society & Natural Resources* 10:485–499.

Popper, K.R. 1959. *The logic of scientific discovery.* New York: Basic Books.

Press, F. 1975. Earthquake prediction. *Scientific American* 232 (May):14–23.

Rayner, S., and E.L. Malone, eds. 1998. *Human choice and climate change.* Columbus, OH: Battelle Press.

Ross, D. 1991. *The origins of American social science.* Cambridge, UK: Cambridge University Press.

Rushefsky, M.E. 1984. The misuse of science in governmental decisionmaking. *Science, Technology, and Human Values* 9(3):47–59.

Simon, H.A. 1982. *Reason in human affairs.* Stanford, CA: Stanford University Press.

Simon, H.A. 1985. Human nature in politics: The dialogue of psychology with political science. *American Political Science Review* 79:293–304.

Toulmin, S. 1961. *Foresight and understanding.* New York: Harper and Row, p. 35.

Wilson, E.O. 1998. *Consilience: The unity of knowledge.* New York: Alfred A. Knopf.

Wyss, M. 1997. Cannot earthquakes be predicted? *Science* 278:487.

Why Predict? Historical Perspectives on Prediction in Earth Science

Naomi Oreskes

Underlying the ten case studies presented in this book is an implicit assumption that scientists have to make predictions. Do they? Before we rush headlong into the scientific, social, and political agenda of scientifically based prediction in aid of public policy, we might ask the question, why predict?

Many people think that it is inherent in the nature of science to make predictions, because prediction is integral to an ideal of scientific method based on testing theories by their consequences. But this kind of prediction—logical prediction—is distinct from the temporal prediction that forms the primary subject of the case studies presented here. Predicting the future—earthquakes, floods, asteroid impacts, climate change—has not traditionally been a major part of the work of earth scientists. On the contrary, for the better part of at least two centuries, most earth scientists eschewed temporal prediction, viewing it as beyond the scope of their science. Times have changed, and earth scientists now routinely attempt to predict the future. But, as the case studies in this volume poignantly demonstrate, these attempts rarely achieve their scientific or societal goals. Why are we making temporal predictions if they are not generally successful? Why have earth scientists now embraced temporal prediction as a goal, when previously they avoided it? Knowing the answers to these questions may affect the way we present our science in the public policy area, and perhaps even the way we do it in the laboratory.

Even logical prediction has come to preeminence in our understanding of science only relatively recently. In the twentieth century, it became conventional wisdom that science works by testing theories through their logical predictions, and therefore that the goal of science *is* to test theories by comparing their predictions to observations. This approach is presumed to apply for all sciences. It is what most people have in mind when they talk about "the scientific method" (e.g., Popper 1959; Hempel 1965). But in the eighteenth and nineteenth centuries, things were viewed differently. To present a complex situation in broad outline, physics and astronomy were understood as predictive sciences, but earth science was not. Explanation was considered to be the principal goal of the earth sciences. Knowledge of the earth was thought to develop not by the testing of theoretical predictions, but by inductive generalizations from observational evidence.

Geology as an Inductive Explanatory Science

Perhaps the most widely read philosopher of earth science in history was the nineteenth-century British polymath William Whewell (1794–1866). Known to historians of science for coining the term *scientist*, Whewell was known to his contemporaries as a brilliant scholar who ranged over the breadth of science, religion, philosophy, architecture, the classics, and history. His *Elementary Treatise on Mechanics* (1819) introduced Leibniz's version of calculus to British audiences, helping to set in motion the mathematization of modern physics and earning Whewell a place in the Royal Society. Ordained as an Anglican minister in 1826, he accepted the Chair of Mineralogy at Cambridge in 1828. Among his geological studies, Whewell pioneered the systematic study of tides, demonstrating that the details of tides in any given place could not be accounted for solely on the basis of a formal mathematical theory. Elected president of the Geological Society of London in 1836, he was acquainted with nearly all the major figures in nineteenth-century British science. In 1837, Whewell published his three-volume *History of the Inductive Sciences,* a comprehensive survey of science from ancient Greece to the present. Three years later, he completed its companion volume, *Philosophy of the Inductive Sciences, Founded Upon Their History.* He had meanwhile been appointed professor of moral philosophy (in 1838), and in 1855 (in a model for all academics) he resigned his chair, having nothing more to say (Elkana 1984).

Whewell studied the history of science because he wanted to know how science worked: "how truths, now universally recognized, have really been discovered" (Elkana 1984, p. xvi). Only by studying how science actually had developed, he argued, could one formulate a useful philosophy of it. And Whewell preached what he practiced. His philosophical conclusion, based on his historical investigations, was that science proceeded inductively. That is to say, facts came first and theoretical concepts emerged from them. Explanations were *induced* by observational facts. Periods of great advance he termed "Inductive Epochs," which became possible when sufficient observational data were present to permit generalization and improved understanding. An important feature of inductive epochs was the achievement of a "consilience of inductions"—the unification of previously unrelated bodies of evidence under one explanatory framework. In Whewell's words, "The consilience of inductions takes place when an induction, obtained from one class of facts, coincides with an induction, obtained from another class" (Whewell 1847). Just as one is more likely to believe the independent testimony of multiple witnesses to an event, so one is more justified in believing an inductive inference when it is independently obtained from more than one set of observational facts. Induction was the act of creative insight that gave meaning to large bodies of observational evidence. This was the basic method of science (Whewell 1847, 1857; see also Laudan 1971; Elkana 1984).[1]

Philosophizing can be an expression of discontent; some people feel compelled to write when the world is not behaving as they think it should. But Whewell's philosophy matches what we know about the scientific world in which he lived. The world of nineteenth-century geology was for the most part a world of inductive science, the goal of most earth scientists to produce explanations of observed facts. The accumulation of particular instances—about rocks, fossils, structural and stratigraphic relations—was the basis from which more general conclusions were derived. Geologists worked from the specific to the general. One consequence of this inductive method, and its accompanying emphasis on explanation, was that prediction, either logical or temporal, played little role. The power and truth of a theory were thought to reside not in its ability to predict the unknown, but in its ability to explain the known (Rudwick 1985; Secord 1986; Oldroyd 1990).

Nowhere is this more evident than in the work of Whewell's brilliant contemporary, Charles Lyell (1797–1875), whose three-volume *Principles of Geology* was crammed with observational detail. Lyell described the details of stratigraphic units, fossil assemblages, volcanic sequences, and structural dislocations in order to understand geological history. His famous injunction—"travel, travel, travel"—reflected his inductive approach to knowledge: one learned about the earth neither through laboratory experiments nor by testing theoretically derived predictions, but by systematically and persistently observing geological processes and their products in as many places as possible. Travel was useful for many reasons, but above all because one accumulated the *extent* of knowledge necessary for reliable inductive insight to emerge. One also encountered phenomena one would never have predicted. Who could have imagined a trilobite? What theory would have predicted the Cambrian explosion of biodiversity? What laboratory experiment could replicate the workings of a volcano? Travel also challenged one's assumptions. In the first edition of *The Principles*, Lyell scoffed at claims of the Fennoscandian uplift—a purported gradual but discernible rise in coastal landscapes (or drop in sea level) in Norway and Sweden—because he could imagine no cause of rapid crustal uplift without accompanying earthquakes, and because geological evidence from around the world pointed to sea level rise, not fall. But then Lyell traveled to Norway, where he heard the accounts of fishermen and saw the physical evidence of raised beaches and displaced shorelines. He was convinced and said so in later editions of his book (Lyell 1830, 1850; see also Oreskes 1999).

This is not to suggest that Lyell had no theoretical point to make. Far from it. *The Principles of Geology* was written to advocate the theoretical position that geological history can be explained by currently observable processes acting in the past at the same rate and intensity as they do at present—what we now call *uniformitarianism*. No major geological catastrophes or other unusual events are required, just the steady accumulation of incremental change. Scholars have argued about which came first for Lyell—his observations or his theory—but there is no question that he tried to demonstrate the latter with reams of the former. Prediction plays no important role in his argument. The point for Lyell was not to predict the world but to explain it.

If Lyell was the most famous geologist of the nineteenth century, the most famous scientist of all was Charles Darwin. When Darwin embarked on the voyage of the H.M.S. *Beagle*, he was a confused young man with few clear ideas, scientific or otherwise. But he had been inculcated in the Cambridge model of natural science by his teachers, geologist Adam Sedgwick and botanist John Henslow, and he took aboard the *Beagle* a copy of Lyell's newly published book (Desmond and Moore 1991). Lyell's influence is evident throughout Darwin's work, but above all in the clear parallel between uniformitarianism and Darwin's insistence on the gradual, incremental, and accumulative nature of evolutionary change. Like Lyell, irrespective of how he came to his explanatory ideas, Darwin supported them by masses of specific, detailed observations (Darwin 1859, 1871; also see Hull 1973; Kohn 1985).

By 1844 Darwin had outlined his theory of evolution by natural selection, but he delayed publishing for fifteen years while he accumulated supporting evidence. Likely he would have continued longer in the same vein, had not Alfred Russel Wallace independently (in 1858) proposed the same idea. Abbreviated as they were in Darwin's view, *The Origin of Species* (1859) and *The Descent of Man* (1871) together constituted nearly one thousand pages, detailing example after example after example. Even so, Darwin worried that his contemporaries would consider his theoretical claims too great a leap—and many (including Whewell) did (Hull 1973). But despite the cavils of Darwin's critics, *The Origin* is an inductive work. Natural selection is an explanatory account of a mass of observational material. It is not a predictive theory. It gives an account of a process by which evolutionary change has occurred, and presumably may continue to occur, but it tells us nothing specific about the future. Darwin no doubt believed that evolution was a continuing process, but how it would continue was unknown, because it was contingent on unpredictable environmental change.

Philosopher of science Karl Popper later criticized Darwin's theory for just this reason: Natural selection can't be disproved, he argued, because it can't be tested at all. By definition, natural selection takes place over time spans to which we do not have access. It is also tautological: the fit survive and reproduce, and we define fitness in terms of survival and reproduction. It is only moderately cranky to argue that natural selection is the survival of the survivors (Popper 1976, p. 171; others have made similar points). Furthermore, any biological feature can be explained *ex post facto* as an adaptation, but this does not prove that it is one (Gould and Lewontin 1979). From these perspectives, Darwin's theory is attacked for being insufficiently deductive, but in his own day he was being criticized by both Whewell and Mill for being insufficiently inductive, proving that among philosophers you just can't win (see Hull 1973, p. viii). Darwin was aware of the problem. "When we descend to details, we can prove that no one species has changed; nor can we prove that the supposed changes are beneficial, which is the groundwork of the theory. Nor can we explain why some species have changed and others have not" (Darwin 1887, cited in Hull 1973, p. 32). Darwin argued instead that his theory had to be understood in the main—for its overall intelligibility as an explanatory account of the whole of nature, despite the impossibility of proving individual instances of it.

Darwin's argument proved correct: natural selection became a celebrated theory because of its explanatory power. The theory made few predictions but lots of sense. Its truth was ultimately understood not in terms of its consequences, but in terms of its ability to account for known features of the natural world. Darwin's theory did not have to make predictions to be convincing. What it had to do was to give a compelling account of the world (Hull 1973; Kohn 1985; Desmond and Moore 1991).

Lyell and Darwin are the most famous examples from the history of earth science; there are many others. The establishment of the geological time scale was the great accomplishment of nineteenth-century geology: it paved the way for Darwin's theory (providing both evidence of life through geologic history and the necessary time frame for gradual evolution); it enabled geologists to give a naturalistic account of earth history; and it caused individuals and communities to rethink the place of humanity on earth (Toulmin and Goodfield 1965; Rudwick 1985; Secord 1986; Oldroyd 1990). But it had little to do with prediction, in either the logical or the temporal sense. The same may be said of virtually all the major accomplishments of eighteenth- and nineteenth-century geology. In Germany, Abraham Gottlob Werner and his students pioneered the study of minerals and mountains, laying the foundations of modern mineralogy and mining geology (Laudan 1987). In France, Georges Cuvier established the reality of extinction, one of the pillars upon which Darwin's theory would later stand (Rudwick 1997). In the United States, G.K. Gilbert demonstrated isostatic rebound through study of the dry beds of Lake Bonneville, while T.C. Chamberlin demonstrated the multiple periods of Pleistocene glaciation though study of glacial deposits and landforms (Pyne 1980; Oreskes 1999). These scientists had diverse attitudes about the interrelation between observation and explanation, and diverse theoretical beliefs and goals. What they shared was an intellectual commitment to experience and observation as the foundation of scientific knowledge about the earth. Confirmation of logical predictions was not required for an explanation to be convincing, for a theory to be judged true.[2]

Why Not Predict?

Why didn't geologists pay more attention to prediction? One answer is obvious. Geologists were deeply concerned with temporal matters, but their concern was the past, not the future. The past could be deciphered by studying its tangible remains, but how could one study the future? The future, by definition, involved the unobservable—the inaccessible—and therefore pushed one beyond the realm of inductive science. Temporal prediction had a religious dimension as well: the future was God's domain. Scientists like Lyell and Darwin believed that their theories were likely to hold true in the future, but specific temporal predictions were considered to be beyond the scope of their study, beyond the scope of their science.

A second answer involves the nature of geology as a science. Prediction was of course central to astronomy, physics, and chemistry. Astronomers had long had the capacity to make accurate temporal predictions of heavenly phenomena, and by the seventeenth century, scientists like Newton and Boyle were extending that capacity to the phenomena of the earth as well. By the start of the nineteenth century, astronomy, physics, and chemistry all had well-developed logically predictive laws concerning the behavior of their objects of study. But this was precisely the point: These sciences were understood to be different from geology (and biology, for that matter) because they dealt with repetitive patterns and events. For them, the future *was* accessible, because the events of interest would occur again—today, tomorrow, next week, next year. The future would soon become the present. But geology was different.

As historian Kenneth Taylor has pointed out, the establishment of geology as an independent science in the eighteenth century involved its distancing from celestial mechanics and astronomy, its definition of a unique and separate set of questions that would define the science of geology *qua* geology. Geology would focus on the earth, literally, on the tangible rocks and mud beneath our feet. Taylor has described the attitude of early geologists as "critically sophisticated analytical empiricism" (Taylor 1998; see also Laudan 1987). Early geologists were not disdainful of theoretical goals, nor were they naive about the relationship of theory to observation and observation to perception. On the contrary, their sensitivity to the problematic nature of scientific inference convinced them to focus on the tangible physical materials accessible to empirical investigation. These materials were all relics of the past. Indeed, it was precisely the inaccessibility of past geological processes that led Lyell to emphasize the need to refer to currently observable processes to explain how rocks and fossils were formed. The present was accessible by observing the geological processes around us, which produced objects similar to those found in the rock record; thus the connection between them could be inferred. But the future was entirely unobservable, and therefore out of bounds. Furthermore, the available relics of the past spoke to singular historical events. The patterns revealed in the rock and fossil record were not repetitive, and therefore not predictable in any meaningful sense. Earth history—like human history—was the unique record of singular events. When notable geological events did occur in the present—earthquakes or volcanic eruptions, for example—they never repeated at sufficiently frequent intervals to permit the development of predictive laws. The nature of geology seemed to distinguish and demarcate it from sciences that had predictive goals.

Taylor writes primarily about continental geologists; for the British another factor was the influence of the intellectual tradition of empiricism. British geologists in the eighteenth and nineteenth centuries worked within the legacy of Bacon, Locke, and Hume, all of whom stressed experience and observation as the basis of knowledge. Hume famously argued that, although we convince ourselves of the reality of causes, and believe their existence gives us warrant to predict the future, we should in fact conclude the reverse: we should expect the unexpected (Hume 1739 [1992]). In the words of philosopher A.J. Ayer, "Our experience leads us to expect the occurrence of unforeseen events; at the most we may believe that they can subsequently be accounted for" (1980, p. 71). While physics or chemistry could rely on repetition, geology could not. Earth history was a succession of unforeseen—and unforeseeable— events. The task of geology was to account for them after the fact.

Empiricism was the dominant Anglo-American philosophy of science in the seventeenth and eighteenth centuries; in the nineteenth century its influence spread to continental Europe with the rise of positivism. In 1830, Auguste Comte (1798–1857) published the first volume of what would become his highly influential, six-volume *Lectures on Positive Philosophy*. Comte, like the British empiricists, believed that positive knowledge (in the sense of "yes, I'm absolutely positive") derived from observation. In Comte's view, the historic triumph of European culture was that it moved through the progressive stages of magical and metaphysical explanation to the stage of scientific explanation based on observation and reasoning. Comte did not deny the role of theories and creative ideas in the development of scientific knowledge; on the contrary, he considered theoretical principles essential to the organization of observational knowledge. But positive philosophers—scientists—kept their principles close to their observations and resisted the impulse to argue final causes and hidden truths.

Positivism was a profoundly influential philosophy, helping to inspire the foundation of the social sciences in its confidence that all problems and questions were susceptible to scientific investigation, the social and political as well as the physical and natural. In Germany, it was taken up by Ernst Mach, whose influence on Einstein is well documented (Holton 1973). In philosophy of science, it reached the apex of its influence in the early twentieth century with the work of the so-called Vienna Circle, or "logical positivists"—Rudolph Carnap, Otto Neurath, Maurice Schlick, and others. But, ironically, as the movement peaked in intellectual sophistication and influence, it was collapsing as a philosophy of science.

From Explanation to Prediction: The Hypothetico-Deductive Method

In the twentieth century, a sea change occurred in how scientists and philosophers thought and talked about scientific knowledge. Our ability to make sense of experience and observation independently of prior theoretical belief was increasingly called into question. Scientists and philosophers began to speak about scientific knowledge in a new way: the hypothetico-deductive method.[3] The shift in thinking can be summarized in the following way: Empiricism is an insufficient philosophy for understanding the development of scientific knowledge because life is a sea of experience and sense perception. How do we decide which experiences and observations to take note of? We must have hypotheses to guide our observations. Without hypotheses, life is just a jumble of sense-perception, a bombardment of competing experiences.

In this view, empiricism appears to be at best a naive or primitive stage through which sciences might pass on their way to intellectual maturity. Once a science is mature, hypotheses come first. But if hypotheses precede observation, and organize it, how do we tell if a given hypothesis is correct? How do we distinguish hypotheses from bias? The answer is by testing. All beliefs have logical consequences, things that would be true if the hypothesis were true. To test a theory or hypothesis, we need only draw out its logical consequences—its logical predictions—and look for them in the natural world or laboratory. If the predictions come true, then we have grounds for believing in the theory. It doesn't matter whether the predicted fact already exists in the world or is something that will soon come to pass. It matters only that the prediction arises as a consequence of our theoretical belief. The hypothetico-deductive model of science thus elevates prediction to an exalted position. Because hypotheses precede observation—and therefore may be generated entirely independently of it—the only way to know if a hypothesis is any good is to test it through its logical predictions.[4]

What caused this shift in thinking? This is a huge question, beyond the scope of this chapter to answer in full. But some reasons are apparent. As alluded to above, two broad traditions have long competed in the philosophy of science, two historically distinct visions of the scientific enterprise. One emphasizes prediction, the other explanation. Physics and chemistry (and, more recently, molecular biology) are highly compatible with the former, and perhaps best explained by it. Geology and biology (at the level of organisms and ecosystems) are better accounted for by the latter. Not surprisingly, then, physicists and chemists have tended to advocate the former, geologists and biologists the latter. As one might expect given competing visions, there has been a history of tension, and at times open conflict, between the proponents of these divergent views. But despite such tensions, these competing visions coexisted—more or less side by side—for the better part of several centuries. Whatever physicists and chemists believed about their sciences, geologists in the main rejected prediction as a central task, and geology developed few predictive laws. Why did one of these two visions become dominant in the twentieth century, to the point of being accepted by earth scientists as well?

One reason is the influence of German philosopher Immanuel Kant (1724–1804). Kant famously questioned our ability to perceive things as they really are and suggested that our perceptions are just that—perceptions, not the things in themselves. Furthermore, our perceptions must be organized by concepts, which leads to the question: which comes first, the conceptions or the perceptions? Kant's answer is that at least some rudimentary concepts must come first, or how would we make any sense of our perceptions? German scientists in the mid- to late nineteenth century were greatly influenced by Kant, and as German experimental science grew in stature and scope—particularly German chemistry and physics—so did the influence of Kant's ideas. (Whewell was also deeply influenced by Kant, leading some philosophers to question in retrospect how inductive his inductive philosophy really was—but that is another story.) With the rise of physics and chemistry in the twentieth century, the philosophy that seemed to fit them rose too.

Physics and chemistry posed profound challenges to a strictly inductive model of science as they expanded their explanatory ambitions. Scientists in those fields confronted abundant evidence of phenomena that were not directly observable. It was not merely that their causes were unobservable—a problem physicists and chemists shared with all scientists—but that even the physical entities influenced by those causes were in many cases unobservable. Eighteenth-century natural philosophers had argued at length about the reality of the ether, through which light was thought to be transmitted, and some refused to believe in it because it was unobservable (Laudan 1984). In the nineteenth century, some resisted belief in atoms for the same reason. But as advances in chemistry confirmed the explanatory power of atomism, and advances in physics made the reality of forces like electromagnetism seem indisputable, resistance to unobservables broke down. Observation alone seemed insufficient to account for the phenomena of the universe. By the end of the nineteenth century, the stage was set for a theory that would challenge the very meaning of observation: Einstein's theory of relativity.

Although Einstein cited the influence of the positivist Ernst Mach, and there are elements of positivism in Einstein's early work, the special theory of relativity undermined the fundamental premise of positivism: the possibility of absolute observation (Holton 1973). One's perception of space and time, Einstein argued, was a function of one's frame of reference. Absolute space and time simply do not exist, and therefore neither does absolute observation. What we see depends upon our frame of reference. Einstein became the stuff of legend, and the fame of his theory of special relativity contributed to a growing twentieth-century skepticism about knowability in general. (If we cannot know time or space absolutely, what can we know absolutely?) But perhaps equally important to epistemologies of scientific knowledge was the success of the celebrated logical prediction of general relativity.

Einstein's theory of general relativity explained gravitational effects not as a force emanating from one body and affecting another, but rather as the influence of massive bodies on the geometry of space-time. If this theory were true, then starlight should be bent in the presence of massive bodies. The light from a distant star, for example, should be bent as it passes near our sun, and this might be detected during an eclipse. In 1919 it was. Einstein's prediction was brilliantly confirmed in a highly publicized expedition organized by the British astronomer Sir Arthur Eddington, a result that Eddington continued to celebrate in popular books in the 1920s and 1930s.

Einstein's success helped to shape the thinking of one of the most influential philosophers of the twentieth century, Sir Karl Popper. No philosopher did more to discredit the Baconian vision of science than Popper, who caricatured inductive science as the "bucket method" of acquiring knowledge—collecting facts like pebbles at the beach. This, he argued, was absurd. True science relied on hypotheses as "searchlights" to illuminate nature; only with hypotheses to guide us do we know which facts to acquire. The importance of facts is not to provide the basis for inductive generalizations, but to provide the means by which we can refute false ideas. Science for Popper was a series of "conjectures and refutations." Scientists propose bold, creative, conjectural ideas, and then they (or their colleagues) go in search of facts that will challenge them. This is what Einstein, with Eddington's help, had done.

Popper (and others) thus inverted two centuries of philosophy of science. In the inductive view, facts come first and hypotheses emerge from them. In the deductive view, hypotheses come first and facts serve as tests of them. Logical prediction becomes essential, for it is the only means by which to test our conjectures. It is the very stuff of scientific progress (Popper 1959; see also Hempel 1965, for a related view).

The rise of the hypothetico-deductive model of science coincided with the consolidation of the sciences as professional disciplines and became enshrined in many twentieth-century textbooks. It also provided the template for the emerging social sciences (Ross 1991). Thus in the mid-twentieth century, one could find scores of explicit pronouncements of scientists declaring that that is what they do. When plate tectonics became the new paradigm of the earth sciences in the 1960s, both scientists and philosophers of science declared that what separated this theory from the older theory of continental drift, debated in the 1920s, was that *this* one made predictions (Wilson 1968; Cox 1973; Frankel 1979; Laudan and Laudan 1989). But continental drift made predictions, too, and a lack of predictive capacity was never raised as a significant argument against it. What was different in the 1960s (among other things) was the way people thought about scientific knowledge. In the 1920s, prediction was not a sine qua non of a geological theory. In the 1960s, for many people it was (Oreskes 1999).

Whatever brought about the change in prevailing views about how science works, the historical point is this: The explanation for this change cannot simply be that now we know how science really works.[5] For this was not how science worked in the nineteenth century, or at least not how a substantial portion of it worked. Today, most scientists unquestioningly accept the importance of logical prediction as a test of the truth of a theory, but from a historical point of view, this is a recent development. Great scientists like Lyell and Darwin thought otherwise.

Contrary to popular belief, there is no necessary connection between science and prediction. It is a particular view of science that privileges prediction. Prediction is a way to test the legitimacy of a theory after it has been generated, but one can also judge a hypothesis by the method by which it was generated in the first place. Scientists may choose to make predictions, they may want to make predictions, they may be asked by others to make predictions. But if we are asked by others to make predictions, we should ask, what for?

Public Policy in the Age of Environmentalism: The Rise of Temporal Prediction

The status of prediction has been further elevated in recent years by a political imperative: the demand for temporal prediction in aid of public policy. As the case studies in this volume show, there is a huge demand for temporal prediction in the earth sciences in response to environmental problems and natural hazards. As more and more people live along America's coastlines, as the population of Southern California grows, as humans increasingly become agents affecting the composition of the earth's oceans and atmosphere, earth scientists are increasingly being asked to predict hurricanes, earthquakes, and the impact of human activities. This has led to an explosive growth in the use of computer models in the earth sciences to generate such temporal predictions. In many cases, these temporal predictions are treated with the same respect that the hypothetico-deductive model of science accords to logical predictions. But this respect is largely misplaced (Oreskes et al. 1994; Oreskes 1997, 1998).

The hypothetico-deductive model presupposes that, if a theory fails its predictive test, then the theory is rejected, or at least must be seriously revisited for necessary modification. This model requires two things. First, temporal predictions can serve as tests of theory only if the time frame of the prediction is short enough to allow us to compare the prediction with events in the natural world. Weather prediction may satisfy this demand, but much prediction in the earth sciences will not. Whatever value we place on logical prediction because of its use in theory testing does not transfer to long-term temporal predictions. Many if not most temporal predictions in the earth sciences are incapable of refuting our theoretical beliefs because the time frame involved is simply too long.

Second, to be of value in theory testing, the predictions involved must be capable of refuting the theory that generated them. This is where prediction in the earth sciences becomes particularly sticky. Nowadays, the majority of predictions in the earth sciences are generated by computer models. There are very few formal predictive laws in earth science. Where such laws exist on their own—for example, Darcy's law, which governs the flow of water through an aquifer—they generate little policy-relevant information. The policy-relevant information discussed in this volume—about acid rain, global climate, floods, hurricanes, nuclear waste—is typically generated by computer models.

These models are a complex amalgam of theoretical or phenomenological laws (and the governing equations and algorithms that represent them), empirical input parameters, and a model conceptualization. When a computer model generates a prediction, of what precisely is the prediction a test? The laws? The input data? The conceptualization? Any part (or several parts) of the model might be in error, and there is no simple way to determine which one it is.

So what use are predictions? A logical prediction can serve as a test that gives us a clue when something is wrong, even if it fails to define what is wrong uniquely. Temporal predictions can serve this role, but only if the time frame is short or the predicted events frequent. Forecasts of events in the far future, or of rare events in the near future, are of scant value in generating scientific knowledge or testing existing scientific belief. They tell us very little about the legitimacy of the knowledge that generated them. Although scientists may be enthusiastic about generating such predictions, this in no way demonstrates their intellectual worth. There can be substantial social rewards for producing temporal predictions. This does not make such predictions bad, but it does make them a different sort of thing. If the value of predictions is primarily political or social rather than epistemic, then we may need to be excruciatingly explicit about the uncertainties in the theory or model that produced them, and acutely alert to the ways in which political pressures may influence us to falsely or selectively portray those uncertainties.[6]

As individuals, most of us intuitively understand uncertainty in minor matters. We don't expect weather forecasts to be perfect, and we know that friends are often late. But, ironically, we may fail to extend our intuitive skepticism to truly important matters. As a society, we seem to have an increasing expectation of accurate predictions about major social and environmental issues, like global warming or the time and place of the next major hurricane. But the bigger the prediction—the more ambitious it is in time, space, or the complexity of the system involved—the more opportunities there are for it to be wrong. If there is a general claim to be made here, it may be this: the more important the prediction, the more likely it is to be wrong.

How are we to judge our science if not by its predictive capability? At present, this question is the subject of active discussion and disagreement, and it is clear that a simple answer will not be quick in coming (see, for example, Ford 1999; Oreskes 2000). What is clear is that there has been a widespread tendency to judge models by their predictive output, and to value certain kinds of models *because* they generate predictions—despite the fact that the predictions may turn out to be false, or in some cases can already be shown to be false. But history shows that there are alternative ways to think about our science. We do not have to focus solely (or even primarily) on predictive outcome as the measure of a model. We can also view models inductively and ask about the information that generated them: the theoretical underpinnings, the quality and quantity of the empirical data base, and the independence of the evidence supporting the model conceptualization. Model evaluation should focus at least as much on model inputs as on predictive output. The inductive model of science—and common sense—tells us that the more information that went into the construction of a model, and the better the quality of that information, the more reason we have to believe that the predictions generated by it are telling us something useful. Lyell and Darwin understood this, and so should we.

Notes

1. Some philosophers have denied that Whewell's philosophy was inductive because he believed that once achieved, successful inductive generalizations should be applicable to as yet unobserved phenomena or even unobservables; because he allows a far more active and creative role for the human mind than implied by Bacon's philosophy; and because, influenced by Kant, he believed some things could be known *a priori*. These beliefs led to a famous and acrimonious public debate over the meaning of induction with John Stuart Mill.

2. This is not to say that no one ever tested a geological theory. Chamberlin's famous method of multiple working hypotheses was based on the idea of testing theories by discriminating between their logical consequences. But in this respect, Chamberlin, who died in 1928, was a harbinger of things to come. In the nineteenth century, the idea of prediction—in either its temporal or its logical sense—did not play a major role in the work of most earth scientists.

3. This is not to imply that the hypothetico-deductive model was invented in the twentieth century. The success of Newton's work in physics had led many eighteenth-century natural philosophers to be convinced of the importance of predictive laws, while in France Cartesian rationalism placed less emphasis on observation, more on ratiocination. At the start of the nineteenth century, Dalton's work in chemistry encouraged a belief in the importance of predictive laws and verifying evidence.

4. It isn't *logically* necessary to assume that hypothesis precedes observation, only that the observational evidence be rationally construable as a logical consequence of the hypothesis. For the hypothetico-deductivist, it doesn't

matter where the hypothesis comes from, only how it is verified (e.g., Reichenbach 1938).

5. Some philosophers, sociologists, and historians have lately criticized the hypothetico-deductive model as not accurately describing what scientists do. Real life is much more messy. Others have pointed out that the method fails in principle as well as in practice, because tests of hypotheses are always underdetermined: there are myriad ways to save a theory that has failed a predictive test by making small adjustments to it (Lakatos 1970). Tests of hypotheses lack the logical force the hypothetico-deductive method implies. Nevertheless, many if not most scientists still adhere to the belief that this *is* the scientific method.

6. Political pressures often cause scientists to downplay uncertainties to permit conclusions and action, but not always: in the case of global warming, there is pressure from some quarters to emphasize the uncertainties of the models. This is not a new phenomenon: In the late 1950s, opponents of a comprehensive ban on the testing of nuclear weapons emphasized the uncertainties of seismic verification in order to inhibit action. This uncertainty became a major justification for the ultimate outcome: the limited test ban treaty, which forbade testing in the atmosphere, in space, or underwater but allowed tests underground, protecting the world from fallout but not from the arms race.

References

Ayer, A.J. 1980. *Hume*. Oxford, UK: Oxford University Press.

Cox, Allan, ed. 1973. *Plate tectonics and geomagnetic reversals*. San Francisco: W.H. Freeman.

Darwin, Charles. 1859 and 1871. *On the origin of species* and *The descent of man*. New York: Modern Library.

Desmond, Adrian, and James Moore. 1991. *Darwin*. New York: Norton.

Elkana, Yehuda, ed. 1984. *Editor's introduction to selected writings on the history of science, by William Whewell*. Chicago: University of Chicago Press.

Ford, David. 1999. Defining the problems of model assessment and quality assurance. In *Quality assurance of environmental models*. Washington, DC: National Center for Statistics and the Environment Web Site: http://www.nrcse.washington.edu/events/qaem.

Frankel, Henry. 1979. Why continental drift theory was accepted by the geological community with the confirmation of Harry Hess's concept of sea-floor spreading. In *Two hundred years of geology in America*, C.J. Schneer, ed. Hanover, NH: University of New England Press, pp. 337–353.

Gould, Stephen J., and Lewontin, R. 1979. The spandrels of San Marco and the panglossian paradigm: A critique of the adaptationist program. *Proceedings of the Royal Society* B205:581–598.

Hempel, Carl G. 1965. *Aspects of scientific explanation and other essays in the philosophy of science*. New York: Free Press.

Holton, Gerald. 1973. *Thematic origins of scientific thought: Kepler to Einstein.* Cambridge, MA: Harvard University Press.

Hull, David L. 1973. *Darwin and his critics: The reception of Darwin's theory of evolution by the scientific community.* Chicago: University of Chicago Press.

Hume, David. 1739 (1992). *Treatise of human nature.* Buffalo, NY: Prometheus Books.

Kohn, David, ed. 1985. *The Darwinian heritage.* Princeton, NJ: Princeton University Press.

Lakatos, Imre. 1970. Falsification and the methodology of scientific research programmes. In *Criticism and the growth of knowledge,* I. Lakatos and A. Musgrave, eds. Cambridge: Cambridge University Press, pp. 91–195.

Laudan, Larry. 1971. William Whewell on the consilience of inductions. *Monist* 55:368–391.

Laudan, Larry. 1984. *Science and values: An essay on the aims of science and their role in scientific debate.* Berkeley: University of California Press.

Laudan, Rachel. 1987. *From mineralogy to geology: The foundations of a science, 1650–1830.* Chicago: University of Chicago Press.

Laudan, Rachel, and Larry Laudan. 1989. Dominance and the disunity of method: Solving the problems of innovation and consensus. *Philosophy of Science* 56:221–237.

Lyell, Charles. 1830 (1990). *Principles of geology* (facsimile of the 1st edition). Chicago: University of Chicago Press.

Lyell, Charles. 1850. *Principles of geology* (8th edition). London: John Murray.

Oldroyd, David R. 1990. *The Highlands controversy: Constructing geological knowledge through fieldwork in nineteenth-century Britain.* Chicago: University of Chicago Press.

Oreskes, Naomi. 1997. Testing models of natural systems: Can it be done? In *Structures and norms in science,* M.L. Dalla Chiara et al., eds. Dordrecht: Kluwer Academic Publishers, pp. 207–217.

Oreskes, Naomi. 1998. Evaluation (not validation) of quantitative models. *Environmental Health Perspectives* 106 (supplement 6):1453–1460.

Oreskes, Naomi. 1999. *The rejection of continental drift: Theory and method in American earth science.* New York: Oxford University Press.

Oreskes, Naomi. 2000. Why believe a computer? Models, measures, and meaning in the natural world. In *The earth around us: Maintaining a livable planet.* Jill S. Schneiderman, ed. San Francisco: W.H. Freeman.

Oreskes, Naomi, Kristin Shrader-Frechette, and Kenneth Belitz. 1994. Verification, validation, and confirmation of numerical models in the earth sciences. *Science* 263:641–646.

Popper, Karl R. 1959 (1968). *The logic of scientific discovery.* New York: Harper Torchbooks.

———. 1976. *Unended quest: An intellectual autobiography.* La Salle, IL: Open Court.

Pyne, Stephen J. 1980. *Grove Karl Gilbert: A great engine of research.* Austin: University of Texas Press.

Reichenbach, Hans. 1938. *Experience and prediction: An analysis of the foundations and structure of scientific knowledge.* Chicago: University of Chicago Press, pp. 3–16.

Ross, Dorothy. 1991. *The origins of American social science.* Cambridge, UK: Cambridge University Press.

Rudwick, M.J.S. 1985. *The great Devonian controversy: The shaping of scientific knowledge among gentlemanly specialists.* Chicago: University of Chicago Press.

Rudwick, M.J.S. 1997. *Georges Cuvier, fossil bones, and geological catastrophes: New translations and interpretations of the primary texts.* Chicago: University of Chicago Press.

Secord, James. 1986. *Controversy in Victorian geology: The Cambrian-Silurian dispute.* Princeton, NJ: Princeton University Press.

Taylor, Kenneth L. 1998. Earth and heaven, 1750–1800: Enlightenment ideas about the relevance to geology of extraterrestrial operations and events. *Earth Sciences History* 17:84–91.

Toulmin, Stephen, and June Goodfield. 1965. *The discovery of time.* Chicago: University of Chicago Press.

Whewell, William. 1847 (1984). The philosophy of the inductive sciences (2nd ed.). In *Selected writings on the history of science*, Yehuda Elkana, ed. Chicago: University of Chicago Press.

Whewell, William. 1857 (1984). The history of the inductive sciences (3rd ed.). In *Selected writings on the history of science*, Yehuda Elkana, ed. Chicago: University of Chicago Press.

Wilson, J. Tuzo. 1968. Static or mobile earth: The current scientific revolution. *Proceedings of the American Philosophical Society* 112:309–320.

Uncertainty, Judgment, and Error in Prediction

Thomas R. Stewart

Every prediction contains an element of irreducible uncertainty. This fundamental fact is not disputed by scientists or by those who use their predictions to inform decisions. However, important *implications* of irreducible uncertainty are rarely discussed and generally not appreciated.

In this chapter, I revisit an old demonstration that shows how actions that are based on predictions lead to two kinds of errors. One is when an event that is predicted does not occur, i.e., a false alarm. The second is when an event occurs but is not predicted, i.e., a surprise. There is an inevitable tradeoff between the two kinds of errors; steps taken to reduce one will increase the other. Often, this results in cycles of policy adjustments intended to reduce one kind of error, then the other, and then the first again, and so on. Reducing both kinds of errors simultaneously, and breaking the back-and-forth cycle of policy change, requires improving the accuracy of predictions, where accuracy is simply defined as the correlation between that which is predicted and that which actually occurs. Since prediction involves human judgment, defined in this context as the synthesis of multiple items of information to produce a single prediction, understanding the judgment process may indicate methods for improving our view of the future. I will briefly explore the role of judgment in the accuracy of predictions.

Uncertainty

Uncertainty in prediction simply means that, given current knowledge, there are multiple possible future states of nature. Within this definition, a number of different types of uncertainty can be identified. Probability is the standard measure of uncertainty, and an important distinction is made between *frequentist* and *subjectivist* views of probabilities (e.g., Morgan and Henrion 1990). The frequentist view is the one taught in most introductory statistics and probability courses. In this view, probabilities are determined by long-run observations of the occurrence of an event. For example, if it rains 90 days out of 1,000 days, the frequentist probability of future rain is $90/1,000 = .09$. In order to apply frequentist probabilities, events have to be well specified, and there must be an empirical record appropriate for estimating probabilities. The data for calculating frequentist probabilities are available for many weather events, but weather is atypical. In the case of earthquakes, global warming, or nuclear waste disposal, because of their relative infrequency, frequentist probabilities are in various degrees not available.

Subjective probability is simply someone's belief that an event will or will not occur. Although subjective probabilities range from 0 to 1, they do not necessarily follow the standard rules of probability theory. Since they are subjective, different people will have different subjective probabilities for the same event. Subjective probabilities are assessed routinely by decision theorists, who have developed elaborate methods for eliciting them (e.g., see Clemen 1990). Some argue that all probabilities are subjective probabilities, because relative frequencies are only sample data of past events that influence subjective probabilities of future events. Others object to the use of subjective probabilities because they are not "objective." Because human judgment invariably plays a role in prediction, it is difficult to discuss uncertainty in any systematic way without considering subjective probabilities.

Uncertainty can also be classified as aleatory or epistemic. Aleatory uncertainty reflects the nature of random processes. For example, even though you know a fair die has six sides, you cannot reduce the uncertainty about what the next roll will show. But you can quantify the uncertainty. For the simple case of the die, the odds are 1 in 6 of any particular face turning up.

Epistemic uncertainty is incomplete knowledge of processes that influence events. Incomplete knowledge results from the sheer complexity of the world, particularly with respect to issues at the interface of science and society. As a result, models (computer or mental) necessarily omit factors that may prove to be important. It is possible to judge the relative level of epistemic uncertainty, e.g., because of the time frames and number of potentially confounding factors, it is higher in nuclear waste disposal and climate prediction than in the prediction of weather and asteroid impacts. Total uncertainty, either subjective or frequentist, is the sum of epistemic and aleatory uncertainty.

Duality of Error

When forecasts and events are frequent enough that data are available for estimating probabilities, uncertainty can be represented pictorially as a scatterplot. The three panels in figure 3.1 illustrate three different levels of uncertainty.

Each point on a scatterplot represents a hypothetical prediction. The predicted value is plotted on the X-axis and the actual event that occurred is plotted on the Y-axis. The scales on the axes are arbitrary. The horizontal scale might represent a forecast of maximum winds in a

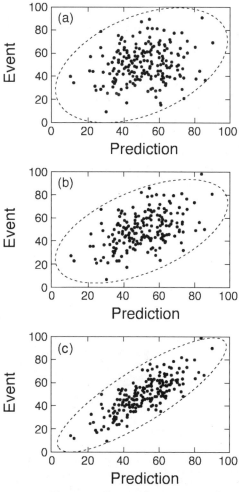

Fig. 3.1 Scatterplots representing various levels of uncertainty.

hurricane at a specific location and time. The vertical axis would then represent the actual observed wind speed at that point and time. Alternately, the horizontal axis might represent a forecast of rainfall amounts or tornadic potential. The vertical axis would then represent the actual observed event. Alternatively, the prediction might be expressed as a probability, such as a forecaster's probability of some magnitude of severe weather. The vertical axis would then be the severity observed, as measured by some criterion, such as wind speed or hail diameter, or a combination of both. Thus, the diagram applies to probabilistic predictions—e.g., 70 percent chance of rain—as well as categorical ones—e.g., tomorrow's temperature will be 77°F.

Figure 3.1a represents a high level of uncertainty. It is drawn such that the correlation coefficient between predictions and events is 0.20. The ellipse drawn around the points represents the uncertainty and is very wide. Figure 3.1b represents a lower level of uncertainty (the correlation is set to .50) and figure 3.1c an even lower level (correlation = .80). Notice how the ellipses become narrower as uncertainty decreases. (In the hypothetical case of no uncertainty, the points fall on a straight line, and the coefficient of correlation is 1.00.) Because the predictions are more highly correlated with the actual event, figure 3.1c obviously represents a more accurate, and therefore more desirable, set of predictions than does figure 3.1a.

The Taylor-Russell Diagram
In figure 3.2, I have placed a horizontal line across a scatterplot. This line represents an arbitrarily chosen boundary between situations requiring action, which are above the line, and situations requiring no

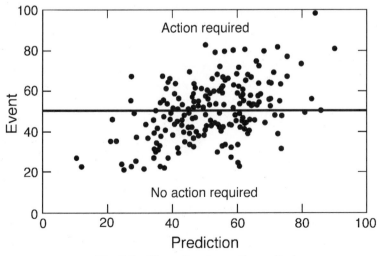

Fig. 3.2   The action/no-action criterion.

action, which are below the line. An example would be hurricane intensity forecasts. If a particular storm's intensity has wind speeds of 100 mph or greater, then the course of action recommended would be evacuation, but below that threshold, it would be no evacuation. This line is called the *criterion*. If the actual event exceeds the criterion, then some sort of preventative or protective action is required.

In figure 3.3a, a vertical line is added to the chart. This is the decision cutoff. In many instances, evacuation decisions being a prime example, because decision makers cannot wait for an event to occur to take action, decisions must be made based on a prediction. Given the expected outcome, based on the criterion, if the predicted event (or probability) is greater than the cutoff, then action is taken (or recommended). If it is less than the cutoff, then no action is taken or recommended. Figure 3.3a is known as a Taylor-Russell diagram, after the authors of the classic paper that first described it (Taylor and Russell 1939). Its original formulation was meant to be applied to decisions based on testing (e.g., admission to college based on SAT scores), but

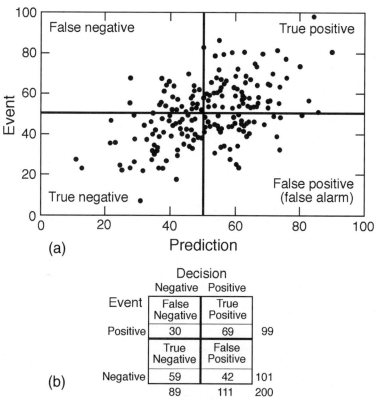

Fig. 3.3   (a) Taylor-Russell diagram with (b) decision table.

Kenneth Hammond (1996) has recognized that it applies to virtually any policy problem where decisions must be made in the face of uncertainty.

As shown in figure 3.3a, the two lines divide the scatterplot into four regions. The regions have been labeled according to standard decision research terminology:

- *True positive*: Appropriate action is taken. For example, people are warned of severe weather, and it actually occurs.

- *False positive*: Inappropriate action is taken. This is often called false alarm. For example, people are warned of severe weather, but none occurs.

- *False negative*: Action should have been taken but wasn't. For example, people were not warned of severe weather, but severe weather occurred.

- *True negative*: No action was taken, and that was appropriate. For example, no warning was issued, and the storm did not become severe.

Figure 3.3b shows the 200 data points in figure 3.3a in terms of a "standard decision table." A decision table has a row for each possible state of nature and a column for each possible decision alternative. The cells represent outcomes, that is, combinations of decisions and states of nature. Entries in the cells may be the frequencies of the outcomes, their probabilities, or their costs and benefits. In the last case, the decision table may also be called a payoff matrix. While the decision table clearly tallies the frequencies of the various outcomes, the Taylor-Russell diagram is required to show how those outcomes are determined by (a) underlying uncertainty, (b) the criterion, and (c) the decision cutoff.

The Tradeoff between False Positives and False Negatives

Suppose figure 3.3b represents the history of 200 decisions to evacuate coastal cities based on forecasts of hurricane winds. Policy makers, and the citizens they represent, might well complain about the number of errors (false negatives and false positives). They might be particularly concerned about the number of false negatives because those cases represent people who were not warned to evacuate when they should have been. That is, the forecast did not reach the decision cutoff required for an evacuation order, but the actual wind speeds did reach the level required to justify evacuation. The result of such a false negative could well be loss of property and life. Consequently, local officials and emergency managers might implement policies designed to reduce the number of false negatives. For example, they might lower their decision threshold for issuing evacuation orders.

Figure 3.4 shows the likely result of policies designed to address false negatives. The decision cutoff is moved to the left, resulting in the outcomes shown in the decision table. False negatives have been virtually eliminated, but that has been accomplished at the expense of greatly increasing the number of false alarms. In our example of evacuation of a city as a hurricane approaches, a false alarm results when evacuation proves unnecessary because the hurricane force does not reach damaging levels.

Confronted with the outcomes shown in figure 3.4b, policy makers are likely to experience political pressure to reduce the number of false alarms, which lead to costly and unnecessary evacuations. Policies designed to reduce the number of false alarms will be put in place. The effect of such policies will be to increase the decision threshold for evacuation decisions, that is, to demand greater certainty prior to taking action, as depicted in figure 3.5.

Notice that a significant reduction in false alarms has been achieved, but at the expense of an increase in the number of false negatives. The

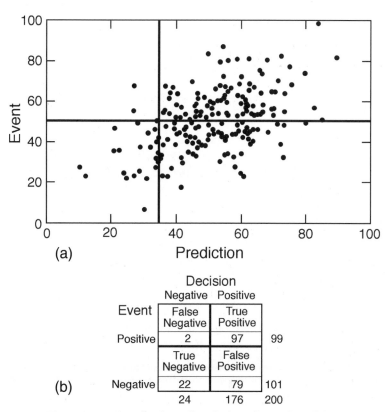

Fig. 3.4 (a) Result of a policy designed to reduce false negatives. (b) Decision table.

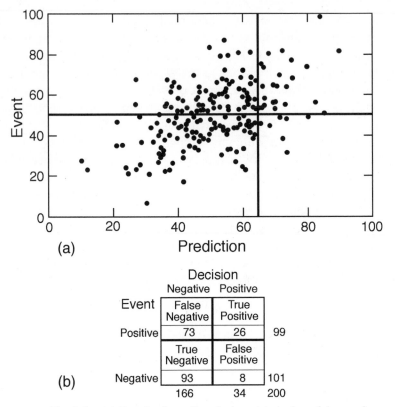

Fig. 3.5   (a) Result of a policy designed to reduce false positives. (b) Decision table.

consequences are predictable: Those who are most concerned about false negatives will mobilize their efforts to have the cutoff moved back to the left, and they will succeed, because they will be more motivated to be politically active than those who are not adversely affected by false negatives.

Thus, the Taylor-Russell diagram illustrates the duality of error and the inevitable tradeoffs between the two kinds of errors. Given a particular level of uncertainty, it is not possible to reduce one kind of error without increasing the other. This is a key feature of many policy problems beyond those involving prediction. For example, many medical issues, such as the age at which women should have regular screening mammograms and policies regarding screening for prostate cancer, have this quality. It can be also found in policies as diverse as affirmative action and airline security. The criminal justice system has struggled with this problem for centuries. How many guilty should go free (false negative) in order to prevent one innocent person from going to prison (false positive)? Of course, all problems involving the use of prediction in the earth sciences involve uncertainty and, therefore, tradeoffs between false positives and false negatives.

Reducing the Uncertainty in Prediction

Hammond (1996) argues that each of the four regions in the Taylor-Russell diagram will develop its own political constituency, and that, in the presence of irreducible uncertainty, the decision cutoff will cycle back and forth over time. Assuming that a prediction is necessary for decision, the only way to break the cycle, and reduce both types of error at the same time, is to reduce the uncertainty in the forecast. This is illustrated in figure 3.6.

When the amount of uncertainty is reduced greatly (here, the correlation has been increased to .90), there is a dramatic reduction in both false negatives and false positives. Hence, all constituencies are happier, conflict is reduced, and there is less pressure for the constant back-and-forth cycling of the decision cutoff. Here is one reason policy makers frequently support research to reduce uncertainties.

An obvious question, then, is how one would go about reducing the uncertainty in the forecast? Or, equivalently, how would one go about increasing the accuracy of forecasts? A related question is when should policy makers seek alternatives to forecasts? Of course, much of the research in the earth sciences is focused, directly or indirectly, on improving predictive accuracy in order to improve the potential for decision making (see chapter 1). Better monitoring technologies provide better information, and better understanding of physical processes can lead to improved methods for combining information to produce predictions. Extensive modeling efforts can systematically incorporate both improved information and improved science into improved products of prediction. But such efforts, as important as they are, can never eliminate all the uncertainty in forecasting, for reasons vividly illustrated by the case studies in this book.

This irreducible uncertainty creates the need for *judgment* (Hammond 1996). As indicated above, people must exercise judgment when they are confronted with several items of information and have to produce a single prediction (in the case of judgments about future events) or diagnosis (in the case of judgments about current events). As Hammond argues, judgment involves a *combination* of intuition and analysis. Uncertainty and judgment are, therefore, common elements in all cases involving scientific input to complex policy issues. Judgment can be a source of inaccuracy in predictions, and a better understanding of the judgment process and its role in the prediction enterprise can contribute to the improvement of predictions, as well as their use by policy

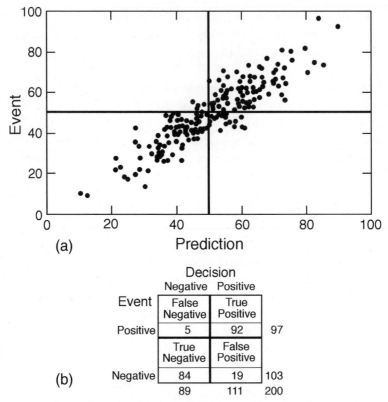

Fig. 3.6   (a) Taylor-Russell diagram with reduced uncertainty.
(b) Decision table.

makers. The next section provides a brief overview of research on judgment.

Research on Judgment

A substantial body of research on human judgment has been generated by multiple disciplines, including psychology, sociology, and economics. The literature is diverse, and any summary of it will reflect the convictions of the summarizer. In my view, the fundamental characteristics of human judgment can be summarized as follows:

1. Humans possess substantial cognitive competence, but human cognitive performance is often poor.

2. Poor performance can be traced to the environment in which decisions are made.

3. The best opportunities for improving performance are in improving the decision environment.

This expresses a view of judgment that is based on the psychological theories of Egon Brunswik (1952, 1956) and Kenneth Hammond (1996). I will elaborate briefly on each with special emphasis on how the decision environment influences judgment.

Human Cognitive Competence Is Substantial, but Human Cognitive Performance Is Often Poor

This requires little elaboration. One needs only to point to the record of human accomplishments in the sciences and the arts, as well as everyday events, such as solving the complex scheduling problems faced by two-career couples with children. All these events point to the potential we have for performing impressive mental feats.

The other side of the coin is that we don't always perform up to our potential. Human error is the cause of many accidents and disasters. There is a large body of research in psychology showing that judgments about uncertainty are suboptimal (Kahneman, Slovic, and Tversky 1982; Plous 1993). Furthermore, poor performance is not limited to laypeople. It has been observed in highly respected professionals, including doctors, accountants, and scientists.

Poor Performance Can Be Traced to the Environment In Which Decisions Are Made

If cognitive competence is so impressive, why is performance often poor? The answer is that we are forced to operate in environments that do not foster optimal performance. By the "environment," I mean the context in which people exercise judgment—that is, the situation or context in which predictions are made. In other words, the problem of poor performance is not completely a problem of limited human ability. It results from a combination of human abilities and properties of the environment.

Three elements constitute the environment for making predictions. One is the nature of the system that is the object of prediction. In the case of weather prediction, that system is the solar/earth/atmosphere/ocean system that produces weather. In the case of earthquake prediction, it is the geological system that produces earthquakes. Those systems are governed by physical laws; scientists have developed various representations of the laws that govern the systems, and those representations may be useful in predicting behavior of the systems. One characteristic of all such systems is uncertainty, both aleatory and epistemic. That is, even if scientists could have perfect information about current conditions, perfect predictions would be impossible. Remember that uncertainty means that, given current knowledge, more than one event is possible. The system that is the focus of predictions is called the *external environment* to distinguish it from the environment for decisions.

The second element that constitutes the environment for prediction is the system that brings information about the external environment to the people making predictions. Generally, this is an information system, which includes instruments, observers, data links, and various displays of data either on paper or computer screens. Information about current conditions is never perfect. Instruments are not perfectly precise, and observations are not continuous in space or time. Therefore, the information systems introduce further uncertainty into predictions. In combination, the uncertainty introduced by the external environment and the uncertainty introduced by the information system make up the *task uncertainty*. Task uncertainty is that component of prediction uncertainty that is beyond the immediate control of the person making a prediction. That is, no matter how intelligent or well trained or experienced people are, when asked to make a prediction, their performance is limited by uncertainty in the external environment and limitations in the information system.

The third component of the environment for making predictions is the procedural, social, and bureaucratic context. Procedures for making forecasts may be specified (e.g., the order in which information is gathered, or the specific calculations or algorithms that are used). The prediction process may be influenced by various policies (e.g., to reduce the number of false alarms). The person making predictions might also be operating under various social conventions, restrictions, or incentives. For example, predictions may be influenced by praise or criticism received for recent successful or unsuccessful predictions. Stress, time pressure, and fatigue are other components of the environment that might influence the accuracy of predictions.

Taken together, the three elements of the environment (external environment, information system, context for making predictions) produce uncertain predictions. From the perspective of the decision maker using predictions, uncertainties reflect the fact that there is more than one possible outcome associated with a specific prediction. Uncertainty in predictions leads to errors and inaccuracy. Among the environmental factors that can affect the uncertainty in predictions, two of the most important are feedback and task uncertainty.[1] These concepts provide powerful tools for explaining why experts working in one domain may make more accurate forecasts than equally qualified experts working in another domain. Each will be discussed briefly below.

Availability of Feedback.

The availability of adequate feedback (knowledge of results) is a fundamental determinant of the quality of cognitive performance. In order to learn from experience, people need sufficient feedback about the accuracy and consequences of their judgments. Weather forecasters receive such feedback for certain daily routine weather events (e.g., temperature and precipitation); they also improve with experience (Roebber and Bosart 1996), and, when task uncertainty is low, their performance is outstanding (Stewart, Roebber, and Bosart 1997). In the case of climate change, by way of contrast, there is very little opportunity for feedback and therefore for learning from experience.

Medicine provides many examples of situations that provide poor feedback. Often, physicians make a diagnosis, and then the patient goes away and gets better. They never find out whether their diagnosis was correct or not. This is obviously a poor environment for learning. In some medical fields, however, knowledge of results is often acquired. One example is anesthesiology, where substantial progress has been made in reducing errors (Gawande 1999).

In the earth sciences, feedback of results is generally acquired for those events that occur relatively regularly, such as severe weather, hurricanes, and floods. For rare events, such as earthquakes, feedback occurs (that is, earthquake detection is easy), but the rarity of the phenomenon limits opportunities for learning. For complex evolutionary processes that occur on large temporal and geographic scales, such as acid rain and global climate change, little or no feedback is possible. Thus, the scale of the event, its familiarity, and the frequency with which it occurs determine whether feedback is potentially available. In cases where feedback is limited or nonexistent, scientists may be reluctant to make predictions, but they will be pressured to do so because policy decisions cannot wait, and policy makers need the judgments of informed experts. In such cases, scientists must rely on "coherence" to make their judgments (Hammond 1996). In other words, the best they can do is make logically coherent judgments that are consistent with what is known about physical and biological processes. Their judgments and predictions should not violate any known natural laws.

Unfortunately, logical coherence and consistency with natural laws leave room for a wide range of predictions. This creates ample opportunities for informed experts to disagree, and such disagreement is common (Mumpower and Stewart 1996). Expert disagreement leaves policy makers in a quandary and often results in opposing, equally credible experts effectively canceling each other out.

Task Uncertainty.

High task uncertainty (resulting from uncertainty in the external environment and uncertainty introduced by the information system) leads to poor cognitive performance. Although it is difficult to make quantitative comparisons of uncertainty among different problem domains, it might be reasonably argued that uncertainty is lower in the case of floods than in the case of an asteroid impact threat or global climate change. In North Dakota, there is experience with floods. There is a historical record on which to base judgments about the flood events and their probabilities as well as their effects. Such is not the case for asteroids or global climate change.

Task uncertainty is important for two reasons. First, it puts an upper bound on the accuracy of forecasts. No forecaster, no matter how wise and well informed, can overcome the inherent uncertainty in the decision environment. Second, for those tasks for which feedback is available, forecasters can learn to cope with the uncertainty in the decision environment. They have learned to make decisions in an uncertain world, and that experience in turn improves judgment.

One of the most important effects of uncertainty is on the reliability (i.e., repeatability or consistency) of forecasts. Paradoxically, task uncertainty can make feedback appear at times erratic, unstable, and arbitrary. In such cases, feedback can actually make a person's judgments less reliable, that is, more erratic, unstable, and arbitrary (Stewart in press; Stewart et al. 1997). As a result, the person will make even more errors than he or she would make without feedback, and the reliability of the resulting predictions may decline.

Assessing Uncertainty.

Since uncertainty has such profound consequences for those who make predictions and for those who use them, the measurement of uncertainty should play a critical role in the use of predictions in decision making. For example, in some cases there is so much uncertainty associated with a particular prediction that the prediction should be ignored. Assessing uncertainty can be extremely difficult, however. In some situations, assessing the uncertainty associated with a prediction will be more technically complex than making the prediction.

When sufficient feedback is available, it is generally a straightforward matter to assess uncertainty. Data on past events and past predictions (if a sufficient number of cases is available and the data can be assumed to generalize to future predictions) yield quantitative estimates of uncertainty (see, for example, Murphy and Daan 1985). Weather forecasting epitomizes these conditions (see chapter 4 in this volume).

When appropriate data are not available, the assessment of uncertainty becomes a judgment. The information used to make a judgment of the amount of uncertainty includes: (a) the amount of disagreement among experts, (b) assessment of whether experts have had an opportunity to learn from experience, (c) the amount of aleatory uncertainty (random but statistically characterizable processes), (d) the amount of epistemic uncertainty (missing information, incomplete understanding of important processes, or even expert convergence on the wrong answer; e.g., see chapter 10, this volume). Unfortunately, given the complexity of these indicators, it would take an expert to make an assessment of the amount of uncertainty in a prediction, and different experts might well, and often do, come to different conclusions. This problem suggests, however, an increasingly important role for scientists: estimating the amount of uncertainty in predictions.

The Best Opportunities for Improving Performance Are in Improving the Environment

Changing the decision environment is the best way to improve human cognitive performance. This is not to deny the role of training and experience, which are obviously critical for performance. But since training and experience are so important, they get a lot of attention, and gains from more training and more experience are likely to be slight. In many cases, however, little attention has been paid to the environment in which decisions are made.

One exception is weather forecasting, where changes to the decision environment resulted in dramatic improvements in forecasting skill during the twentieth century. Although weather forecasters certainly know more about atmospheric processes at the beginning of this century than they did at the beginning of the last, a large portion of the increase in prediction skill can be attributed to the availability, in the weather forecasting environment, of both better information from improved instrumentation (e.g., satellites, radar) and aid in using that information (guidance from weather forecasting models).

Determining exactly how to change a particular decision environment requires a detailed analysis of that environment. Such an analysis addresses questions about the kind and amount of information available, how that information is organized, what the rewards or penalties are for good and bad decisions, and requirements for justifying decisions, as well as many other factors. Analysis of the environment for judgment and decision making is the first step in improving the judgmental component of prediction. It can also be useful in identifying potential improvements in the judgments and decisions of those who use predictions.

Conclusion

The duality of error, the tradeoffs between different kinds of errors, and the resulting conflicts among various political players are major impediments to the use of scientific predictions in policy decisions. These impediments are a result of the uncertainty inherent in any prediction. Some of that uncertainty may be reduced through improvement in information systems, understanding of physical processes, and models for processing information. It may further be reduced by taking action that can support the judgment process—especially by improving the decision environment. Even so, some uncertainty will remain in every prediction. Although uncertainty can never be eliminated, we should make use of every available tool for assessing it, reducing it, and coping effectively with what remains.

Notes

1. Many have been studied, but no coherent, comprehensive "theory of the task" has been developed. Cognitive continuum theory, developed by Hammond (1996), is one ambitious attempt.

References

Brunswik, E. 1952. *The conceptual framework of psychology*. Chicago: University of Chicago Press.

Brunswik, E. 1956. *Perception and the representative design of psychological experiments* (2nd ed.). Berkeley: University of California Press.

Clemen, R.T. 1990. *Making hard decisions*. Boston: PWS-Kent.

Gawande, A. 1999. When doctors make mistakes. *The New Yorker* 74(Feb. 1): 40–55.

Hammond, K. R. 1996. *Human judgment and social policy: Irreducible uncertainty, inevitable error, unavoidable injustice*. New York: Oxford University Press.

Kahneman, D., P. Slovic, and A. Tversky, eds. 1982. *Judgment under uncertainty: Heuristics and biases*. New York: Cambridge University Press.

Morgan, M.G., and M. Henrion. 1990. *Uncertainty: A guide to dealing with uncertainty in quantitative risk and policy analysis*. New York: Cambridge University Press.

Mumpower, J.L., and T.R. Stewart. 1996. Expert judgment and expert disagreement. *Thinking and Reasoning* 2: 191–211.

Murphy, A.H., and H. Daan. 1985. Forecast evaluation. In *Probability, statistics, and decision making in the atmospheric sciences*, A.H. Murphy and R.W. Katz, eds. Boulder, CO: Westview Press, pp. 379–437.

Plous, S. 1993. *The psychology of judgment and decision making*. New York: McGraw-Hill.

Roebber, P.J., and Bosart, L.F. 1996. The contributions of education and experience to forecast skill. *Weather and Forecasting* 11: 21–40.

Stewart, T.R. In Press. Improving reliability of judgmental forecasts. In *Principles of forecasting: A handbook for researchers and practitioners*, J.S. Armstrong, ed. Norwell, MA: Kluwer Academic Publishers.

Stewart, T.R., and Lusk, C.M. 1994. Seven components of judgmental forecasting skill: Implications for research and the improvement of forecasts. *Journal of Forecasting* 13: 579–599.

Stewart, T.R., W.R. Moninger, K.F. Heideman, and P. Reagan-Cirincione. 1992. Effects of improved information on the components of skill in weather forecasting. *Organizational Behavior and Human Decision Processes* 53: 107–134.

Stewart, T.R., P.J. Roebber, and L.F. Bosart. 1997. The importance of the task in analyzing expert judgment. *Organizational Behavior and Human Decision Processes* 69(3): 205–219.

Taylor, H.C., and J.T. Russell. 1939. The relationship of validity coefficients to the practical effectiveness of tests in selection: Discussion and tables. *Journal of Applied Psychology* 23: 565–578.

PART TWO

Disasters Waiting to Happen: Predicting Natural Hazards

■ t might seem that the relation between scientific predictions and natural hazards is just a matter of common sense: predict the hazard and evacuate the area. But accurately predicting natural hazards is a tough scientific assignment. And ordering an evacuation is the kind of thing that can get a mayor unelected (especially if the science turns out to be wrong). Things without legs or wheels, of course, cannot be easily evacuated—so perhaps we should worry less about predicting floods with science, and more about controlling them with dams? Less obviously, different types of decision makers may be looking for different types of predictions. A perhaps apocryphal story attributes to former American Airlines president Robert Crandall the following quote: "I need accurate predictions of good flying weather." Next time your flight gets cancelled due to a blizzard, you'll know who to blame: the weather forecaster.

The National Weather Service generates about 10 million weather predictions each year. Not surprisingly, the weather service has gotten pretty good at it—but not necessarily for the reasons that one might think. Scientists have gotten pretty good at flood predictions, too. So why do the damage costs from floods and other weather-related hazards keep increasing? The costs of asteroid impacts, on the other hand, have remained at about zero, but they could, if we're unlucky, rapidly rise to the equivalent of the entire economic output of the world. We think we're pretty good at predicting asteroids, but given that we've had no major impacts and only one significant public prediction—which was wrong—there's not much basis for evaluation. Evaluating earthquake prediction, on the other hand, is easy: we don't know how to do it. So, instead, we try to design buildings that can resist earthquake shaking. Still, one might ask: Why do people insist on living in floodplains and near fault zones? If people didn't move to places like the Mississippi Valley and California, we wouldn't have to worry about predicting floods and earthquakes. (And if people didn't live on earth, we wouldn't have to worry about predicting asteroid impacts.)

Short-Term Weather Prediction: An Orchestra in Need of a Conductor

William H. Hooke and Roger A. Pielke, Jr.

In 1863, in the midst of the Civil War, an aspiring weather forecaster, Francis Capen, approached President Lincoln proposing to supply weather forecasts to the Union Army. President Lincoln was not convinced:

> It seems to me that Mr. Capen knows nothing about the weather, in advance. He told me three days ago that it would not rain again till the 30th of April or 1st of May. It is raining now [April 28th] and has been for ten hours. I can not spare any more time to Mr. Capen. (Whitnah 1961, pp. 14–15)

Since that time, the role of weather forecasts in the United States, and indeed around the world, has changed dramatically.[1] Weather forecasts are today a fundamental component of modern life. Forecasts of routine weather are used by individuals to plan what to wear and to schedule activities. They are an essential element in daily decisions in a range of sectors that span the economy: from electric utilities to aviation to insurance (Pielke 1997). Forecasts, and related warnings, of pending extreme weather, like tornadoes or floods, are a centerpiece of emergency response. Weather forecasting has been shown to have predictive "skill" and in many cases provides great benefits to users (Katz and Murphy 1997).[2] Indeed, by way of contrast with other predictive sciences discussed in this volume, the ability of scientists and policy makers to document long-term improvements in skill and related value of weather predictions is unique. At the end of the twentieth century, the greatest challenge facing the weather forecasting community, and those who depend on its products, is to improve the linkage of advances in predictive knowledge with the decision needs of users.[3]

The exchange between Capen and Lincoln well captures the relationship of meteorologists, the public, and the federal government. Meteorologists have sought federal support of their science, promising in exchange benefits to society. For its part, the public, and its elected representatives, have evaluated weather forecasts based on their vast experience with them. For the most part, in recent decades the outcome has been more favorable for the weather community than it was for Mr. Capen. In recent decades the U.S. government has spent tens of billions of dollars on the nation's forecasting enterprise. This includes support for research in federal agencies, operational forecasting, and data gathering (e.g., with radars and satellites) and processing. At the same time, the private sector has developed a robust industry that distributes weather information. Weather predictions are provided to a wide range of interested parties that includes individuals, the media, government agencies, and private corporations. These various players and perspectives have come together in the modern weather prediction system that we see today.[4]

This chapter has three purposes. First, it provides an overview of weather prediction in the United States. Second, it uses three cases—hurricane, tornado, and flood forecasting—to illustrate the challenges to achieving greater benefits from weather forecasts. Third, it discusses at a general level steps needed to overcome those challenges and concludes with a vision for the future of weather services as the nation enters the twenty-first century.

Weather Prediction in the United States

Weather forecasts are the result of measurements and other kinds of information that characterize the state of the atmosphere; computer models based on scientific understanding and statistical relationships; and human judgment. Weather forecasts have been produced for many decades, and thus a considerable body of theory and experience can be used to assess their accuracy. In the early development of chaos theory, Lorenz showed in the 1960s that weather prediction is inherently limited because small errors in knowledge of the current state of the atmosphere propagate into large differences in future states, a phenomenon popularly called the butterfly effect. Because of this inherent uncertainty, scientists estimate that the fundamental limit of weather prediction is about two weeks. Because of the uncertainties associated with predicting the weather, forecasts are inherently *probabilistic*, meaning that numerous events could occur, given a particular forecast. Forecasts are sometimes, but not always, expressed probabilistically (e.g., 70 percent chance of rain).

Institutional Bases

Weather prediction is characterized by some interesting and unique attributes. First, there is the sheer number of short-term weather predictions issued. Within the United States every day, over one hundred National Weather Service (NWS) local forecast offices produce for public consumption a wide range of products, totaling some *24,000 predictions per day or ten million predictions each year*. Second, the historical improvement in short-term weather prediction has been directly related to supporting research (AMS 1991). There is a long track record of forecast improvements, and there is a high degree of community consensus on how to achieve continuing forecast improvements in the near term (NRC 1998). By contrast, examples such as earthquake prediction or beach erosion modeling show that the historical relationship between research and forecast skill is not always obvious, nor is there necessarily agreement among researchers on how best to proceed (see chapters 7 and 8).

Given the large number of short-term predictions and societal contexts available for study, a third attribute of weather prediction is that the opportunities for *evaluating* the societal benefits of improved forecasts are diverse, plentiful, and often very public (e.g., Katz and Murphy 1997). By contrast, the decades or more that it takes for events such as major asteroid impacts and global change to occur make evaluation of the forecasts, much less their actual value, impossible (see chapter 16). Fourth, decision makers find weather information addictive; i.e., as the quality of weather forecasts and their utility have advanced, user demands for information have expanded commensurately (NRC 1999). This in part results from a well-understood tendency of organizations to "systematically gather more information than they use" (Feldman and March 1981). But another reason for this addiction is that, historically, those making weather-sensitive decisions have not had reliable forecasts. As a result they've tended to make decisions that minimize the maximum losses they could face should the weather prove unfavorable, rather than seek to optimize their decisions to take advantage of favorable opportunities. As the accuracy of forecasts has improved, and the time horizon has lengthened, it has been gradually possible to shift to a more aggressive approach to decisions.

Historically, weather prediction has been the province of national governments and international bodies such as the World Meteorological Organization. Modern-day meteorology dates from the invention of the telegraph, which provided the means for scientists to construct a comprehensive "picture" of the weather at regional and global scales. This allowed the tracking of weather systems over time. The modern functional equivalents to the telegraph are technologies such as satellites and radars, which allow for the tracking of weather by synoptic observation. High-speed computing in the second half of the twentieth century has enabled the prediction of the evolution of weather patterns and specific phenomena from hours to days in advance.

The federal government has organized the nation's weather prediction responsibilities into two related areas. First,"operations" are the basis for production and dissemination of official forecasts and warnings. (Operational services are also divided between public-sector predictions, both civilian and military, and private-sector, value-added dissemination and prediction services.) Annually, federal operational weather services in 1999 spent more than $2.2 billion (table 4.1). Related technological development activities—radars, satellites, etc.— account for another $500 million a year (NRC 1998). Private-sector expenditures, including those for the operations of broadcast meteorologists, total more than $1 billion annually (Hererra 1999). Second, research, systems development, and technology development and implementation are supported to improve the skill of predictions. These activities are in some cases tightly coupled to operational efforts, while others have a weaker connection. Research is carried out within both federal laboratories and universities. Research and development, whether in federal laboratories or universities, are largely supported by the federal government and amount to about $500 million each year (OFCM 1998). Thus, on an annual basis resources devoted to the nation's weather forecasting system total more than $4 billion dollars.[5] This system supports many weather-related decisions. Estimates suggest that $1 trillion of the nation's $7 trillion economy is weather sensitive (NRC 1998).

TABLE 4.1

Federal budget for meteorological operations and supporting research, FY 1999 (in thousands of dollars).

Agency	Operations	% of Total	Supporting Research	% of Total	Total	% of Total
Agriculture	$ 12,600	0.6	$ 15,500	4.0	$ 28,100	1.1
Commerce	$1,303,450	59.0	$ 70,768	18.1	$1,374,218	52.9
Defense	$ 438,228	19.9	$ 87,013	22.3	$ 525,241	20.2
Interior	$ 800	0.0	0	0.0	$ 800	0.0
Transportation	$ 448,648	20.4	$ 13,955	3.6	$ 462,603	17.9
EPA	0	0.0	$ 5,700	1.5	$ 5,700	0.2
NASA	$ 2,963	0.1	$197,095	50.5	$ 200,058	7.7
NRC	$ 110	0.0	0	0.0	$ 110	0.0
TOTAL	$2,206,799	100.0	$390,031	100.0	$2,596,830	100.0

As weather prediction capabilities have matured in recent years, a broader spectrum of weather-sensitive industrial sectors has increasingly incorporated weather forecasts on all time scales into its decision making (Pielke 1997). Box 4.1 shows examples of weather impacts on some industries and how improving forecasts could reduce those impacts. In response, the value-added private sector has also matured, building on the government's provision of observations and global models to create specialized weather services for different business interests and other uses. Because of the changing nature of technology and user demands, public and private responsibilities are frequently debated, leading to an uneasy relationship (see e.g., Leavitt 1997).

For many decision makers, alternatives to weather prediction (see, e.g., chapter 14) include a wide variety of practices to make human activities as robust as possible with respect to the variability of weather conditions. For example, South Florida has adopted a building code for wind that is among the strongest in the nation. It has arguably limited that region's total hurricane impacts (Pielke and Pielke 1997). Farmers have the option of planting a mix of crops to ensure some production, whether the summer growing season proves hot or cold, dry or wet, or somewhere in between. More recently, agribusiness, utilities, the recreation sector, and other weather-sensitive industries have turned to weather derivatives and other financial instruments to hedge their seasonal and longer-term weather risks (Salpukas 1999).

The federal role in weather prediction is well codified and clearly stated (for an overview, see U.S. Congress 1979). The nation's policy goals for weather prediction are captured in the 1890 law creating the National Weather Service, which assigned to the NWS responsibility for:[6]

> forecasting of the weather, the issue of storm warnings, the display of weather and flood signals for the benefit of agriculture, commerce, and navigation, the gauging and reporting of rivers, . . . the reporting of temperature and rainfall conditions, . . . the distribution of meteorological information. (15 USC Sec. 313)

Subsequent laws direct the NWS to:

> furnish such weather reports, forecasts, warnings, and advices as may be required to promote the safety and efficiency of air navigation in the United States and above the high seas . . . [and] study fully and thoroughly the internal structure of thunderstorms, hurricanes, cyclones, and other severe atmospheric disturbances . . . with a view to establishing methods by which the characteristics of particular thunderstorms may be forecast.

BOX 4.1

The effects of weather and the potential value of improved weather information for different industrial sectors:

Oil and gas exploration and production

Improved forecasts of tropical weather conditions (wind, waves) could reduce delays in drilling operations at a cost of up to $250,000 per rig per day (there are several thousand rigs in the Gulf of Mexico).

Improved hurricane track predictions could reduce days of production shutdown, each day of which costs the industry and the U.S. Treasury a combined $15,000,000.

Vegetable processing

Improved temperature and precipitation forecasts could lead to greater efficiency in chemical spraying (e.g., pesticides), which costs $10–$15 per acre per application for hundreds of thousands of acres.

On a national scale the annual cost of lost production to the vegetable processing industry, primarily due to weather, is $42,500,000.

Insurance

A single hurricane can lead to more than $80,000,000,000 in damages.

Weather-related catastrophes have led to more than $48,000,000,000 in property insurance claims over the period 1989–93.

Rail transportation

It costs $2,000 per hour to stop a train; a single tornado warning covering fifteen miles of track for fifteen minutes can lead to seven stopped trains.

Most weather-related derailments cost between $1,000,000 and $5,000,000.

Electric power

Improved thunderstorm forecasts could save a single utility $200,000 annually in reduced outage time.

"Good quantitative precipitation forecasts" could save a single utility $2,000,000 over five years.

Improved temperature forecasts could save "hundreds of millions annually nationwide for the utility sector."

Aviation

Every avoided cancellation saves $40,000; every avoided diverted flight saves $150,000.

For the sixteen members of the Air Transport Association, delays and cancellations cost $269,000,000 annually.

The federal government has also expressed its sense that:

> a reliable and comprehensive national weather information system responsive to the needs of national security; agriculture, transportation, and other affected sectors; and individual citizens must be maintained through a strong central National Weather Service that can work closely with the private sector, other Federal and State government agencies, and the weather services of other nations.

In short, weather predictions, and improved use of weather predictions, directly support the following policy goals: safer communities in which to live and work; job growth and economic prosperity; and protection of national security. The clarity, strength, and consistency of this mission is arguably a critical factor in the successes of the weather prediction system.

The Forecast System as a Symphony Orchestra

One of the most important criteria for evaluating weather forecast products is accuracy, defined as the difference between what is forecast and what actually occurs. The more accurate a forecast is, the greater its potential value to a decision maker. Scientists have invested considerable effort in evaluating the accuracy of forecasts, although some have suggested that additional efforts are needed (Doswell and Brooks 1998). The results have shown a long-term improvement in the ability of forecasters to predict the weather, as measured by objective criteria (e.g., NRC 1999). According to the American Meteorological Society (AMS 1991, p. 1273):

> The notable improvement in forecast accuracy that has been achieved since the 1950s is a direct outgrowth of technological developments, basic and applied research, and the application of new knowledge and methods by weather forecasters. High-speed computers, meteorological satellites, and weather radars are tools that have played major roles in improving weather forecasts.

Of course, weather forecasts are far from perfect, but their accuracy (or relative inaccuracy) can be quantified reliably. This track record sets weather forecasting apart from all other predictive earth sciences.

Predictions are produced in the environment of a broader *prediction process*, which includes the production of forecasts, but also communication of forecast information and the incorporation of that information in user decisions. The process might be thought of as a symphony orchestra in which the different sections must work together harmoniously to produce music (Drucker 1993). The analogue to music in the forecast process is effective decision making with respect to weather. Often, some mistakenly ascribe a linear relation to the three subprocesses, i.e.:

$$\text{predict} \rightarrow \text{communicate} \rightarrow \text{use}$$

These three subprocesses are instead better thought of as occurring in parallel, with significant feedbacks and interrelations between them. Table 4.2 illustrates the elements of the forecast process and the outcomes associated with each element.

An accurate forecast is insufficient for effective decision making (Katz and Murphy 1997).[7] More generally, success in any one of the three subprocesses does not necessarily result in benefits to society (e.g., Pielke, 1999, 2000; Doswell and Brooks 1998; Roebber and Bosart 1996; Vislocky, Fritsch, and DiRienzo 1995). A technically skillful forecast that is miscommunicated or misused can actually result in costs to society. Similarly, effective communication and use of a misleading forecast can lead to decisions with undesirable outcomes. For the process to work effectively, success is necessary in all three elements of the forecast process: prediction, communication, and use. Further, success requires healthy connections between the elements; they cannot be considered in isolation, i.e., with the tasks of prediction, communication, and use delegated to isolated or poorly connected parties. Integration of evaluation methods is a necessity. At the interfaces of the elements lie several questions that ought to be asked and answered in a healthy forecast process:

TABLE 4.2

Evaluation methods for elements of the prediction process.

Element of prediction process	Outcome	Criteria of evaluation	Methods of evaluation (example reference)
Prediction	Forecast products	Skill, quality, etc. (Murphy 1997)	Verification
Communication	Guidance	Information transfer (Sorensen 1993)	Survey, interview, etc.
Use	Decisions	Value	Prescriptive/descriptive decision studies, etc. (Stewart 1997; Wilks 1997; Changnon 1997)

Between Prediction and Use

- What ought to be predicted?
- How are predictions actually used?

Between Prediction and Communication

- What does the prediction mean in operational terms?
- How reliable is the prediction, and how is uncertainty conveyed?

Between Use and Communication

- What information is needed by the decision maker?
- What content or form of communication leads to the desired response?

Because the forecast process is composed of multiple elements, no single measure captures the societal "goodness" of a forecast process. Instead, multiple measures are needed to evaluate the technical, communication, and decision dimensions of forecasts. Table 4.2 also summarizes some of those measures. Typically, policy makers have focused attention on the economics of forecasts in order to determine a bottom-line assessment of value, while social scientists have studied the communication process (e.g., warnings) and physical scientists have evaluated forecasts according to technical criteria like skill scores and "critical success indexes." These different foci are clearly important and necessary; however, the segregation of evaluation tasks has meant that no one is responsible for evaluation of the entire forecast process. The result is that we try to improve the system by working on its components while ignoring critical interactions visible only from a more comprehensive perspective.

Consequently, when policy makers or other users of weather forecasts ask the general question "What is the value of an improved forecast?" and expect to get an aggregate answer in dollars or lives, they ask the wrong question. They ought instead to ask "What changes to the existing forecast process (predict, communicate, use) can we expect to lead to better outcomes?" and expect the answer to be contextual, multidimensional, and subjective. Consider the following examples in hurricane, tornado, and flood forecasting.[8]

Case: Hurricane Forecast Improvements

Forecasts of hurricane motion have become increasingly accurate in recent decades, but it is not clear how those improvements have led to improved decision making. Figure 4.1 shows the improvement in hurricane track predictions for 24-, 48-, and 72-hour forecasts. But at the same time, the actual length of coastline warned per storm by the

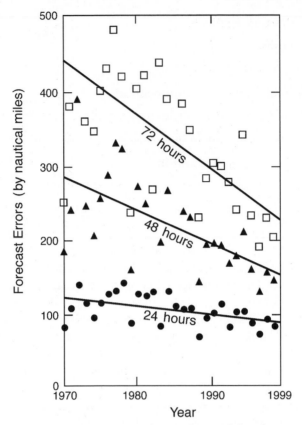

Fig. 4.1   Improvement in accuracy of hurricane track forecasts, 1970–98. Circles are 24-hour forecasts, triangles are 48-hour forecasts, and squares are 72-hour forecasts. (Data courtesy of Tropical Prediction Center and National Hurricane Center)

National Hurricane Center (NHC) has increased from less than 300 nautical miles (nm) in the late 1960s to about 400 nm over the past ten years, as shown in figure 4.2. This at first seems counterintuitive because if the hurricane track is better known, it would seem that areas believed not to be at risk (because of improved track predictions) would not need to be warned. The length of coastline warned is important because it dictates how many people will be ordered to evacuate and will otherwise choose to prepare for the storm.

According to Jarrell and DeMaria (1999), "The increase is somewhat surprising, because, since 1970 . . . [errors in] official NHC track fore-

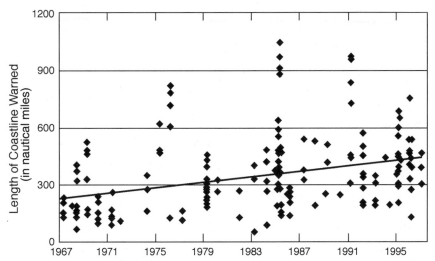

Fig. 4.2   Scattergram and linear trend line of the length of the U.S. coastline included in hurricane warnings 1967–97. (Data courtesy of Tropical Prediction Center, National Hurricane Center, and NESDIS/CIRA, as shown in Jarrell and DeMaria 1999)

casts have been decreasing at about 1% per year." Jarrell and DeMaria speculate that the improvement in track forecasts has translated into longer lead times for evacuation decisions. Lead time, from the time the first warning is issued to the time that the storm's center crosses the coast, has increased from about 18 hours to 24 hours. Increased lead time results in more miles of coastline warned because of the larger uncertainty associated with predictions made for longer periods, as shown in figure 4.1. This could be an example of the forecast process working in a healthy manner—i.e., for many decision makers an increase in lead time could be a worthwhile tradeoff with miles of coastline warned.

But there are other possible explanations for the upward trend in miles of coastline warned, including (a) the desire of emergency managers (and elected officials) to shift accountability to other sources by requesting that NHC warnings be extended to cover their communities; (b) a desire throughout the evacuation decision process to avoid the error of a strike on an unwarned population (thus, translating the forecast improvement into lower risk); and (c) the fact that more and more people inhabit the coast, meaning that evacuation times are much greater, making necessary longer lead times and lengths of coastline warned. Unfortunately, despite these hypotheses, it has not been con-vincingly demonstrated why the length of coastline warned per storm has increased during a period of decreasing forecast errors and whether this increase confers a net benefit or cost on society. In this instance, no one has assessed whether the orchestra is in fact making music.

Given the large costs involved with overwarning, both in unnecessary preparations including evacuations and in potential public disgust with false alarms, causing them to ignore subsequent warnings (see Dow and Cutter 1998), it would seem to be in the best interests of forecasters, pol-icy officials, and the general public to obtain a greater understanding of the use of hurricane forecasts. The hurricane research community has made a convincing case that it is well positioned to make dramatic advances in the science of forecasting (Marks, Shay, and Prospectus Development Team #5 1996)—but for forecasts to be effectively used by decision makers, and thus to be of significant benefit to society, there must at the same time be advances in the scientific understanding of how hurricane forecasts are used in the decision-making process.

Case: Tornado Verification Statistics

On Monday, May 3, 1999, more than seventy tornadoes tore through Oklahoma and Kansas, killing forty-six people, injuring scores more, and resulting in more than $1 billion in damage. The National Weather Ser-vice called the outbreak one of the most severe in this country in the past fifty years. According to Doswell (1999), one of the nation's experts on tornadoes, two important factors kept loss of life from being even higher. One was luck—in that the storm did not strike in the middle of the night or cross a busy interstate crowded with rush-hour traffic. The second was that the nation's severe-weather infrastructure performed admirably in a region where people take official warnings seriously and are prepared to respond to them. The event was but one of many that recently have underscored the importance of the nation's forecasting system.

But with increasing importance to the nation comes more visibility and responsibilities, and ultimately greater demands from the public to improve the forecast process. Future improvements in the forecast process, perhaps defined in this case by some measure of lives lost related to tornadoes, depend on knowing the reasons behind past suc-cesses and failures.

Figure 4.3 shows that the nation has gone from an average annual loss of life to tornadoes of more than 300 per year in the 1920s to less than 100 per year in the 1990s (Brooks 1999). The data are even more striking when considered in the context of the nation's population

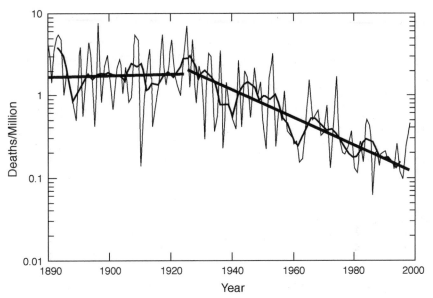

Fig. 4.3   Decrease in incidence of tornado deaths per unit of population, 1880–1998. (Source: Doswell and Brooks 1998)

growth. From 1880 to 1924 the death rate from tornados was about 1.8 per million; by 1997 it had declined to 0.14 per million. Considering that losses have decreased steadily since the 1920s, yet official tornado warnings date only to 1948, Brooks admits:

> I cannot explain the decrease. Lots of things contribute— improved forecasts and warnings, communication of warnings, better housing, the movement of people from rural areas to urban areas, less time being spent outdoors, etc. Before we, in the meteorological and preparedness communities pat ourselves on the back, it is important to note that lightning deaths also show a decreasing trend over this time period, but we do little forecasting or preparedness work on that problem, in comparison to tornadoes.

Others have criticized how the National Weather Service evaluates its severe weather forecasts. (See, for example, Pettit 1999, Brooks 1999.)

If the economic and other impacts associated with tornadoes increase in absolute terms, demands from the public and policy makers for the weather community to do something will also increase. In fact, impacts associated with most types of extreme weather are expected to increase, simply because of the growing population of the nation (see

Kunkel, Pielke, and Changnon 1999). The weather prediction system will be able to respond to such demands more effectively if the people who manage it better understand the relationship between its products and societal outcomes. The National Weather Service could conduct research needed to gain that understanding, or, perhaps more appropriately, an independent entity could oversee the needed studies (cf. chapter 12). The bottom line is that if the nation expects decreased impacts from tornadoes, then it must better understand the relationship of weather forecasts and societal outcomes. Once again we see that little attention has been paid to the overall performance of the "orchestra."

Case: Red River Floods

In April 1997, the Red River of the North, which flows north along the North Dakota–Minnesota border, experienced extreme flooding (this event is also discussed in chapter 5). Damages related to the event have been estimated at $1–2 billion, with most damage occurring in Grand Forks, North Dakota, and East Grand Forks, Minnesota. Through the spring the NWS had been predicting a flood crest (i.e., a maximum river level) of forty-nine feet at East Grand Forks, which would have been a record, but the actual crest was fifty-four feet. This discrepancy contributed to decisions that arguably exacerbated the damages, such as failure to remove personal property from the areas at risk or "sacrificing" parts of the community to allow more water to pass through the town. In the aftermath of the event, considerable attention was focused by policy makers on flood predictions and their role in decisions that preceded the peak flooding (see Pielke 1999 for an in-depth discussion of this case).

Figure 4.4 shows historical overestimates and underestimates of recent Red River flood crests. By historical standards, the forecast of the 1997 flood crest was not unusually inaccurate. Given that the forecast was for a record event, i.e., one for which there was no experience, the forecast was arguably much better than many issued in previous years. In interviews conducted in May 1997 with various decision makers in the Red River of the North basin, it is clear that different people interpreted the NWS prediction in different ways. Some viewed the flood crest forecasts as a maximum, i.e., a value that would not be exceeded. For example, on April 8, 1997, the *Grand Forks Herald* reported that "[NWS] experts are still forecasting a maximum 49-foot crest for the Red at East Grand Forks." Others viewed the prediction as exact, i.e., that the crest would be forty-nine feet. Still others viewed the outlook as somewhat uncertain; examples of the degree of uncertainty ascribed to the prediction by various decision makers ranged from one to six feet.

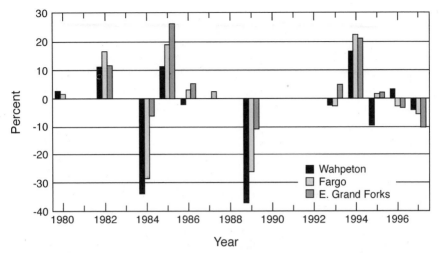

Fig. 4.4   Prediction error at Wahpeton, Fargo, and East Grand Forks, North Dakota, as a percentage. (Source: North Central River Forecast Center)

While scholars have long recognized that communication involves both *sending* and *receiving* information, little attention has been paid to the manner in which flood forecasts are interpreted by decision makers, and subsequently, to the role of this information in the forecast process. In short, it is apparent that the use and value of existing flood forecasts are not well understood, much less the potential usefulness and value that might be attained through "improving" the overall forecast process. Other recent experience suggests that this circumstance may be fairly common in flood forecasting in general (e.g., see chapter 5) and in other areas of prediction as well (e.g., see chapters 6, 8, and 12).

All three of these cases show that the challenge of more effective use of forecasts cannot be solved by simply providing more information, such as by improving the accuracy of existing products or developing new products, e.g., probabilistic forecasts. If decision makers have difficulty using existing products, these difficulties will not go away simply through providing more or "better" information. Indeed, several recent studies of judgment suggest that as the amount of information available increases, the judgment process may actually become less reliable due to information overload, especially in contexts of high uncertainty and high risk such as the forecasting of extreme events like hurricanes, tornadoes, and floods (e.g., see Stewart et al. 1992). More attention must be paid to how forecasts are issued, who actually receives what information, and with what effect.

From Research to Societal Benefit

What stands in the way of achieving improvements in the overall forecast process? The most common answer is that improvements are constrained by limited resources—resources for research, for computers, for forecasters, for training and education, and so on. More research, faster computers, more (and better trained) forecasters can demonstrably lead to better predictive capabilities (e.g., NRC 1999), but better predictive capabilities are at best a necessary but insufficient condition for improving the overall process. It is as if an orchestra's conductor were to focus only on the string section to the exclusion of the horns and percussion. The situation was aptly summarized by a colleague who said, "As forecasters, the gap between what we know and what we communicate to the public has never been larger." But if that view is correct, and society is not fully making use of the knowledge and resources that it already has, policy makers may rightly ask what would be the point of adding more resources to the system, unless at the same time we improve the forecasting process itself.

Twenty years ago, the administrator of the National Oceanic and Atmospheric Administration (NOAA) wrote to the president of the National Academy of Sciences to ask that the academy conduct a study to assess how advances in science and technology could best lead to an improved forecasting system for the nation. In its report (NRC 1980), the academy concluded that there was a need for new observing systems, improved computer systems, and better communication systems. But the report also noted, "The rate of progress toward better services is not limited by technology"—and, by extension, not by resources for technology. Instead, the constraint was an "inadequate mechanism for transferring weather and hydrological information and knowledge of applications to specific users." That statement clearly went beyond matters of *forecast production* to the broader issue of the *forecast process*. It seems clear that—even twenty years after the academy report—the weather community has yet to fully and systematically address the challenge of understanding the factors necessary and sufficient for advances in science and technology to most effectively contribute benefits to society.

The nation has invested considerable resources in the development of understanding and technologies to meet the expected demands of its citizens for improved forecasts. These include the modernization of the National Weather Service and the U.S. Weather Research Program. Regrettably, many of the fruits of those investments have not been transferred via an effective forecast process into useful products for decision makers; hence, benefits to the nation are much less than they might be. Examples include data from satellites and radars that are not fully or

effectively used (Dabberdt and Hales 1998), techniques for the manipulation of data that are understood but not used (Schlatter et al. 1999),
and knowledge of human judgment that is not incorporated into the
development of useful products (Pielke 1997). At the core of such problems is a mismatch between the weather forecasting system's capability
of producing scientific and technological advances and of translating
those advances into useful information. A challenge facing the nation is
thus to implement an improved forecasting process that takes full
advantage of our continuing national investment in observations,
research, and technology. Without enabling an improved forecast
process—spanning research to societal benefit—the nation will not fully
benefit from its ongoing investments, and may thus fail to meet society's
growing expectations of weather prediction.

As the United States moves into the twenty-first century, the nation
expects greater accuracy, timeliness, and reliability in weather forecasts,
as well as an increased number of useful products (NRC 1998). In an
improved forecasting process, such products could play an increasingly
important role in both public- and private-sector decision making. Furthermore, an improved forecast process could provide expanding
opportunities to better protect life and property, stimulate economic
activity, enhance national competitiveness, and contribute to environmental management (NRC 1998).

Because the nation has not fully benefited from its weather investments,
many scientists are concerned that funders will react by reducing support
for research. But this would be a mistake and would run contrary to the
need to view weather forecasting as a prediction process. It would be much
as if the owner of a grocery store with a backlog of bread on its shelves were
to address the problem by telling farmers to stop planting wheat. The issue
is not like regulating flow through a pipeline, but rather like managing
numerous parallel processes to form a coherent whole—like trying to get a
symphony orchestra to produce music. Thus, a central challenge facing the
community is to identify opportunities to improve the existing forecast
process and to recommend alternative courses of action necessary and sufficient to open the way to more effective and efficient capitalization of the
ongoing investment in weather forecast research and technology.

A limitation more fundamental than that of resources is that of leadership. If the forecast process is indeed like a symphony orchestra, then
it suffers from the lack of a conductor. No one has assumed responsibility for the task of improving the forecast process. A committee of the
National Academy of Sciences has arrived at a consistent conclusion in
a broad review of atmospheric research: "No one sets the priorities; no
one fashions the agenda" (NRC 1998, p. 58). While many participants in

the nation's forecasting system agree that the process *should* be improved (by someone), the community has not organized itself to systematically evaluate the existing process and implement improvements. There are candidate conductors, such as the National Weather Service, which issues forecasts, and the U.S. Weather Research Program, which seeks to better understand the science of weather, including impacts, forecasts, and use. But until responsibility to improve the process is assumed, it is likely that a gap will continue to exist, if not broaden, between knowledge of weather forecasting and its effective use. The recommendation of specific steps to improve the forecast process goes beyond the scope of this chapter, but some steps are being taken. The National Research Council, for instance, has begun to study the issue, which marks a significant departure from its traditional focus on natural science research isolated from societal application. However, without recognition that it is the process that needs attention, not simply improvement in forecast products, even the best-intentioned advice is likely to fall short of achieving the nation's potential in outcomes related to weather prediction.

The Future of Weather Services

The following exchange on the subject of hurricane forecasts took place between a member of Congress and a National Weather Service official during a 1993 congressional hearing on the performance of the agency in several instances: Hurricane Andrew in 1992, the 1993 "Superstorm" East Coast blizzard, and the Midwest floods of 1993 (U.S. Congress 1993):

> MEMBER OF CONGRESS: Have we ever undertaken a study that . . . would try to analyze . . . a cost benefit analysis of whether we . . . are actually dollars ahead in terms of the ability to predict whether or not hurricanes are going to hit, when they are going to hit, where they are going to hit, how they are going to hit, how hard they are going to hit? . . . Has it been worth [the investment in prediction]?

> NWS OFFICIAL: Okay. First, you respond for protection of life . . . second, with a warning, you can indeed protect your property so you have much less loss . . .

> MEMBER OF CONGRESS: I understand that. But have we ever done an analysis that would compare the amount of money

that has been spent preparing for hurricanes that have not occurred against the amount of money that would be saved or that is saved in preparing for the one that does occur? . . .

NWS OFFICIAL: . . . Okay. It's a tremendous savings there . . .

MEMBER OF CONGRESS: Wait, excuse me. I am sorry, but my question is: Has that sort of analysis ever been attempted or performed?

NWS OFFICIAL: I don't think it has been done in detail.

The exchange between the member of Congress and the NWS official illustrates the significance of understanding the weather forecast process. The member of Congress wanted to know the relationship between funds appropriated to hurricane forecasting and the outcomes that result, with outcomes defined societally, not technically. Presumably, such information could be used by Congress to help set priorities for weather spending versus the myriad other items on its agenda. Knowledge of the effectiveness of the forecasting process is thus squarely in the public interest.

But the answer of the NWS official is representative of the broader circumstance: policy makers lack information that would allow for systematic, comprehensive answers to questions about the forecast process: "Somewhat surprisingly . . . relatively little attention has been devoted to determining the economic [or other societal] benefits of existing weather forecasting systems or the incremental benefits of improvements in such systems" (Katz and Murphy 1997). As a result, all users and potential users of forecast information suffer. Debate and discussion of the health of the weather forecasting process depends more on unverified anecdotes, optimistic assumptions, and simple politics than on reliable information and systematic analysis (cf. GAO 1996, 1997).

If leadership in the weather prediction community comes forward and implements steps to better understand the forecasting process as an integrated system—as compared to a simple understanding of the discrete elements of that process focused on products—then the following sort of exchange might be expected to occur between a member of Congress and an NWS official at some point in the near future, to the benefit of policy makers, the weather community, and an entire nation that has grown dependent on its weather forecast system:

MEMBER OF CONGRESS: Once again, your agency has come before this committee with a request for a budget increase in order to improve forecasts. Now you realize that each agency under this committee's jurisdiction is asking for an increase. Fiscal realities dictate that budget increases will be difficult. Have you any information that would lead this committee to expect that the increased investment you have requested will yield a societally beneficial outcome?

NWS OFFICIAL: I'm glad you asked that question. In fact, over the past several years we have completed a number of studies on the forecast process that cover the interrelated aspects of prediction, communication, and the use of forecast information in decision making. In addition to these retrospective evaluations, we have conducted several forward-looking analyses that, while preliminary, suggest a number of contexts in which decision making can improve, with the following potential benefits expected from this year's request . . .

Notes

1. This chapter focuses on weather forecasting in the United States. We use the terms *forecast* and *prediction* interchangeably throughout this chapter.
2. *Skill* is defined as a prediction's improvement over a baseline measure of performance.
3. The first of ten recommendations in a 1999 National Research Council report on the future of the National Weather Service was that the organization, and its parent agency, NOAA, "should more aggressively support and capitalize on advances in science and technology to increase the value of weather and related environmental information to society" (NRC 1999, p. 1).
4. Throughout this chapter the term *forecast system* is used to refer to the institutions, infrastructure, and resources that are involved with the production, communication, and use of weather predictions.
5. NASA's Earth Observing System is not usually associated with weather prediction; however, its data products could be used for that purpose (NRC 1999), in which case the total would increase by more than $1 billion.
6. See chapter 5 for a brief history of the National Weather Service. See Whitnah (1961) for an in-depth history.
7. There are also obviously situations in which a forecast is unnecessary for effective decision making. Understanding when and when not to rely on predictions is an essential aspect of the effective use of prediction. See chapter 14.

8. These examples focus on the prediction of extreme weather. A similar case can be made for the importance of prediction of "routinely disruptive" weather, see Pielke 1997 for discussion.

References

AMS (American Meteorological Society). 1991. Weather forecasting: A policy statement of the American Meteorological Society as adopted by the Council on 13 January 1991. *Bulletin of the American Meteorological Society* 72:1273–1276.

Brooks, H.E. 1999. Observed record—All tornados and deaths. In *Tornado-related items*, unpublished document on Internet, www.nssl.noaa.gov/~brooks.

Changnon, S.A. 1997. *Assessment of uses and values of the new climate forecasts*. Boulder, CO: University Corporation for Atmospheric Research.

Dabberdt, W., and J. Hales. 1998. *Nowcasting and predictions for urban zones*. Report of the Prospectus Development Team #10 of the U.S. Weather Research Program. Online at uswrp.mmm.ucar.edu/uswrp/PDT/PDT10.html.

Doswell, C.A., III. 1999. *Tornados: Some hard realities*, unpublished copyrighted manuscript on Internet at www.wildstar.net/~doswell/Tornado_essay.html.

Doswell, C.A., III, and H.E. Brooks. 1998. Budget cutting and the value of weather services. *Weather and Forecasting* 13:206–212.

Dow, K., and S.L. Cutter. 1998. Crying wolf: Repeat responses to hurricane evacuation orders. *Coastal Management* 26:237–252.

Drucker, P.F. 1993. *Post-capitalist society*. New York: HarperCollins.

Emanuel, D., E. Kalnay, and Prospectus Development Team #7. 1996. *Observations in aid of numerical weather prediction for North America*. Report of the Prospectus Development Team #7 of the U.S. Weather Research Program. Online at uswrp.mmm.ucar.edu/uswrp/PDT/PDT7.html.

Feldman, M.S., and J.G. March. 1981. Information in organizations as signal and sign. *Administrative Science Quarterly* 26:171–186.

GAO (General Accounting Office). 1996. *NWS has not demonstrated that new processing system will improve mission effectiveness*. Report no. GAO/AIMD-96-29. Washington, DC: GAO.

GAO (General Accounting Office). 1997. *National Weather Service: Closure of regional offices not supported by risk analysis*. Report no. GAO/AIMD-97-133. Washington, DC: GAO.

Glantz, M.H., and L.F. Tarleton. 1991. *Mesoscale research initiative: Societal aspects*. Report of the workshop December 10–11, 1990. Boulder, CO: Environmental and Societal Impacts Group, National Center for Atmospheric Research.

Grand Forks Herald. 1997. *Come hell and high water: The incredible story of the 1997 Red River flood*. Grand Forks Herald, Box 6008, Grand Forks, ND 58206-6008.

Herrera, S. 1999. Weather wise. *Forbes* (June 14). Online at www.forbes.com/forbes/99/0614/6312090a.htm.

Jarrell, J.D., and M. DeMaria. 1999. An examination of strategies to reduce the size of hurricane warning areas. In *Twenty-third conference on hurricanes and tropical meteorology,* vol. 1. Boston: American Meteorological Society, pp. 50–52.

Katz, R.W., and A.H. Murphy, eds. 1997. *Economic value of weather and climate forecasts.* Cambridge, UK: Cambridge University Press.

Kunkel, K.E., R.A. Pielke, Jr., and S.A. Changnon. 1999. Temporal fluctuations in weather and climate extremes that cause economic and human health impacts: A review. *Bulletin of the American Meteorological Society* 80(6):1077–1098.

Leavitt, M.S. 1997. Testimony of the Commercial Weather Services Association before the Subcommittee on Energy and Environment, U.S. House of Representatives, April 9. Downloaded from www.house.gov/science/leavitt_4-9.html.

Marks, F., L. Shay, and Prospectus Development Team #5. 1996. *Landfalling tropical cyclones: Forecast problems and association research opportunities.* Report of Prospectus Development Team #5 of the U.S. Weather Research Program. Online at uswrp.mmm.ucar.edu/uswrp/PDT/PDT5.html.

Murphy, A.H. 1997. Forecast verification. In *Economic value of weather and climate forecasts,* R.W. Katz and A.H. Murphy, eds. Cambridge, England: Cambridge University Press, pp.19–74.

NRC (National Research Council). 1980. *Technological and scientific opportunities for improved weather and hydrological services in the coming decade.* Washington, D.C.: National Academy Press.

NRC (National Research Council). 1998. *The atmospheric sciences: Entering the twenty-first century.* Washington, DC: National Academy Press.

NRC (National Research Council). 1999. *A vision for the National Weather Service: Road map for the future.* Washington, DC: National Research Council.

OFCM (Office of the Federal Coordinator for Meteorology). 1998. *The federal plan for meteorological services and supporting research for fiscal year 1999.* FCM-P1-1998. Washington, DC: OFCM.

Pettit, P. 1999. *A review of National Weather Service severe weather warning statistics, 1994–1997.* Unpublished document on the Internet, www.weatherconsultant.com/Verification.html.

Pielke, R.A. Jr. 1999. Who decides? Forecasts and responsibilities in the 1997 Red River flood. *Applied Behavioral Science Review* 7:1–19.

Pielke, R.A., Jr. 2000. Policy responses to the 1997/1998 El Niño: Implications for forecast value and the future of climate services. In *The 1997/1998 El Niño in the United States,* S. Changnon, ed. Oxford, England: Oxford University Press.

Pielke, R.A., Jr., ed. 1997. *Workshop on the Social and Economic Impacts of Weather.* Proceedings of workshop held April 2–4, 1997, in Boulder, Colorado. Boulder, CO: Environmental and Societal Impacts Group, National Center for Atmospheric Research.

Pielke, Jr., R.A., and R.A. Pielke, Sr. 1997. *Hurricanes: Their nature and impacts on society.* Chichester, England: John Wiley and Sons.

Roebber, P.J., and L.F. Bosart. 1996. The complex relationship between forecast skill and forecast value: A real-world comparison. *Weather Forecasting* 11: 544–559.

Salpukas, A. 1999. Firing up an idea machine: Enron is encouraging the entrepreneurs within. *New York Times,* Sunday, June 27, Section 3, p.1.

Schlatter, T.W., F.H. Carr, R.H. Langland, R.E. Carbone, N.A. Crook, R.W. Daley, J.C. Derber, and S.L. Mullen. 1999. *A five-year plan for research related to the assimilation of meteorological data.* Report of USWRP Workshop on Data Assimilation, December 9–11, 1998. Internet: uswrp. mmm.ucar.edu/uswrp/reports/five_year_plan/title.html.

Sorenson, J.H. 1993. Warning systems and public warning response. Paper prepared for January workshop *Socioeconomic Aspects of Disaster in Latin America.* Oak Ridge, TN: Oak Ridge National Laboratory, pp. 1–15.

Stewart, T.R. 1997. Forecast value: Descriptive decision studies. In *Economic value of weather and climate forecasts,* R.W. Katz and A.H. Murphy, eds. Cambridge, England: Cambridge University Press, pp. 147–182.

Stewart, T.R., W.R. Moninger, K.F. Keideman, and P. Reagan-Cicerone. 1992. Effects of improved information on the components of skill in weather forecasting. *Organizational Behavior and Human Decision Processes* 53:107–134.

U.S. Congress. 1979. *Atmospheric Services and Research and a NOAA Organic Act.* Report prepared for Subcommittee on Natural Resources and Environment of the Committee on Science and Technology, U.S. House of Representatives, 96th Cxongress.

U.S. Congress. 1993. *NOAA's response to weather hazards—Has nature gone mad?* Hearing before Subcommittee on Space of the Committee on Science, Space, and Technology, 103rd Congress, September 14. Washington, DC: U.S. Government Printing Office.

Vislocky, R.L., J.M. Fritsch, and S.N. DiRienzo. 1995. Operational omission and misuse of numerical precipitation probabilities. *Bulletin of the American Meteoriligical Society* 76:49–52.

Whitnah, D.R. 1961. *A history of the United States Weather Bureau.* Urbana: University of Illinois Press.

Wilks, D.S. 1997. Forecast value: Prescriptive decision studies. In *Economic Value of Weather and Climate Forecasts,* R.W. Katz and A.H. Murphy, eds. Cambridge, England: Cambridge University Press, pp. 109–146.

Flood Prediction: Immersed in the Quagmire of National Flood Mitigation Strategy

Stanley A. Changnon

The goal of United States flood policy is to minimize loss of lives and damages to property.[1] Floods in the United States caused over $30 billion in damages between 1993 and 1997—more than any other natural hazard (Pielke 1999a). Loss of life from floods ranks third behind temperature extremes and lightning (Changnon et al. 1996). Damaging floods also create long-lasting effects on a region's economic productivity and on victims' physical and mental health. The magnitude of the flood problem suggests that flood predictions should rank high in the nation's flood mitigation policy, but, as this chapter shows, this is not the case. How and why this situation exists can be understood only within the broader context of national flood mitigation policy.

This chapter explores flood prediction in four sections. The first introduces the production and use of flood forecasts from both scientific and institutional perspectives. The second briefly reviews the evolution of flood prediction capabilities from the late 1800s to the present. Section three assesses the use of flood predictions in two recent major floods. The last section examines the significance of flood predictions in the context of national flood mitigation policy, with an eye to where we go from here.

The Production of Flood Forecasts

A prediction of the timing, location, and magnitude of a specific flood is called a flood forecast. The production of a sound flood forecast demands expertise in meteorology, hydrology, and hydraulics. Meteorology is the science of weather. Hydrology is the science "dealing with the properties, distribution, and circulation of water on the surface of the land, below the surface, and in the atmosphere"; hydraulics, on the other hand, treats "the mechanical properties of water in motion" (FIFMTF 1992, pp. 6–11). Flood experts working in public institutions integrate these three areas of expertise to provide information useful to decision makers concerned with floods.

Institutions

The National Weather Service (NWS) routinely issues flood forecasts for about four thousand locations in the United States (figure 5.1). The NWS, which resides in the National Oceanic and Atmospheric Administration (NOAA) of the Department of Commerce, has responsibility for "forecasting of weather, the issue of storm warnings, the display of weather and flood signals for the benefit of agriculture, commerce, and navigation, the gauging and reporting of rivers. . ." (15 USC Sec. 313). In the development of flood forecasts, the NWS relies on a number of other federal agencies to achieve its mission, particularly the U.S. Geological Survey (USGS) and the Army Corps of Engineers (COE; the Corps).

Within the NWS, the Office of Hydrology has responsibility for "the collection and processing of hydrologic data for river and flood forecasts" (Stallings and Wenzel 1995, p. 458). The NWS also is home to

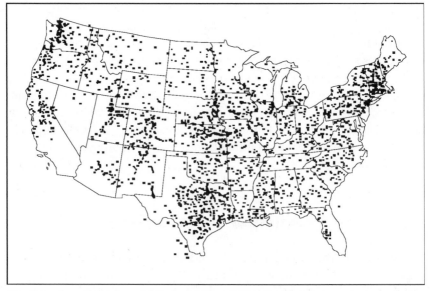

Fig. 5.1   Locations for which the National Weather Service routinely issues flood forecasts. (Source: NOAA 1994)

five regional hydrologists who have expertise on the flood threat in different parts of the nation. Figure 5.2 shows the organization of the NWS Office of Hydrology as of the mid 1990s.

In order to monitor the nation's river systems, the NWS operates thirteen River Forecast Centers (RFCs) across the nation, each centered on at least one major river basin (figure 5.3). The RFCs provide flood information to NWS Forecast Offices, where the information is passed on to city, county, and state officials as well as to the public. About three thousand communities are served by the thirteen RFCs, each of which is staffed by about sixteen people (Stallings and Wenzel 1995). In periods of normal river flows, RFCs produce streamflow forecasts for barge operations as well as water management decisions for applications such as electric power generation. RFCs also work with local communities to help them better prepare for floods and flash floods.

Process

Following disastrous flooding associated with Hurricane Camille in 1969, the National Weather Service established a "systematic operational and organizational approach" to flash flood guidance (Zevin 1994, p. 1268). From that beginning the modern river forecast system has evolved. The process of the development of a flood forecast for public dissemination typically involves three steps: (1) "[using] observations (precipitation, temperature, etc.) to estimate the net amount of water

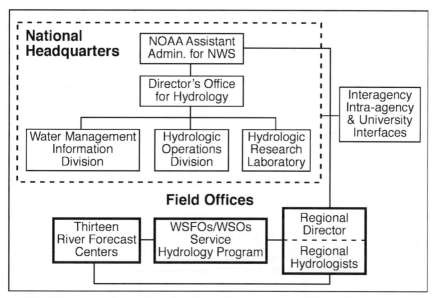

Fig. 5.2   Flow chart of National Weather Service's Office of Hydrology.

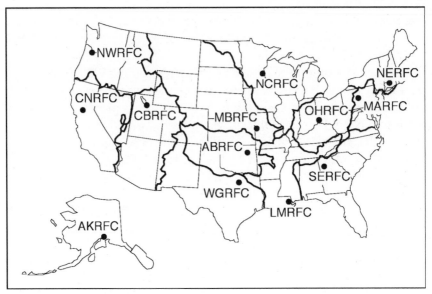

Fig. 5.3   Locations of the thirteen National Weather Service River Forecast
Centers.

entering the [river] basin from rainfall and/or snowmelt," (2) "[convert-
ing] the net input of water (from rainfall or snowmelt) into a volume
that enters the stream (runoff)," and (3) "[calculating] the volume rate
of water (discharge) that flows from a point in the stream to points farther downstream" (DOC 1994, p. 4-4).

Hydrometeorological Data.

Streamflow data is an essential component of flood forecasting. The stream-
gauging program of the U.S. Geological Survey is the largest national program responsible for the collection and dissemination of streamflow data.[2]
This effort is supported by more than six hundred federal, state, and local
agencies and organizations, and in 1994 only 549 stations (of more than
7,000 total) were funded solely by the USGS. The Natural Resources Conservation Service, Salt River Project, Tennessee Valley Authority, and
NOAA are important contributors to the collection of streamflow data. The
USGS opened its first stream-gauging station in 1889 on the Rio Grande
River and by 1994 operated a total of 7,292 throughout the country. Of these
stations, about 3,000 contribute data used by the NWS to forecast floods. In
addition to flood forecasting, the stream-gauging program collects data that
are used in water resources planning, hydrologic research, and the operation of water resources projects (such as dams). In recent years the stream-
gauging program has been scaled back due to increasingly tight budgets.

Streamflow data are also collected through the NWS hydrologic network of about eleven thousand private citizens. Some of these volunteers report river and rainfall information to their local Weather Service Office or Weather Service Forecast Office on a daily basis. This information is tabulated and communicated to the appropriate RFC for use in river and flood forecasting.

The spatial and temporal records of streamflow in the United States expanded throughout the twentieth century for a number of reasons. Early in this century, irrigation and hydroelectric power needs created demand for streamflow information. The extensive drought of the early 1930s, followed by extreme flooding later in that decade in the Ohio and Potomac River basins, created a demand to better understand extreme low- and high-water flows on major rivers. Also, the continental United States contains more than 846,000 drainage basins that are between one and two square miles in size (FIFMTF 1986, pp. 6–8), and in the 1950s, the establishment of the Federal Interstate Highway system created a need to estimate potential flood flows at the many locations where interstate highways pass over small streams and drainages. Finally, the establishment of the National Flood Insurance Program in 1968 created demand for data on floodplain mapping and flood frequency. In short, the stream-gauging program of today was developed incrementally over the past one hundred years and today provides valuable data for flood forecasting and other water resources–related uses.

Data collected through the stream-gauging program includes the "stage" of a river—that is, "the height of the water surface above a reference elevation" (Wahl, Thomas, and Hirsch 1995, p. 13). Stage data are useful, for example, in mapping floodplains. However, for flood forecasting, a river's "discharge" is also important. "Discharge is defined as the volume of flow passing a specified point in a given interval of time and includes the volume of water and any sediment or other solids that may be dissolved or mixed with the water" (Wahl et al. 1995, p. 13). Discharge is typically expressed in cubic feet (or cubic meters) per second and is often determined through direct measurement or through a statistical relation with the stage level.[3] Discharge is a fundamental input to hydraulic models. The relationship between river stage and discharge for a specific location, determined empirically, is known as the rating curve. Using a rating curve, one can easily convert between stage and discharge.

National Weather Service River Forecast System.

The River Forecast Centers use a set of hydrologic models called the National Weather Service River Forecast System (NWSRFS) to turn raw data into flood forecasts. The NWSRFS provides the hydrologist with a number of different procedures and models to characterize snowmelt, rainfall and runoff, runoff distribution, and channel routing in order to develop a flood forecast (Larson et al. 1995). Hydrometeorological data are fed into hydrological models in order to produce flood forecasts. In areas where snow has accumulated, the forecast must combine outputs of a snow model, a soil moisture model, and a flow model. A snow model takes precipitation and temperature data as input and sends its output (melt+rain) to a soil moisture model. In addition to the snow model output, the soil moisture model also uses as input data on soil type, slope, land use, season, vegetation, and other parameters. The output of a soil moisture model is a calculation of runoff from a particular drainage basin. Runoff is described using the observed relationship between precipitation (or melt) and discharge over time for a particular drainage basin. Finally, a flow model is used to combine runoff from different headwater basins where they flow together.

Existing models—and thus flood forecasts—would benefit from improvements in quantitative precipitation forecasting (QPF), a technique that predicts "the amount of . . . precipitation that will fall over a particular area over a given period" (Schwein 1996, p. 1). Such improvements have been "steady, albeit slow" (Krzysztofowicz 1995, p. 1143). The goal of using QPF in the flood forecasting process is to provide a longer lead time for flash flood (i.e., floods that occur within a twelve-hour period or less on the upstream branches of a river basin) and river stage forecasts. Of course, the potential success of this tool depends in large part on the ability to forecast reliably the amount of precipitation that will fall over a particular watershed. To date, effective use of QPF has not been fully achieved. Experiments with it have had mixed results—some flood forecasts improved, but others worsened. Studies of the use of QPF show benefits to decisions associated with flood warning, reservoir control, and commercial navigation (Krzysztofowicz 1995).

Other potential contributions to improved production of flood forecasts include the use of advanced radar technologies to identify in real time (as well as to forecast) areas of heavy rainfall, development of more accurate rainfall-runoff models, and improved collection of field data (Krzysztofowicz 1995).

Using Flood Forecasts

The production of a flood forecast is only one step toward its effective use by decision makers. As a congressional report clearly stated more than thirty years ago, "It cannot be overemphasized that the mere supply of information as to where the water has reached and when, does not necessarily lead decision makers to avoid the flood threat" (TFFFCP 1966, p. 14). Indeed, a flood forecast becomes effective only through a process of hearing, understanding, believing, personalizing, responding, and confirming (Sorensen and Mileti 1988). A decision by an individual to act in response to the information about a flood, or any other extreme event for that matter, is conditioned by a wide range of factors including perceived risk, education, and time before impact. The use of a forecast must be understood in the broad context of a process that includes information production, communication, and response. Social scientists have a well-developed understanding of this context (e.g., see Drabek 1986). Application of social science knowledge is thus a key element of successful flood forecasts.

The NWS produces and issues a number of flood-related products: flood warning, flood statement, river summary, river statement, flash flood watch, flash flood warning, flash flood statement, severe thunderstorm warning, severe weather statement, and special weather statements. Once a forecast or warning is produced, it is typically distributed to a range of federal, state, and local agencies, including the Federal Emergency Management Agency (FEMA), the Corps of Engineers, and state and local emergency management agencies. Similarly, print, television, and radio outlets receive NWS weather products and are a critical part of the communication process. NOAA also broadcasts its weather products on its Weather Radio. The World Wide Web provides a wide range of information in text and graphic formats on government and commercial sites.

Thus, information abounds. But a forecast is useful only if a particular user incorporates it into the decision process. Consequently, local NWS field offices are responsible for interacting with the decision makers in their communities. A meeting of emergency managers and National Weather Service officials found a number of opportunities to improve the warning process by actions to ". . . simplify and standardize warning and forecast products, better support local and state informed decision making for severe weather and flood response, and enhance the training of emergency managers on severe weather and flood preparedness and response . . ."(Adams 1995, p. v).

Evolution of Flood Prediction Capabilities and Related Policy Issues, 1870–1990

To understand today's flood policy issues requires a brief sojourn into the history of the nation's development and the ensuing struggle of humans against the peril of flooding. The nation's population grew from 7 million in 1800 to 50 million by 1870, and much of that occurred within the flood-prone Mississippi River basin (Changnon, 1999). By the middle of the nineteenth century, national policy makers recognized flooding as a key problem. By 1860, Congress recognized *flood control* as a means to reduce losses, but federal law forbade government involvement in flood control (Hoyt and Langbein 1953). Afterwards, Congress enacted flood control as "navigation works"—thus circumventing the legal prohibition. For the ensuing fifty years, this primarily meant levee construction and stream channelization (Morrill 1897). However, flood losses continued to grow, and the federal government began the post-flood-relief assistance payments that some see as negative incentives for flood mitigation. In this historical context, flood prediction can be seen as one of a multitude of flood mitigation activities that fall within the broader, complex issue of national water policy.

The Early Decades of Flood Prediction, 1871–1930

Flood prediction represented the nation's first systematic efforts to predict a natural hazard. In 1871 Congress directed the secretary of war to establish stream-measuring instruments on the Mississippi River.

The systematic collection of climatic data, which had begun in earnest in the 1870s, provided by the 1890s a database from which scientists could begin to relate magnitudes of rainfall and snowmelt in the basin to the floods along the Mississippi River (Morrill 1897). Flood forecasts were empirical; that is, they were based on the status of what had already happened (rain amounts and river stages). This approach was necessary because there was no scientific basis for quantitative hydrology. And, as the flood and rainfall database grew, and more rain gauges and river gauges were installed, knowledge of flood conditions expanded and forecasting improved.

In 1889, a devastating flood killed 2,209 people in Johnstown, Pennsylvania. Partly in response to that event, Congress reorganized the nation's weather infrastructure into a newly named Weather Bureau whose responsibilities included "flood warnings and the gauging and reporting of rivers" (Office of Hydrology 1997). In 1895 Congress assigned to the USGS the role of "water resource monitoring for the nation," which included measuring the volume of flow on major streams. Thereafter, the USGS and the Weather Bureau worked together to measure river flows and share data. At the close of the nineteenth century, flood forecasting was conducted on few rivers other than the

Mississippi. But, in 1903, severe floods on the Kansas River brought public outcries for action, and ensuing congressional pressures on the Weather Bureau resulted in further extension of flood forecast services. In 1910 the growth of irrigation projects in the West led Congress to assign to the Weather Bureau the task of flood forecasting for the Bureau of Reclamation (Weather Bureau 1961), thus bringing another agency into flood prediction.

Recent Decades, 1931–90

Numerous devastating floods in the 1920s and 1930s fueled actions by the Roosevelt administration; the average annual flood losses during this period were 190 lives lost and $120 million (current dollars) in damages. The early New Deal government actively addressed the flood prediction needs of the nation, leading to major research and operational forecasting advances. The science of weather forecasting developed rapidly during the 1930s, partly in response to the demands created by the growing use of aviation for carrying the U.S. mail and passengers. One result was the Weather Bureau's initiation in 1937 of flood forecasts on two basins in the East, the flood-prone Susquehanna and Potomac Rivers (Weather Bureau 1941). These experimental efforts were based on "expected rainfall," as well as the time-honored assessment of what had already fallen. In the 1930s, the collaboration of hydrologists and meteorologists led to an important scientific development: the "unit hydrograph," an easy-to-use empirical relationship between basin rainfall and runoff.

The Weather Bureau first assessed the economic value of its flood forecast system in 1941. Analysis of floods during 1924–40 showed that total flood losses were $825 million and estimated savings due to flood prediction were $247 million (a fourth of the total possible losses of $1,062 million). Further, the cost of the river forecast service was only $3 million, providing a benefit-cost ratio of 75:1 (Weather Bureau 1941). In 1948, Bernard (1948) found a benefit-cost ratio (savings in flood losses to cost of forecasting) of 31:1.

Major research and operational activities of the late 1940s and ensuing decades up to 1990 fell within three areas. First came a major operational change in the nation's flood prediction effort, the establishment of regional River Forecast Centers, beginning in the 1940s. Second, the post–World War II increase in flood-related research quickly paid dividends. Storm rainfall-runoff relationships were improved, and by 1950 quantitative precipitation forecasts for the following twenty-four-hour period for various basins became standard input to river forecasting. The first weather radars designed specifically for rainfall monitoring were installed nationwide by the Weather Bureau between 1957 and 1962. This new technology increased monitoring capabilities and allowed the rain gauge network to be used more effectively to quantify heavy rainfall over a basin. The 1960s also saw the beginning of the computer age in river and flood forecasting endeavors, allowing faster issuance of flood forecasts.

The third area of major progress concerned flash flooding, theretofore largely ignored. Deadly flash floods in 1951 and 1952 ushered in greater scientific attention. Radar and rain gauge observers, along with general weather forecasts of potential precipitation, could now be used to provide near real-time storm data. Interest in flash floods raised a related issue, the dissemination of warnings, which by 1960 was thought to be a weakness in flood forecasting efforts. State and local agencies became involved along with the media in disseminating flood warnings to the public.

This growing interest continued to be fueled by huge loss of life to flash floods in South Dakota (239 killed in 1972), in Colorado (136 dead in 1976), and in Pennsylvania (79 dead in 1977). The situation called for action. Forecasters admitted that existing forecasting systems were inadequate for flash floods (e.g., Fox and Hurst 1976). The numerical hydrologic models used at the thirteen River Forecast Centers were considered outmoded and still too dependent on the old empirical approaches. Related developments included scientific criticism of quantitative precipitation forecasts as inadequate for flash flood forecasting (Georgakakos and Hudlow 1984), a 1978 NWS Flash Flood Program plan, and a National Academy of Sciences recommendation to improve QPFs for flash flood forecasting (Panel on Precipitation Processes 1980). Again, flood prediction policy was disaster driven.

Improvements occurred, and by 1981 the NOAA administrator stated, "There is no reason for a single loss of life due to flooding. Recent advances in flood prediction now make such loss of life unnecessary" (Byrne 1982). Nevertheless, during the early 1980s flash floods caused $2 billion (current dollars) in damages annually and more than 200 deaths a year. Scientists continued to challenge the new flash flood warning systems of the Weather Service, indicating that for several reasons they did not actually improve warnings. Government interest in the flash flood problem waned in the 1980s—some saw the problem largely solved, and others simply lost interest.

Major Advances in Flood Policy, 1870–1990

Review of flood prediction activities and related policy development from 1870 to 1990 reveals that federal agencies responsible for flood prediction (including data collection and warning) established programmatic goals internally and formulated those goals around their mandated missions, which shifted in response to flood events. Six key factors stand out as affecting the major advances in flood prediction:

1. After most flood disasters, agencies used windows of opportunity to obtain more funds for staffing, research, and equipment.

2. Flood forecasting activities began after the nation had launched major efforts to control flood damages through other means, originally by levees and later by reservoirs. Over time, several federal agencies joined the nation's flood control and mitigation efforts. Thus, flood prediction remained just one part of a broad, evolving, and very complex flood mitigation effort.

3. The Weather Bureau saw flood forecasting as only a part of its central mission, which was to improve forecasting of *all* weather conditions. Further, disciplinary differences hindered progress: there often was confusion and debate over who was, or who should be, responsible for flood forecasting—hydrologists or meteorologists.

4. By the time *flood prediction based on rainfall forecasts* became a useful tool in flood protection in the 1950s, flood loss mitigation involved many federal, state, and local agencies with multiple approaches for addressing the nation's complex flood problem. This situation created confusion in the policy debate about what should be done.

5. When all parties recognized the flash flood forecasting problem as a national problem in the 1970s, significant action followed, and major progress was made in forecasting and warning dissemination.

6. Advocates of nonstructural approaches to the nation's flood problem, active since the 1950s, did not embrace flood prediction because they saw the public's belief that flood prediction equaled flood protection as counterproductive to achieving land-use changes in flood-prone areas (White 1994).

Between 1870 and 1990, flood prediction research and operations evolved significantly, leading to major improvements: Research yielded continually improving hydrological and meteorological understanding of floods; ever better instruments measured and transmitted flood and rainfall data; and government policies changed to allow and exploit these advances. By 1990, flood predictions and warnings had vastly improved.

Assessment of Predictions of Recent Floods

Assessments of the predictions of recent major floods provide insights into the scientific status of flood forecasting and the problems in the forecasting-warning system.

Midwestern Flood of 1993

This massive summer flood developed in the north (Minnesota and Wisconsin) and gradually migrated south (Iowa, Kansas, Missouri, and Illinois) as the summer wore on. The culmination was a unique hydrologic event—the simultaneous arrival of twin flood peaks, one on the Missouri River and one on the Mississippi River, at their confluence near St. Louis on August 1, the first known occurrence of such an event. The floodwaters broke all-time flow records on fifteen hundred river miles in eight states (Interagency FMR Committee 1994). The flood took fifty-two lives, mostly in flash flood events, and cost $18 billion in losses and responses, making this the second most damaging weather event in the nation's history (Changnon 1996d).

The flood condition pronouncements of the National Weather Service and those of the Corps of Engineers during the summer of 1993 were often not only inaccurate but also different, and they were therefore seen as controversial (Interagency FMR Committee 1994; Changnon 1996a). NWS flood predictions, which often significantly underestimated future river levels and the dates when the crests would occur, and the Corps' optimistic and incorrect public statements issued in June, which asserted that the flood would end soon, helped lead to underestimates of the flood's seriousness. These incorrect predictions of an early ending of the flood had detrimental effects on the decision making of those seriously affected by the flood, including the barge industry (Changnon 1996b). The forecasts certainly affected strategies used to fight the flood and to seek alternatives in shipping and other flood-impacted endeavors.

A key factor in the NWS's forecasting problem was the use of hydrologic models—later deemed to be "antiques"—that were unable to handle the effects of levee breaks or to integrate forecasts of future precipitation (NOAA 1994). Nor did the models account for many changes made to the river channels in the pertinent basins. Moreover, unfolding experience soon proved that the historical record did not cover all flood possibilities (Koellner 1996). Post-flood assessments revealed that there had been a serious lack of interest and commitment by the nation's weather agency to flood forecasting on the Mississippi River system (Changnon 1996c). The NWS's investigation of its own activities during the flood revealed that the prediction problems were compounded because many flow-monitoring facilities were damaged, lost, or unable to function during the flood (NOAA 1994). Furthermore, precipitation measurements were seldom adequate for detecting and measuring flash-flood rain events. However, the few new NWS radars operating in 1993 proved invaluable for measuring heavy local rains (NOAA 1994). A major loss of life was prevented in Wisconsin by the automatic flood detection system and the alert actions of NWS flood forecasters (Changnon 1996d).

After the flood ended, members of Congress from the affected states began calling for special flood studies, various interest groups put forth widely different views in congressional hearings, and bills were offered to correct various ills of the nation's flood mitigation program. But *none* addressed the forecasting failures and needs for improved flood prediction. Given the predictive errors, one might ask why more complaints, lawsuits, or other forms of inquiry were not raised. One answer seems to rest in the record height of the flood, a situation that acted to provide, among many of those severely impacted, an understanding of the event's uniqueness and a sense of forgiveness for the errors of the predictions. Further, NOAA deflected criticisms of its prediction problems by reporting that the ongoing modernization of the NWS would handle the 1993 problems in future floods (NOAA 1994). At the national level, a presidentially mandated federal assessment of the flood surprisingly indicated that flood prediction was not a key area needing improvement (Interagency FMR Committee 1994).

Ohio River Flood of 1997

Beginning on February 28, 1997, six to twelve inches of rain fell in a period of two days over most of Kentucky and southern portions of Ohio and Indiana along an axis parallel to the Ohio River. This downpour resulted in numerous flash floods in the tributaries of the Ohio River and a near-record flood on the Ohio from Cincinnati to Cairo, Illinois. The crests on several tributaries set all-time records; the flooding on the Ohio at Cincinnati was the worst since 1964, while farther down-river it was the worst since 1937. Only 7 percent of flooded structures were covered by flood insurance.

The NWS issued sixty-two flash-flood warnings for tributary basins in Kentucky and Ohio. However, forecasters admitted that they underestimated the rain potential and thus the crest for several tributaries (*Cincinnati Inquirer*, March 3, 1997). The flood forecast problems were also subsequently blamed on the lack of timely rainfall data from the large storm area, and on the lack of streamflow data—the USGS had removed some automatic river gauge monitoring stations in 1994 (the year after the 1993 flood had shown their importance) as a result of federal cost-cutting measures (*Cincinnati Inquirer*, March 5, 1997). One result was that some people were caught without warning—a factor contributing to the deaths of some of the fourteen drowned while driving on rural roads and the five drowned when floodwaters swept away their residences. This event revealed that the benefits expected from a modernized NWS having more accurate and timely data for its forecasts were not available in 1997 (Fread et al. 1995).

Red River of the North Flood of 1997

The Red River of the North flood of 1997 has already been recounted in the previous chapter. It bears reiterating, however, that in that case local and state officials in Minnesota and North Dakota blamed poor NWS forecasts for the inadequate local preparations and the massive damages that ensued (Pielke 1999b). As in the cases of the two floods described above, a major difficulty for forecasters was that this was a record event beyond past experience—a problem that vexes experience-based models. Another difficulty for users was that the forecasts did not properly express the uncertainty of the forecasts. Local decision makers assumed that the predicted values were accurate. Further, with hindsight, it was found that the COE's crest forecasts were more nearly correct than those of the NWS, but the two agencies neither exchanged nor discussed their forecasts. The NWS forecasters also blamed flow impediments caused by new bridges that affected flows but were not a part of their models. In summary, this flood had an enormous lead time, and users had every reason to expect that the forecast process would reduce losses, but, for Grand Forks, it did not.

Collectively, these three recent floods reveal that the forecasting was imperfect in the areas of extreme flooding. In all three cases, predicted crests were lower than the actual final levels. Further, the lessons of the 1993 floods (NOAA 1994), combined with the NWS modernization program, did not lead to expected improvement in flood predictions and their use during the two 1997 events.

Policy Outcomes

Scientific Outcomes

There are two issues to be addressed in order to gain improvements in flood predictions. One relates to *scientific research,* and the other relates to *implementation of existing and well-developed scientific methods and operational systems.* Both involve policy decisions and funding.

A number of federal research programs are focused on flood forecasting. The scientists involved include atmospheric scientists, hydrologists, and social scientists. Much of the hydrologic knowledge necessary for useful predictions exists, but basin flood models across the nation need improvement to make use of the science. Forecasters also lack near real-time data on streamflow and rainfall during floods. Moreover, research on flash-flood and coastal flood forecasting receives greater support from Congress than does research on riverine floods. As a result, flood peaks in recent riverine floods were missed, reflecting many needs including closer cooperation between forecasters and the users of predictions, and development of improved formats for presenting predictions (Pielke 1999b). Necessary research to enhance long-range (months and seasons ahead) flood prediction is underway.

There are other untapped opportunities for improvement. For example, assessment of the 1993 midwestern flood revealed the lack of high-performance operational hydrologic basin models that could rapidly integrate continuing inputs of water data, soil moisture data, and precipitation forecasts to generate more accurate flood predictions (Changnon 1996a). The science and technology existed to accomplish such modeling well before 1993. If skilled staff with adequate resources were devoted to such tasks, the models and their predictions could be of high quality, as an Iowa field test demonstrated in March 1997 (Braatz et al. 1998). As shown by the 1993 and 1997 floods, allied improvements need to focus on continuing assessments of basins to detect land-use changes that affect flooding, and subsequent updating of model inputs to reflect these changes. There is also a need to implement a system for enhanced collection of stream and river conditions, now more easily accomplished through remote sensing and data transmittal via satellite systems. A special problem in flood flow monitoring is that responsibility is diffused among several federal agencies and some states, which has resulted in station closures in tough budget times. In short, the technologies exist to monitor and model relevant conditions in near real-time, but they need to be maintained and implemented, which requires resources and staffing (Larson et al. 1995).

Policy-Making Issues: Key Questions

What can improving flood predictions do for society? Presumably, it would help reduce loss of lives and property. But where does flood prediction rank in the array of national flood mitigation options? And why? Are the real or perceived benefits of improved forecasts and warnings sufficiently important to call for major policy attention and action? Evidence suggests that the case has not been made for the economic and lifesaving benefits of improved flood predictions.

For many decades the nation's flood policies were based on structural solutions for flood protection, which, aided by flood predictions, were expected to reduce society's vulnerability to flooding. The nation has vastly improved its flood forecasting and warning systems over the past 125 years and has spent billions of dollars to build flood control structures, but flood losses continue to grow. Moreover, historically unprecedented events continue to confound prediction and response capabilities. The populace at risk tends not to purchase flood insurance, and government aid to victims of floods might reduce individuals' acceptance of personal responsibility for flood risks. It seems that the principal cause of the national flood problem is faulty human choices about where to live, and about land-use and construction practices. The net effect is that society is making itself ever more vulnerable to flood damage. As Wright (1996) states, the key to improving flood mitigation is for individuals to assume responsibility for their locational decisions.

Given this situation, do further improvements of the nation's flood forecast and warning system make policy sense? Are improved flood warnings only a perceived panacea, acting indirectly to lure the populace to vulnerable locales where they believe they are safe? One possible reason for less than top-level attention to the flood prediction option is that flood forecasting and warning systems, like levees, can be viewed as working counter to the achievement of the preferred goal of mitigation through changing land use in flood-prone areas. Gilbert

White, known to support nonstructural solutions to the nation's flood problem, recently stated that "better refined forecasts and more efficient emergency operations will be important but they will not necessarily reduce damages and they neglect the measures that might assure sound use of hazardous areas" (White 1994).

Prediction is only a part of national flood mitigation policy, and floods are only part of our overall national water policy. For two hundred years the nation has developed water policies for broad issues like water development and water allocation, and flood policy has been but a subset of those broader issues. Hence, regardless of the objectives of flood policies in addressing what society needs and wants, those policies cannot be separated from the objectives of other water policies that protect water quality, seek water for irrigation, develop public water supplies, promote fisheries and wildlife, and enhance navigation. Further, the objectives of various water policies and priorities often conflict with one another, technically and politically. (For example, dams often force water managers to make trade-offs between different objectives, such as power generation, flood mitigation, fisheries management, and recreation.) Hence, it is no wonder that the flood prediction issue is difficult to address.

Credible studies are needed to better define the losses, as well as the loss reduction already achieved, and to estimate what more could be gained by improvements in (a) forecast lead time, (b) forecast accuracy, (c) emergency preparation, and (d) effective emergency management. Is the nation's policy for handling the flood problem internally consistent and well conceived? *No*. It is an assortment of conflicting options coupled to extemporaneous post-flood actions that collectively create uncertainty for the populace and top decision makers as to what to do and where to place emphasis. A prestigious interagency assessment of the record 1993 midwestern flood issued several major recommendations for future policy actions, and none of those addressed flood prediction (Interagency FMR Committee 1994).

Given the lack of a strong national focus on flood prediction, the current policies, programs, and power of NOAA, the lead agency for flood prediction, become fundamental factors for assessing the future of flood prediction. But NOAA is facing serious budget problems and has a history of varying, often inadequate, attention to flood prediction research and operational endeavors. A plan exists to address this problem but is yet to be implemented (Fread et al. 1995). Many technical problems would be solved by the ongoing modernization and restructuring of the Weather Service. However, hydrologic modeling-forecast integration improvements have been slow to develop in the 1990s, with only a one-month demonstration project done in 1997. A recent assessment of the agency's flood forecasting activities from 1970 to 1994 (Zevin 1994) revealed a focus on flash floods but a neglect of the larger-scale, more damaging regional floods of the 1990s. With respect to the *use* of forecasts, the National Weather Service has never comprehensively evaluated how decision makers use flood forecasts and how the wording and mode of conveying the forecast affect the responses taken.

To summarize, the missions of the nation's water agencies have included a strong focus on floods, whereas flood forecasting at the NWS has never been the top-priority issue. From its early stages, the national effort in flood prediction has been a multiagency process involving the NWS and its ancestors, the Corps of Engineers, the Bureau of Reclamation, the U.S. Geological Survey, and the U.S. Department of Agriculture. The NWS continues to assemble river and weather data, interpret them, and issue public flood warnings; and the others play key supporting roles in data collection, research, and forecasting for their specific missions. This multiagency involvement in flood mitigation affects flood prediction policies, sometimes for the common good, but other times resulting in confusion and lack of focus on the flood prediction issue. No one is in charge, and no reliable information exists on the benefits to be derived from better prediction and more effective communication of predictions. Predictions have been more or less unconsciously placed within the arsenal of flood mitigation tactics—partly in reaction to specific, catastrophic flood events—but their utility remains poorly understood.

Conclusions

Given the uncertainties surrounding the nation's flood mitigation policies, it is not surprising that the policies relating to flood prediction and warning, as part of the nation's flood mitigation arsenal, are less than clear. Further, one could argue that the nation's flood prediction-warning system has been quite successful, as reflected by a 50 percent reduction in lives lost since the 1980s, and is close to achieving the forecasting skill and warning systems needed, at least for saving human lives. Has prediction helped reduce losses to a point where little further reduction can be achieved? Evidence about lives lost in floods reveals that many victims had received warnings and were aware of the flood but made bad decisions such as driving down a flooded highway or entering a flooded basement. There is concern that flood predictions encourage individuals to build and reside in flood-prone regions; that is, skilled flood predictions are considered by some to act contrary to the objective of promoting responsible public behavior.

Can improved predictions still make a significant difference in property loss reduction, the other main policy objective? No credible studies exist to answer this key question. However, the erroneous predictions in recent floods led to massive property losses that possibly could have been avoided. Of equal importance is attention to the needs of users of forecasts and presentation of flood predictions in improved formats that present the risks in more understandable fashion and that will result in more rational actions.

Ironically, most of what is needed to improve flood predictions could be achieved with attention and funding; the technical knowledge by and large exists, and the scientific questions are well framed (FIFMTF 1994). The flood prediction issue is immersed in the quagmire of the nation's flood mitigation policy. A federal policy commitment to improve flood predictions in the context of an overall national flood policy has not been made. Prediction will need recognition and elevation as a critical issue if the potential gains that better flood predictions can provide are to be achieved.

Acknowledgments

The documents and assistance of Frank Richards and Juliann Meyer are deeply appreciated, as are the valuable comments and additions of Roger A. Pielke, Jr.

Notes

1. See the National Weather Service's mission statement and authorizing legislation at www.nws.noaa.gov.
2. The following discussion of the USGS stream-gauging program is based on the comprehensive overview presented in Wahl et al. (1995).

3. Discharge is calculated as the product of a river's cross-section and velocity, both of which must be measured. The USGS makes approximately sixty
thousand discharge measurements each year.

References

Adams, C. 1995. *Building better partnerships*, Report of the National Weather
Service Emergency Management Forum, cosponsored by the National
Emergency Management Association and the Federal Emergency Manage-
ment Agency.

Bernard, M. 1948. Flood forecasts that reduce losses. *Engineering News
Record* (December 9).

Braatz, D.T., J.B. Halquist, M. DeWeese, L. Larson, and J. Ingram. 1998.
*The advanced hydrologic prediction system: Moving beyond the
demonstration phase.* Chanhassen, MN: NWS North Central River
Forecast Center.

Byrne, J. 1982. Ceremony at Johnstown opens campaign to curb flood-related
deaths. *Bulletin of the American Meteorological Society.* 63(8): 942–943.

Changnon, S.A. 1996a. Defining the flood: A chronology of key events. In *The great
flood of 1993*, S.A. Changnon, ed. Boulder, CO: Westview Press, pp. 3–28.

Changnon, S.A. 1996b. Impacts on transportation systems: Stalled barges,
blocked railroads, and closed highways. In *The great flood of 1993*, S.A.
Changnon, ed. Boulder, CO: Westview Press, pp. 183–204.

Changnon, S.A. 1996c. The lessons from the flood. In *The great flood of 1993*,
S.A. Changnon, ed. Boulder, CO: Westview Press, pp. 300–320.

Changnon, S.A. 1996d. Losers and winners: A summary of the flood's impacts.
In *The great flood of 1993*, S.A. Changnon, ed. Boulder, CO: Westview
Press, pp. 276–299.

Changnon, S.A. 1999. The historical struggle with floods on the Mississippi
River basin: Impacts of recent floods and lessons for future flood management and policy. *Water International* 23(1):263–271.

Changnon, S.A., D. Changnon, E.R. Fosse, D.C. Hoganson, R.J. Roth, Sr., and
T. Totsch. 1996. *Impacts and responses of the weather insurance industry to recent weather extremes.* Final report to the University Corporation
for Atmospheric Research, CRR 41. Boulder, CO: University Corporation
for Atmospheric Research.

DOC (Department of Commerce). 1994. *Natural disaster survey report: The
great flood of 1993.* Washington, DC: Department of Commerce.

Drabek, T. 1986. *Human system responses to disaster.* New York:
Springer-Verlag.

FIFMTF (Federal Interagency Floodplain Management Task Force). 1986. *A
unified national program for floodplain management.* Washington, DC:
Government Printing Office.

FIFMTF (Federal Interagency Floodplain Management Task Force). 1992. *Floodplain management in the United States: An assessment report, Volume 2: Full report.* Washington, DC: L.R. Johnston Associates.

FIFMTF (Federal Interagency Floodplain Management Task Force). 1994. *A unified national program for floodplain management 1994*, FEMA #248. Washington, DC: Government Printing Office.

Fox, W.E., and W. Hurst. 1976. Community self-help river forecast procedures. *Proceedings of the Mississippi Water Resources Conference.* State College: Mississippi State University.

Fread, D.L., R. Shedd, G. Smith, R. Farnsworth, C. Hoffeditz, L. Wenzel, S. Wiele, J. Smith, and G. Day. 1995. Modernization in the National Weather Service River and Flood Program. *Weather and Forecasting* 10: 477–484.

Georgakakos, K.P., and M.D. Hudlow. 1984. Quantitative precipitation forecast techniques for use in hydrologic forecasting. *Bulletin of the American Meteorological Society.* 65:1186–1200.

Hazen, A. 1938. *Flood flows: A study of frequencies and magnitudes.* New York: John Wiley & Sons.

Hoyt, W.G., and W.B. Langbein. 1953. *Floods.* Princeton, NJ: Princeton University Press.

Interagency FMR (Floodplain Management Review) Committee. 1994. *Sharing the challenge: Floodplain management into the 21st century.* Washington, DC: Floodplain Management Task Force.

Koellner, W.H. 1996. The flood's hydrology. In *The great flood of 1993*, S.A. Changnon, ed. Boulder, CO: Westview Press, pp. 68–100.

Krzysztofowicz, R. 1995. Recent advances associated with flood forecast and warning systems. *U.S. National Report to International Union of Geodesy and Geophysics, 1991–1994*, Washington, D.C: Joseph Henry Press, pp. 1139–1147.

Larson, L.W., R. Ferral, E. Strem, A. Morin, B. Armstrong, T. Carroll, M. Hudlow, L. Wenzel, G. Schaefer, and D. Johnson. 1995. Operational responsibilities of the NWS River and Flood Program. *Weather and Forecasting* 10: 465–476.

Linsley, R.K. 1951. The hydrologic cycle and its relation to meteorology—River forecasting. *Compendium of Meteorology.* Boston: American Meteorological Society, pp. 1048–1056.

Morrill, P. 1897. *Floods of the Mississippi River.* Weather Bureau no. 143. Washington, DC: Department of Agriculture.

NOAA (National Oceanic and Atmospheric Administration). 1994. *Natural disaster survey report—The great flood of 1993.* Washington, DC: Department of Commerce.

Office of Hydrology. 1997. *Operations of the NWS Hydrologic Services Program.* Washington, DC: NOAA, National Weather Service, pp. 1–23.

Panel on Precipitation Processes. 1980. *Atmospheric precipitation: Prediction and research problems.* Washington, DC: National Research Council.

Pielke, R.A., Jr. 1999a. Nine fallacies of floods. *Climatic Change* 42(2):413–438.

Pielke, R.A., Jr. 1999b. Who decides? Forecast and responsibilities in the 1997 Red River flood. *Applied Behavioral Science Review* 7:83–106.

Pielke, R.A., Jr., and M.W. Downton. 1999. U.S. trends in streamflow and precipitation: Using societal impact data to resolve an apparent paradox. *Bulletin of the American Meteorology Society* 80:1435–1436.

Pielke, R.A., Jr., J. Kimpel, and Prospectus Development Team #6. 1997. Societal aspects of weather: Report of the Sixth Prospectus Development Team of the U.S. Weather Research program to NOAA and NSF. *Bulletin of the American Meteorological Society* 78(5):867–876.

Schwein, N.O. 1996. *The effect of quantitative precipitation forecasts on river forecasts.* NOAA Technical Memorandum NWS CR-110. Kansas City, MO: National Weather Service Office.

Sorensen, J.H., and D.S. Mileti. 1988. Warning and evacuation: Answering some basic questions. *Industrial Case Quarterly* 5(1):33–61.

Stallings, E.A. 1997. *The benefits of hydrologic forecasting.* Elliott City, MD: EASPE, Inc.

Stallings, E.A., and L.A. Wenzel. 1995. Organization of the River and Flood Program in the National Weather Service. *Weather and Forecasting* 10:457–464.

TFFFCP (Task Force on Federal Flood Control Policy). 1966. *A unified national program for managing flood losses,* Report no. 67-663. Washington, DC: U.S. Government Printing Office.

Wahl, K.L., W.O. Thomas, Jr., and R.M. Hirsch. 1995. *Stream-gauging program of the U.S. Geological Survey.* U.S. Geological Survey Circular 1123. Reston, Va.

Weather Bureau. 1941. *The river and flood forecasting service of the Weather Bureau.* Washington, DC: Department of Commerce.

Weather Bureau. 1961. Hydrology. *Topics* (October): 169–173.

White, G.F. 1994. A perspective of reducing losses from natural hazards. *Bulletin of the American Meteorological Society.* 75:1237–1240.

Wright, J.M. 1996. The effects of the flood on national policy: Some achievements, major challenges remain. In *The Great Flood of 1993,* S.A. Changnon, ed. Boulder, CO: Westview Press, pp. 245–275.

Zevin, S.F. 1994. Steps toward an integrated approach to hydrometeorological services. *Bulletin of the American Meteorological Society.* 75: 1267–1276.

The Asteroid/Comet Impact Hazard: Homo Sapiens as Dinosaur?

Clark R. Chapman

The earth is subjected to a continuing "celestial rain" of cosmic debris. Nearly all of it is fine dust or larger objects that burn up in the atmosphere and are of little consequence to our environment. Occasionally, pieces of larger, stronger objects survive as meteorites. Every few years a meteorite penetrates a roof, strikes a car, or otherwise annoys us; such events are newsworthy because of their extraodinary nature. Yet the size distribution of cosmic debris has no large-size cutoff (comets of unlimited size could conceivably approach the earth), unlike other terrestrial natural hazards. Very rarely (every few 100,000 years or so, or one chance in several thousand during a human lifetime) a comet or asteroid more than a mile wide strikes the earth with serious global environmental consequences that could well threaten the future of civilization as we know it (Chapman and Morrison 1994). Even more rarely (every 100 million years or so, but it could happen tomorrow) a cosmic projectile five to ten miles across strikes, with consequences so terrible that species (like the dinosaurs 65 million years ago) are threatened with extinction. It is plausible that an even larger object, perhaps larger than twenty-five miles across, will strike the earth during our sun's lifetime, which could virtually sterilize the surface of our planet.[1] The impact hazard has often been discussed in the context of establishing an international telescopic Spaceguard Survey to find threatening asteroids, especially those larger than one kilometer in diameter that dominate the hazard (Harris 1999).

This "impact hazard" by near-earth objects (NEOs) is strikingly different from most natural hazards in two ways: (1) the potential consequences of a major impact exceed any other known natural or human-made hazard (including nuclear war), and (2) the probability of a major impact occurring in a politically relevant timescale (e.g., during our lifetimes) is extremely low. However, multiplying the probability of a major impact by its consequences yields an annualized death rate similar to that of some of the traditional natural hazards discussed in this volume (hundreds of deaths per year worldwide, primarily due to near-earth asteroids [NEAs] one to three kilometers in diameter, which have a one-in-a-thousand chance of striking during any century, producing a global climatic catastrophe that would threaten civilization: see Morrison, Chapman, and Slovic 1994). Cosmic impacts, therefore, are the ultimate high-consequence, low-probability hazard.

In the context of prediction, a feature of this hazard is that scientific awareness has evolved from nearly zero two decades ago (when a serious impact would certainly not have been predicted) to one in which three quarters of potential impacts could, by about the year 2010, be predicted so exactly that effective, technologically feasible mitigation measures could be applied. For the remaining one quarter, the prediction would either be too late to mount more than partial countermeasures or might even occur as a wholly unforecast "act of God." The public, including civilian and military officials, were once wholly unaware of the impact hazard. Now it is a common cultural touchstone, largely due to an "impact scare" in March 1998 followed by release of two major impact-disaster movies.

Current technology can, in principle, provide precise predictions about when and where asteroids that have been discovered might impact. (Astronomers have been applauded for centuries for accurate predictions of sunrise, sunset, and eclipses, and engineers routinely guide spacecraft to distant planets.) Technical approaches to mitigation (e.g., deflection of the oncoming object or evacuation of ground zero), while preliminary in development and very expensive, are inherently simpler than for many hazards. Lead times before impact will likely be measured in decades rather than months or years. And for those potential impactors that haven't been discovered (so far the great majority, although that may change if the Spaceguard Survey is fully mounted), the probability of impact is a simple cosmic lottery: our inherent chances of losing are readily calculated and well known (Chapman and Morrison 1994). There are, of course, uncertainties in the environmental consequences of impact—just as there are for other phenomena like storms, global warming, the ozone hole, and El Niño—but the sudden, exceptional drama of an impact event itself would overwhelm the perceived importance of such uncertainties. (Even the 1979 reentry and impact of the relatively tiny Skylab caused public anxiety.)

However, the impact hazard does raise major issues about prediction. Those issues concern how individuals, society, and political institutions may prepare for and respond to predictions of such a horrific but unlikely disaster as the impact of a body about a kilometer in size. Given the difficulty people have in conceptualizing an event never witnessed during modern times, and the challenge of "rationally" responding to very low probability events, policy makers may focus on uncertain or false impact predictions (or exhibit excessive alarm about actual but innocuous impacts). In such cases, the predictions will be the consequential events, not the impacts.

The 1997 XF11 Affair

An erroneous prediction of a possible impact was issued on March 11, 1998. Supposedly credible astronomers made public assertions that there was a "small" (e.g., more than 0.1 percent) chance that the earth would be hit in the year 2028 by a mile-wide asteroid (1997 XF_{11}, hereafter XF11). This first apparently "official" prediction of a significant chance of a dangerous asteroid impact during our lifetimes dominated news around the globe. A day later, headlines expressed relief as newly found observations were reported to yield a miss distance twenty times farther away than originally reported and zero chance of impact. Follow-up coverage was often noncritical toward the astronomers who made the predictions, yet a cynical undercurrent persisted about the initial hasty and erroneous warning. National Public Radio commentator Daniel Schorr asked on March 22, discussing the asteroid event as well as another premature announcement about terrorists, "Couldn't they have waited a few days before scaring us half to death?" Behind the scenes, a complex, often bitter exchange of e-mail ensued between asteroid researchers trying to assess data while dealing with a media frenzy. Meanwhile, astronomers continued to research impact predictions, and by spring 1999, when two other cases analogous to XF11 occurred, a constructive attempt was made to devise a generic approach for responsibly dealing with such predictions in the future.

Discovery of 1997 XF11
Using the Spacewatch telescope on Kitt Peak, Arizona, Jim Scotti discovered 1997 XF11 on December 6, 1997. It was the brightest NEO discovered by Spacewatch—a program focused on smaller and fainter NEOs—in more than a year. Follow-up observations by Japanese amateur astronomers permitted calculation of an approximate orbit for the asteroid with a MOID (minimum orbital intersection distance with earth) of close to zero, making it the 108th known potentially hazardous asteroid, or PHA.[2] In general, an asteroid's orbit has a different size,

ellipticity, and orientation than the earth's orbit; most such orbits do not intersect the earth's orbit, so that a collision could never occur, barring an exceptional and predictable change in the asteroid's orbit (e.g., by passage of another planet). If an orbit comes as close as 0.05 astronomical unit (AU) to intersecting the earth's orbit, then small changes in the asteroid's path—perhaps induced by the earth's gravity itself during close passages—could make the orbits intersect. Even then, the chance of a collision is very small, because the asteroid and the earth would have to be at the intersection point at the same time. Because of the consequences of any collision, however, even small probabilities must be taken seriously. For this reason, once an asteroid is discovered, the astronomer's first task is to take sufficient measurements of the asteroid's position to calculate its orbit and determine if it has a small MOID, and whether that MOID might decrease to zero in the next century. If so, more observations may be needed to be sure that the position of the asteroid as it moves around its orbit is never located at the intersection point with the earth's orbit during the fraction of an hour (each year) that it takes the earth to pass through the intersection point.

Brian Marsden, who directs the International Astronomical Union's Minor Planet Center (MPC) in Cambridge, Massachusetts, maintains a list of PHAs on his Web site. His classification of XF11 as a PHA, a few weeks after discovery, placed it higher on observers' priority lists; but as new measurements trickled in, astronomers remained generally unaware of any near-term close approach of XF11 to earth, for reasons discussed below. By February, Marsden's private orbital calculations and extrapolations indicated to him that XF11 would pass by the earth roughly twice as far away as the moon (0.005 AU) in October 2028.

As new observations are added, orbits can be recalculated, resulting in improvements with smaller uncertainties. In early March, Peter Shelus of the University of Texas submitted new positions of XF11 to Marsden. Marsden did not post Shelus's positions on the Internet—his practice was to embargo orbital data for a month—but he did use them in a new calculation of XF11's potentially close pass in 2028. Marsden was amazed that the nominal miss distance jumped down to just 46,000 km (29,000 miles, or 25,000 miles above the earth's surface); from Marsden's decades of experience, this seemed to him to be a remarkably close approach.

Announcement of Impact Possibility

Without checking with any outside scientists, Marsden published an International Astronomical Union Circular (IAUC #6837) during the early afternoon of March 11, announcing that XF11 would pass just

0.00031 AU from the earth on October 26, 2028, thirty years hence. As an indication of the estimate's reliability, Marsden wrote that an approach within 0.002 AU—a distance closer than the moon—was "virtually certain." (On April 18 in IAUC #6879, Marsden explained that the "virtually certain" statement was in error and that miss distances could have been 10 to 15 times larger.)

The IAU Circulars—distributed by e-mail, posted on the Internet, and mailed in postcard format to observatories worldwide—are the chief way that astronomers learn about rapidly changing events in the heavens, such as supernovae, x-ray bursters, and comets, which require observational follow-up that can't wait for later publication in scientific journals. The IAUCs, which began as telegrams many decades ago, are an official service of the International Astronomical Union; Marsden edits and publishes them. His stated reason for mentioning the close approach on an IAUC was to motivate new observations of XF11, including searches for any unreported past observations, in order to refine its orbit and the prediction.

Marsden knew that his words would be read by many people besides the targeted astronomers. Subscribers to the IAUCs include science journalists and amateur scientists, among others.

Unusual predictions had previously caused telephone inquiries from reporters. Accordingly, Marsden prepared a kind of press release (a Press Information Sheet, or PIS) prior to releasing the IAUC. The PIS reported that "the chance of an actual collision is small, but one is not entirely out of the question," describing the miss distance as ranging from "scarcely closer than the moon," on the one hand, to "significantly closer than 30,000 miles." Marsden then described the "splendid sight" the object would make in Europe's evening skies during closest approach in 2028. He explicitly asked colleagues to search for prediscovery observations during several particular years. Within half an hour of the posting of the IAUC, astronomer Stephen Maran, press officer of the American Astronomical Society (AAS), obtained from Marsden the already prepared PIS and without further investigation e-mailed it to his list of science journalists.

According to Maran, both the *Dallas Morning News* and the *New York Times* began writing stories based on the IAUC even before receiving the associated PIS. Many news media learn about important stories from a story list put out on the *New York Times* news wire long before publication of the *Times'* first edition. The *Washington Post* was informed, prior to receiving the PIS, by an amateur astronomer "news tipster."

The Swift-Tuttle Precursor

A 1992 event (Steel 1995) provides context for appreciating how Marsden's XF11 announcement was received by some of his colleagues. As astronomers searched for the returning long-period comet Swift-Tuttle, Marsden worked on observers' positional data, just as he later did for XF11. While talking with *Boston Globe* science reporter David Chandler, he made a back-of-the-envelope calculation that the comet would have a 1-in-10,000 chance of colliding with earth during a close approach early in the twenty-second century. Marsden repeated his estimate to the *New York Times*. Don Yeomans, of the Jet Propulsion Laboratory (JPL), later showed that the earth was actually safe from the comet. A public argument between Marsden and Yeomans was highlighted by *Newsweek* in a cover article entitled "Doomsday Science."

Initial Reactions to XF11

Although Marsden never stated a quantitative impact probability for XF11, his words had a straightforward meaning for astronomers. From the quoted miss distance and estimated error of many times the miss distance, a simple calculation indicated a probability of impact that was at least 0.1 percent. Alan Harris, of JPL, was among those to draw this simple conclusion, and he e-mailed it to nine colleagues within two hours of the IAUC's posting. Undoubtedly, many others made the same elementary calculation (Goldman 1998).

Since XF11's observed brightness meant that its size was about one or two kilometers (roughly that of a civilization-threatening impactor), a one-in-a-thousand chance of civilization's demise occurring on a specific date just thirty years hence deserved the public attention it received. Richard Binzel, of MIT, considered the threat to measure 3.5 on his proposed 5-point impact hazard index (Binzel 1997; see below)—higher than he had imagined would ever be registered on the scale. Still, the chance of one of the approximately thousand so-far-undiscovered asteroids as large as XF11 colliding with earth in the next thirty years is about 0.01 percent, only ten times less than the initially calculated collision probability of XF11. In other words, such an impact was not all *that* unlikely. What was remarkable was that such a threatening object had been found when the Spaceguard Survey was barely underway.

Don Yeomans and Paul Chodas at JPL approached Marsden's announcement with understandable skepticism, given the Swift-Tuttle history. They immediately sought Shelus's original XF11 data from Marsden, in order to make an independent assessment. While waiting about two hours for Marsden to comply, they calculated an effectively zero probability of impact from just the data available through early February. Within fifteen minutes of receiving the later data from Marsden, Chodas reaffirmed his previous calculation that the chances for collision in 2028 were zero ("That's zero, folks," Yeomans emphasized in an e-mail sent to numerous colleagues that evening, five hours after the IAUC posting).

Yeomans and Chodas immediately emphasized a vital point apparently overlooked by Marsden. The uncertainty of an asteroid's position at any given time can be displayed as an ellipse. In the case of XF11, the error ellipse was extremely narrow, more nearly a line than an ellipse (more than a thousand times longer than wide), and even though the earth was "near" the center of the ellipse, it still lay *outside* the ellipse (figure 6.1). Indeed, when Chodas attempted to calculate the impact probability, it was less than one chance in ten to the 300th power (sensibly zero to anyone but a mathematical purist).

Two of Marsden's colleagues called on him late on March 11 to distribute, via a new IAUC, the Yeomans-Chodas probability = zero calculation so as to forestall the developing media frenzy. But Marsden rejected the suggestion, arguing that an IAUC—which is a technical document—shouldn't be used to correct text in a press release. By early afternoon of March 12, with Marsden still sticking by his original story, Eleanor Helin (of JPL's Near Earth Asteroid Tracking program, an element of Spaceguard) reported that she had located prediscovery observations of XF11 on films taken at Mt. Palomar observatory in 1990. The multiyear time baseline would permit a much more accurate calculation of the 2028 encounter circumstances. Later in the afternoon, the Helin data were used by both Yeomans-Chodas and Marsden in new calculations (see figure 6.2). Still later (too late for evening newscasts), Marsden finally issued a new IAUC, and the JPL group reported to the press essentially the same thing: the asteroid will miss the earth by nearly a million kilometers, about twice the distance between the earth and the moon.

"No Impact" Aftermath

The following morning's newspaper headlines told the story of the earth's escape from cosmic doom. The *New York Post* was emphatic: "Kiss Your Asteroid Goodbye!" Another *Post* story headlined "NASA Needs a 'Crash' Course in Math" and continued that the "doomsday figures were way off [the] mark."

Headlines in the March 23 issues of *Time* and *Newsweek* illustrated the tone of media reaction to what one labeled a "cosmic false alarm": "Oops, Never Mind!" and "For a day, it looked like we could all be toast as

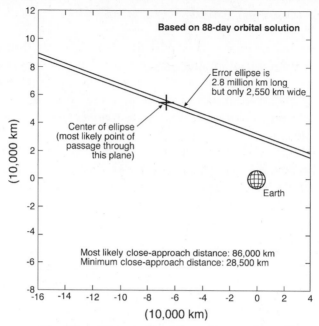

Fig. 6.1a   Near-earth section of uncertainty ellipse in earth target plane on October 26, 2028, for 1997 XF11's passage, calculated from positions obtained December 1997–March 1998 (magnification of figure 6.1b). The asteroid cannot impact earth.

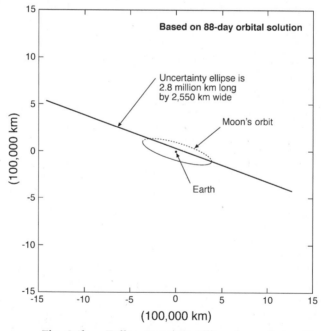

Fig. 6.1b   Full uncertainty ellipse in earth target plane on October 26, 2028, based on positions obtained December 1997–March 1998. Asteroid could pass well beyond the moon's orbit.

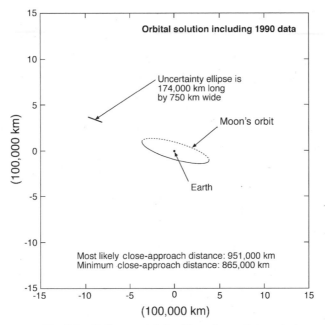

Fig. 6.2 Full uncertainty ellipse in earth target plane on October 26, 2028, calculated with older 1990 positions included. New ellipse falls within the uncertainty ellipse shown in figure 6.1b, but it is now clear that the asteroid will pass far beyond the moon's orbit. From Chodas and Yeomans (1999). The area within the ellipse represents the path of the asteroid at a 99.7 percent certainty level.

an asteroid hurled through space. Then astronomers double-checked their figures." Many reports noted the fortuitous promotion of the Hollywood movies *Deep Impact* and *Armageddon,* due to premier within the following few months. At least one conspiracy-obsessed reporter persistently attempted to uncover illicit connections between Marsden and the Dreamworks producers of *Deep Impact.* While the media generally handled the coverage accurately and benignly, astronomers surely lost some of their vaunted reputation for unassailable predictions, thanks to the XF11 affair.

Several analyses of the interactions between science and science journalists have cited the XF11 affair as a case where science essentially "worked," rather than a case of a wholly erroneous prediction that never should have been made in the first place (cf. Gladstone 1998). Gardner (1998) and Gordon (1998) exemplify the myth that "science worked" in the XF11 case and that, as originally reported just after the scare (Browne 1998), "old photos" of XF11 saved the day. As Muinonen

(1999) later reported, data obtained as early as December 22, 1997, and made available somewhat later were sufficient to calculate a 2028 impact probability of less than 10^{-42}; much smaller probabilities are derived when data through February are properly analyzed. The 1990 Mt. Palomar data weren't needed to rule out a 2028 impact.

Chaos and Keyholes

If XF11 were to pass through one of a few tiny places, "keyholes," along the immense length of the 2028 error ellipse (prior to its shortening by the 1990 data) in 2028, it would hit the earth in a later year. Chodas (1999) concurs with Marsden (1999) that there was about one chance in a hundred thousand that XF11 could have passed through such a keyhole—that is, until the 1990 observations eliminated such possibilities. However, that discovery of tiny chances of impact in 2037 and 2040 hardly justifies the original scare about the zero-probability 2028 impact. Recently, there has been more thorough analysis of keyholes, special resonances, and other phenomena that describe the attributes of the celestial dynamical "chaos" that governs the motions of small bodies that pass through the gravity fields of large bodies like earth. After a close pass, the trajectory of the small body becomes genuinely chaotic and inherently unpredictable over even rather short periods of time. Subsequent encounters can no longer be described by neat "error ellipses," yet Monte Carlo statistical simulations can still be run to estimate impact probabilities. In essence, after a future close approach with the earth (e.g., closer than the moon), an asteroid begins to rejoin the "background" of undiscovered objects—until precise observations are made to determine its new trajectory. In that sense it once again becomes neither more, nor less, dangerous than any other similar asteroid that has not yet been discovered.

1999 AN10 and 1998 OX4

On January 13, 1999, the LINEAR program (a major element of Spaceguard, operated by MIT's Lincoln Laboratory, with U.S. Air Force and NASA funding) discovered a mile-wide earth-approaching asteroid designated 1999 AN10. Andrea Milani, of the University of Pisa, and his colleagues showed that there was approximately a one-in-a-billion chance of an AN10 earth impact in the middle or late part of the twenty-first century. Rather than publicizing this unremarkable result (a thousand times less likely to hit than XF11's might-have-been scenario), Milani sought a peer review. In late March 1999, he sent his manuscript, drafted for a professional journal, to several international colleagues requesting a reply within two weeks. Upon receiving no complaints, he posted his paper (Milani, Chesley, and Valsecchi 1999)—but with no fanfare—on his Web site.

Milani intended his actions to demonstrate a responsible way to handle impact calculations. Indeed, officials of the IAU—including IAU general secretary Johannes Andersen—were kept informed by Milani, and they considered his effort a forerunner of procedures that might be routinely adopted by the IAU for such matters. Unfortunately, in mid-April, Benny Peiser, the moderator of the *CCNet Digest* Internet forum, discovered the unheralded paper on Milani's Web site and charged "cover-up." Journalists picked up the story of AN10, and once again, although without XF11's page-one hype, newspapers around the world reported the slight possibility of an impact with an asteroid. A chance-in-a-billion of impact would hardly seem to be news in a world fraught with war, disease, and accidents, but obviously the idea of catastrophic impacts, however unlikely, arouses great public interest. It is now known that AN10 cannot possibly hit the earth during the next century.

Milani arranged a more focused challenge for attendees of the international IMPACT workshop (sponsored by the IAU, NASA, the European Space Agency, and other entities) held in Turin, Italy, the first week of June 1999. On the meeting's first day, Milani claimed to have discovered yet another asteroid with a non-zero chance of an earth impact on a particular date in the next century. He asked that the group devise clear recommended guidelines, which he promised to follow. Should he just release the information, or should he submit his analysis for peer review? What specific criteria should he use to decide? The challenge helped inspire the drafting of recommended procedures. On the final day of the meeting, Milani was able to apply those procedures; by the recommended criteria, the object—a recently discovered PHA named 1998 OX4—was both too small (a few hundred meters in diameter) and too unlikely to hit (less than a one-in-a-million chance) to be worthy of peer review (or news media interest, for that matter). It also scored as a zero in Binzel's new 10-point impact hazard scale, which I describe below.

Astronomers studying the behavior of earth-approaching asteroids increasingly realize that impact probability is in no way constant. In reality, it is represented by numerous "spikes" of increased probability for individual asteroids, which occur on specific future dates when their MOIDs are nearly zero and the asteroids may be near an intersection point with the earth. Our knowledge, today, of whether a particular asteroid has a chance of hitting on such a date is rendered uncertain by two factors. First, observational errors in the asteroid's orbit and position along its orbit can always be improved by additional telescopic positional measurements. Second, should there be a near-term (i.e., interim) close pass by the earth, then the chaotic uncertainty introduced by earth's gravity may make it impossible to calculate future encounters accurately.

Indeed, 1998 OX4's location is highly uncertain. While its orbit is well determined, its position in its orbit is very poorly known: it could be at virtually any orbital longitude. However, Milani, Chesley, Boattini, and Valsecchi (1999) have developed a way to learn for certain whether OX4 will hit during the next century, without requiring an impractical search around 360 degrees of sky to recover it. They have calculated the very few places where OX4 could be in its orbit that would result in collision. So if astronomers look in just those few places and don't find OX4 there, then we can know that we are safe from its minuscule threat.

The studies of 1999 AN10 and 1998 OX4 have greatly increased astronomers' understanding of the nature of NEO encounters and of how to assess impact probabilities. At the same time, Milani's pedagogic use of these cases to explore ways of handling public release of impact predictions is helping entities like the IAU and NASA to develop protocols for assimilating and reporting such results.

Issues Raised by the XF11 Affair

The 1997 XF11 impact scare raises issues involving the affected communities—asteroid astronomers, institutional and public officials, the news media, and the public—concerning both their internal workings and their mutual interactions. This section evaluates several such issues, focusing on the astronomical community, of which I am a part.

Astronomical Issues
Urgency vs. Peer Review.
A fundamental dilemma within the predictive sciences concerns the balance between being careful and being timely. One wishes not to issue a hazard warning without good basis; yet if one waits too long, it might be too late. Because astronomers usually work for universities, where the only pressure to conclude a research project is "publish-or-perish," the IAU Circulars were created to handle matters of comparative urgency. Some things change in the sky and require observational follow-up quickly, compared with the year or more it takes to publish a paper. Marsden was operating within this tradition when he called for

observations of XF11 in his March 11 IAUC. Although the IAUCs are a form of publication, they are issued quickly, and generally without peer review. Of course, they rarely deal with phenomena of such popular and practical importance as a catastrophic impact.

As in the Swift-Tuttle case, Marsden was criticized because of the *public* announcement of a sensational, unchecked result that proved wrong. A lesson for astronomers surely is that the normal, out-of-the-limelight, ivory-tower protocols that work for informal communications among themselves have very different effects if the topic treats potential danger and thus is inherently sensational. One could readily understand Marsden's haste had the prediction demanded an immediate response. But with thirty years to go, there was no public-policy urgency. In an astronomical context, it is always desirable to try to follow up interesting NEOs quickly, since they may become much fainter within hours or days. (The problem is compounded by the inflexible tradition of assigning observing time on telescopes typically half a year in advance.) However, there was less urgency for XF11 because any new observations of it during spring 1998 (before it became unobservable near the sun) would reduce errors little; moreover, as a relatively bright asteroid, it could readily be reobserved in later months or years.

More careful procedures should be adopted for communications among astronomers about inherently sensational topics. An informal approach might have worked. For example, Marsden could have omitted mentioning the 2028 encounter in the IAUC and simply sought additional observations for this "very interesting" asteroid. Or he could have just telephoned the several most likely observers who could make additional observations. Marsden might answer that there are many potentially interesting asteroids and he needed to highlight the reason for raising XF11's priority. Such an approach trusts the community to withhold predictions until they are checked. In the future a more formal approach may be needed to prevent the release of hasty, and inevitably public, announcements.

Technical Confusion and Human Error.

Specialists who calculate orbits normally have weeks or even years to do their work. The perceived urgency of the XF11 case resulted in unusually hasty, error-prone work, done under a media spotlight. It is human to err, and scientists operating outside of their normal routines are likely to err, as they did. In such situations, researchers should realize that errors of interpretation or judgment might cause mistakes *much larger* than formal, strictly technical, levels of uncertainty. Until requirements of "checking one's math" and peer review are met, there must be restraint in believing and interpreting one's preliminary results, let alone broadcasting them to the world.

Consider Marsden's original report (that XF11 was "virtually certain" to be closer than the moon), which was seriously erroneous. Marsden first responded that his calculations were technically correct, while admitting that his words had been "unfortunate." Later he confessed to initial misunderstandings of his technical calculations. What triggered Marsden's excitement—that the center of the error ellipse happened to fall near the earth—was actually accidental and insignificant. Moreover, various uncertainties that could have affected anybody's calculations were not fully evaluated. The data sets available for orbital analysis are not uniform, as they come from various amateur and professional observers, are subject to various systematic errors (e.g., reference star positions), and are distributed peculiarly in time (e.g., due to the difficulties of observing during the two weeks when the moon is bright). It can be misleading to apply formal statistics to evaluate uncertainties in orbits calculated from such disparate data. Scientists routinely report "formal" estimates of errors, occasionally multiplied by an arbitrary factor (like 3) to account for these unknown errors. In the case of XF11, several calculators (in addition to Marsden) reported errors that were too small.

The real disagreement, however, was over the possibilities for earth impact, which is not the same as the possibility that the asteroid would pass *near* the earth. The "That's zero, folks" conclusion of Yeomans was, in some ways, as misleading as Marsden's assertion of the "small" chance of impact. For one thing, it was too soon for Yeomans and Chodas to be sure they hadn't made a mistake; indeed, they were laboring under false impressions, thanks to an error in their plotting software. Also, the word "zero," when used by a mathematical scientist, implies precision that was premature, no matter what the formal errors were. Furthermore, prior to Chodas's new calculation based on Helin's 1990 data, other members of the very small community of asteroid orbit experts had made their own estimates of collision probabilities.

Karri Muinonen, of the University of Helsinki, reported a *maximum* probability of impact in the range between two and four in a million. The fact that no other expert had yet settled on a zero probability should have cautioned that it was premature to announce zero to the media. NASA official Carl Pilcher used more appropriate common English words in simply stating to the media that the asteroid was "not going to hit."

Asteroid astronomers need to appreciate what many more practical scientists have long known: they must report uncertainties that take into account factors beyond just the formal uncertainties that they normally discuss with each other. A public official needs to know what the real chances are of the predicted event happening, and that includes the chance, which is uncomfortable for a researcher to deal with, that he or she has made an error. For the impact hazard, where probabilities of only one in a million must be taken seriously because of the potential consequences, one must jump into an unfamiliar frame of mind about uncertainties. Human beings are familiar with estimating odds involving flips of a coin, or even 1 percent, but the small odds that typify impact probabilities are outside the realm of experience. The chances for computer errors, illness of the researcher, or a host of usually minor technical problems are far larger than the infinitesimal chances of impacts, so it is irresponsible for a researcher to insist, for example, on the validity of technical odds that are smaller than, say, one in a billion. Uncertainty calculations must include the possibility of human error.

Centralized vs. Distributed Evaluation.
Marsden's scientific competitors have long argued that he should make all data immediately available, so that his work may be checked and so that the data can be rapidly used for other purposes. NASA, which has recently funded part of the MPC's operations, has tried to mandate such an open policy. The fact that the critical data that motivated Marsden's March 11 announcement were not yet public brought this matter to a head.

Several NEO researchers, primarily observers who provide Marsden with positions, have defended Marsden's March 11 IAUC and his unique role as guardian of the data. Through him, they have seen their individual observations integrated into something important (asteroid and comet orbits). Several of these observers have prominently advocated on the Internet that Marsden should be praised, rather than criticized, for the XF11 announcement because the higher public visibility may yield increased funding of the observers' programs.

However, if the data were available to all experts, independent verifications of potential impacts would be easier. Marsden's colleagues have different, independent computer programs for calculating orbits and impact probabilities. As exemplified by Chodas's quick work upon receiving the full XF11 data, analytic calculations are fast (and the more thorough Monte Carlo simulations take only a matter of hours to a day or so). Whether such checks and balances would actually have come into play had the data been available is another matter. For example, Chodas's calculations made a few hours before receiving the full XF11 data, which were more than sufficient to rule out an impact, could have been done weeks earlier but simply weren't.

As a result of the XF11 affair, Marsden is making a larger fraction of NEO data available much more rapidly than before. Also, other researchers are now making routine calculations of future close encounters. But it remains to be determined whether formal (and funded) responsibility will stay with a single person or instead be distributed within the international community, as is now feasible, thanks to the Internet. The IAU, assisted by Milani's handling of the AN10 and OX4 cases, is beginning to formalize some voluntary internal procedures for vetting a potentially sensational prediction by a timely peer review before announcement. While voluntary, there is little doubt that there would be strong pressure from peers and funding agencies to adhere to such policies.

Issues Involving Interface with the Public
Secrecy vs. Peer Review.

In Internet chats and newspaper editorials after the XF11 affair, some commentators worried about "secrecy" and urged that scientists *immediately* make impact calculations public. This view must be taken seriously, for it resonates with expectations of democratic openness in our society. Moreover, the realities of the Information Age are forcing revision of traditional methods of ensuring credibility of scientific results.

The world has long been exposed to the uncensored babble of anyone who wants to say something and has access to a printing press. Yet society, and the scientific community, have developed ways to sort out the wheat from the chaff. With the advent of the Internet, the babble has become a roar and new methods must be developed. Traditionally, the scientific community has "certified" reputable results by having them checked, peer-reviewed, and published in technical journals. Also, by self-regulation (or by editorial policies of journals like *Nature*), scientists are discouraged from going to the media before the publication date of the journal. Violators of this tradition often meet with skepticism. For example, just days before the 1997 XF11 affair, there was Internet chat of a claim by unknown foreign scientists that the asteroid Icarus might impact earth in the year 2006, but the false story generated few headlines around the world, presumably because the source was suspect and because such "official" sources as the MPC did not endorse the prediction.

Particularly loud complaints were made about guidelines suggested by NASA, in the immediate aftermath of XF11, that its officials be informed, say, twenty-four hours before public announcement of an impact prediction. Scientific organizations have long required peer review before publishing research results. And there is an understandable institutional desire as well as potential public benefit for NASA officials to be given a heads-up about a public announcement before the press calls on them unawares; indeed, it has long been standard NASA operating procedure that its officials be informed prior to press releases based on NASA data. One can also appreciate distrust by a public wary of government cover-ups, especially internationally, so the competing interests need to be balanced. It will be far better to negotiate these issues in advance than to rely on arbitrary, ad hoc responses to an evolving crisis. Other cases in this volume describe effective means for analyzing predictions in a manner that balances scientific credibility with public openness. Approaches like the Nuclear Waste Technical Review Board (see chapter 10) or the California Earthquake Prediction Evaluation Council (chapter 7) should be evaluated by the space science community.

Media Relationships.

The relationship between scientists and journalists typically involves fractious mutual disrespect (Hartz and Chappell 1998). Thus many astronomers, in the wake of the XF11 media frenzy, believed that journalists were responsible for the sensationalism. While media accounts of XF11 included the usual journalistic errors, the fundamental fault lay with the scientists. Given the statements in Marsden's PIS, and given that Marsden was as reputable and official a source on this topic as anyone in the world, the subsequent headline treatment was fully warranted. Even at small odds, the threatened impact in 2028 was clearly sensational on its face. The public policy mistake was not the error in prediction, but the failure to take simple steps (checking with colleagues), which would have demonstrated the error before the announcement.

Uncertain Orbits and the Hazard Scale.

Several asteroid experts have been concerned about the evolution from first discovery of an NEO to the eventually certain prediction that it will not (or unluckily will) hit the earth, and how to evaluate and report such evolving results. Bowell and Muinonen (1992) first presented a might-be scenario. This was amplified by Chodas and Yeomans (1997). Initial observations of an NEO permit only crude estimates of its orbit, which might admit of a tiny chance of earth impact in the future. Subsequent observations will shrink the size of the error ellipse (unless there is a close approach and the earth's gravitational field introduces a chaotic element into the asteroid's trajectory); the smaller ellipse almost always excludes the earth. In the highly unlikely case that the smaller ellipse included the earth, the unexpected scenario of a very close approach or actual impact would have to be addressed.

1997 XF11 presents an excellent example of such an evolving situation, beginning with Marsden's late December 1997 listing of XF11 as a PHA and concluding with the March 12 evaluation of 1990 observations that ruled out any impact in 2028. Had anyone checked the data and made proper calculations, it would have been known from the outset that the earth was *never* within the error ellipse in 2028. According to Chodas and Yeomans (1998), the highest-impact probability their algorithm ever would have calculated subsequent to December 24, as data trickled in, was 2×10^{-30}, based on the sixty-five observations available as of January 10. Muinonen (1999), using the Monte Carlo technique, finds even smaller collision probabilities based on the early data alone.

Some NEO observers argue that the answer to any uncertainty in the potential close approach by an NEO is acquisition of new data. They argue that debates about what analysis of pre-March data on XF11 might have shown are moot: the correct thing to do, they say, is always to get more data and make sure. Unfortunately, there are unintended but inevitable consequences of using a prediction to motivate astronomers to obtain more data: for example, scaring the world half to death. In future cases, it may be difficult to obtain timely data—the vagaries of weather and rapid motion of an NEO may delay successful observing for months. So it is vital to assess uncertainties of calculations based on partial data sets as knowledge of the orbit of a new PHA is gradually improved. A lesson that *should* be learned from the XF11 affair—a lesson obscured by the common media story that the prediscovery data were responsible for correcting Marsden's original prediction—is that if the choice is to pay observers to make follow-up observations *or* to pay for better orbital calculations using existing data, the latter might often be more effective and more timely.

How do we best make public the discovery of a new NEO that can't yet be ruled out as potentially impacting the earth in the foreseeable future? In some cases ambiguities may persist for months or even years, depending on asteroid observability. The example of 1998 OX4 is one with *very* low (but non-zero) probabilities—less than one in a million—of hits on January 20, 2044, and again on January 20, 2046. In the time between initial discovery and resolution of the ambiguities, how should such a prediction be presented to the public?

Richard Binzel, of MIT, has attempted to condense the detailed complexities facing astronomers by employing a simple numerical scale, like the Richter scale used to measure earthquakes or the scale used to characterize hurricanes. He proposed (Binzel 1997; Verschuur 1998) a five-point scale, and associated evaluation of the appropriate level of concern, that is based on two vital measures of a specific impact threat:

(a) probability of impact and (b) size of the impactor, hence impact energy and resulting damage. Clearly, we are more concerned if the impact probability is high or if the threatening object is large—a few km across as distinct from a meteoroid that will burn up harmlessly in the upper atmosphere. Another feature of a potential impact, not included in Binzel's scale, is how soon the predicted impact will occur. Obviously, the seriousness and urgency of responding to a predicted impact depends on whether it will occur in a year, a decade, or a millennium.

In consultation with colleagues and science journalists, Binzel has modified his scale and made it explicitly applicable to evaluating predictions of specific impacts that might occur during the twenty-first century. It is now a ten-point scale (figure 6.3), and its graphical presentation emphasizes that zero and one are the only numbers likely to be used. The boundary between 0 and 1 is defined to be 1/10 the annual chance of impact by a kilometer-sized or larger asteroid, or one chance in a million (a common rule-of-thumb threshold for significance in the field of risk assessment; cf. Okrent 1987).

The higher numbers are defined primarily to serve a pedagogic purpose, as users learn about the impact hazard by trying out "what if"

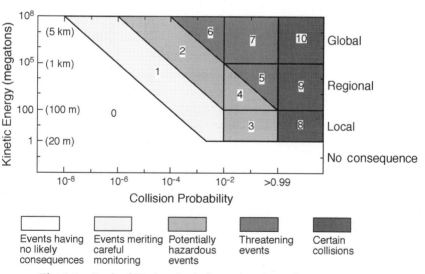

Fig. 6.3   Revised Torino Scale for rating the seriousness of concern for impact predictions that may be made for specific dates in the twenty-first century. Associated with each number, 0 through 10, is a one- or two-sentence description of the category. (Courtesy R. Binzel)

scenarios. The scale was endorsed by the Turin IMPACT workshop in June 1999. Whether it will be adopted by official bodies and actually used by astronomers and science journalists remains to be seen. Furthermore, just when and how it should be used remains problematical, as illustrated by Binzel's judgment that XF11 was an astonishingly high 3.5 on his original 5-point scale, due to unquestioning application of the erroneous estimate of 0.1 percent chance of impact.

NASA and Scientific Community Reactions.

NASA has walked a tightrope with regard to the impact hazard. Initially, NASA, perhaps scared of the "giggle factor" that suffused news reports of the impact hazard in the early 1990s, responded minimally to calls by Congress that it study the hazard and propose a program to deal with it. A previous associate administrator for space science regarded the impact hazard as lacking *scientific* importance; other parts of NASA were happy to leave "planetary defense" to the Defense Department. After the XF11 scare, NASA's stance shifted as asteroid impacts became the second most common subject of public communications directed to NASA administrator Daniel Goldin (after the "Face on Mars"—an area on the Martian surface that looks, on some satellite images, like a human face). In his May 1998 testimony to the House Subcommittee on Space and Aeronautics, Carl Pilcher, director of NASA's Solar System Exploration program, committed NASA to achieving the goals of the Spaceguard Survey—discovery of 90 percent of earth-approaching asteroids larger than one kilometer by 2009. NASA claims that it has doubled its previous $1.5 million annual funding of NEO search programs; that amount probably needs to be doubled again if Spaceguard goals are to be met (Chapman 1998). Harris (1999) believes that the current discovery rate lags behind what is needed to meet the goal by a factor of about eight; achieving this probably requires constructing one or more two-meter wide-field telescopes, which is unlikely given NASA's present budget.

Pilcher has expressed more concern about smaller impactors than the kilometer-scale objects being sought by Spaceguard. Those, of course, are more likely to occur on the watch of politicians and public officials, despite the fact that once per century, several-megaton impacts threaten only about 1 percent of the total expected mortality of all natural hazards (floods, earthquakes, etc.) of similar intensity (Morrison et al. 1994). In fact, NASA is even more concerned about impact predictions, since the XF11 event caught NASA officials unprepared, and since predictions, erroneous and otherwise, will occur with much greater frequency than the one in a hundred thousand years between actual impacts of large bodies.

Commentary and Conclusions

Implications of the Spaceguard Survey

Inevitably, the hazards community will be confronted by the impact hazard, especially as more potential impactors are discovered. Roughly 1,600 NEAs of greater than one kilometer in diameter are thought to exist (Harris 1999). About 290 of these—plus ten earth-approaching short-period comets—have been discovered. Thus, the vast majority of asteroids large enough to potentially threaten civilization (and virtually none of the comets) have not yet been discovered. Should the Spaceguard Survey be undertaken at full strength, that situation will change over the next decade to one in which *most* of the potential impactors will have been surveyed, and it will have been reliably determined whether or not (almost certainly not, but if so, then when and where) any of them will impact earth during the twenty-first century. Even if the Spaceguard Survey is not formally undertaken, rapid advances in astronomical instrumentation—including that available to amateur astronomers—combined with the augmented interest in the topic are likely to accomplish its objectives in the next several decades.

During this time of augmented discovery, there could be many "false alarms," especially because of the inevitable period of uncertainty between first discovery and eventual determination that an impact is impossible. Even though the odds (one in a thousand) are strongly against finding even one such large object in an orbit that will actually strike earth in this century, some asteroids will be found that, for a period of days or conceivably years, will appear to have a non-zero (even if very low) probability of collision with the earth. Such a scenario is even more likely for the many smaller objects that Spaceguard will inevitably discover in the course of searching for kilometer-sized bodies. Indeed, the size of such objects may be uncertain for some time, and several factors (observational biases as well as psychological factors) may cause small objects to be reported as being much bigger, hence more dangerous, than they really are.

Preliminary attempts to establish regular communication and coordination between the potential discoverers of NEOs (professional and amateur astronomers in the United States, Japan, France, Russia, China, and other countries, as well as Air Force observers) and those who calculate orbits and collision probabilities are in their infancy. In the current, uncoordinated environment, competitive pressures actually work against efficient attainment of Spaceguard goals. Observers feel rewarded for the most discoveries, which is accomplished by "skimming the cream" of asteroids near the opposition point (directly opposite the sun in the sky, where asteroids are brightest) and ignoring the sky along the earth's orbit (where similar-sized asteroids are fainter, but which includes a larger fraction of bodies that might actually hit the earth). Furthermore, the diverse telescopic resources being applied to Spaceguard cannot be used optimally if the leading program—currently, the U.S. Air Force effort—doesn't leave some of the sky for them to search in. Analysis of the total system, sharing information about observing plans and actual sky coverage, and active coordination are among the first steps toward rectifying this situation, and those are goals of NASA's newly formed NEO Program Office at JPL. Perhaps the incentives for participation in the Spaceguard Survey can be restructured to channel efforts more constructively toward the project's goals.

Communications and coordination between such astronomers and those public or military officials who might be called on to respond to potentially threatening impacts remains wholly ad hoc. Fortunately, it is much more likely that a threatening object will be found decades, rather than hours or weeks, before impact (as exemplified by XF11's thirty-year advance warning). But, especially if sizes are overestimated or if exaggerated attention is given to small objects, imminent impacts could well be thrust into the laps of unprepared high government officials. The fall of Skylab in 1979 and the 1996 crash of a failed Russian Mars probe exemplify how even small space debris can create a sensation. Exploding meteors, with yields approaching a megaton, enter the upper atmosphere occasionally, and are spectacular even if they do little or no damage on the ground. The White House was alerted in 1994 when a modest atmospheric impact (between tens and hundreds of kilotons) was detected by downward-looking surveillance satellites (McCord, Morriss, and Schmidt 1995). Indeed, with widening media interest in the topic, even modest impacts sometimes generate major media reports (such common fireball or bolide events are dramatic, even if they are unthreatening). Bolides or actual ground impacts by meteorites could even be misinterpreted as aggressive military activities (Chyba, van der Wink, and Hennet 1998).

We should plan for action by public and military officials in the face of a genuine near miss, with attendant media and public reactions. For instance, there is roughly a 5 percent chance that an object at least one kilometer in diameter really will, sometime in this century, pass as close to the earth as Marsden originally reported for XF11. And one or two times a decade, an object large enough to cause a Tunguska-like event passes within the same distance.[3] While astronomers might consider these comfortably safe encounters, political and military leaders might take a different view.

Should Society Respond to the Impact Hazard?

Several years ago, *Fortune* magazine published a cost-benefit analysis demonstrating that spending hundreds of millions of dollars a year on planetary defense was worthwhile. Some advocates of planetary defense have argued, in effect, that billions should be spent annually. Gerrard and Barber (1997) show how one could defend even an annual expenditure of $16 to $32 billion, thousands of times what is now being spent. However, others argue that the probability of impact is so low that few public funds should be expended on this hazard—instead, efforts should focus on other hazards that *certainly* will happen (even if with more modest consequences) during the next years.

It is difficult to think clearly about how to deal with this unusual hazard. A common mistake is to think that there is no reason to deal with it during our lifetimes (or for a politician to think it is irrelevant to his or her "watch") because there is a long waiting time until the next impact. The chances of a civilization-threatening impact are extremely low during a politician's term, but they are no different from chances during any other similar time interval, whether during the next decade or several hundred thousand years from now. Similarly, people build on floodplains, perhaps thinking, "The hundred-year flood occurred here a few years ago, so I need not worry." It is a faulty argument. A valid argument is that the *chances* of a flood (or an impact) are so low that we choose to do nothing (or little) to avoid the consequences and, instead, spend our funds to mitigate lesser but more certain hazards.

Another choice is to let future generations, with their no-doubt superior knowledge and technology, deal with the problem. But that choice avoids responsibility for consequences that could happen on our watch. Future generations cannot deal with an impact that happens during the next thirty years, but such an impact could have a catastrophic effect on future generations. To be sure, the odds are so extremely small that such an impact will happen during our lifetimes that it may well be justifiable to do little or nothing about it. The choice is necessarily subjective, but the issues need to be engaged. It seems to me that, at the very least, a conscious decision should be made by society about whether to deal with the impact hazard or not.

Implications for Predictive Science and Policy

Astronomy has, traditionally, been the predictive science of unexcelled precision. Of course, some astronomical predictions are problematic (e.g., predictions of the date of the next "great" Leonid meteor shower, of solar flares, or of how bright a comet may become). Still other aspects of astronomy are so uncertain as to preclude prediction. The impact hazard, since it is based on the same kind of physics that successfully delivers spacecraft to planetary encounters, would seem to be fairly foolproof, but the 1997 XF11 affair illustrates the pressures and interests that can complicate the application of the laws of physics and lead to unreliable predictions.

Public officials rarely have to deal with astronomical predictions. Powerful solar flares, which can disrupt radio transmissions and satellite operations, are the chief exception. The impact hazard may be another. Experience with the recent history of impact science suggests that whenever the public is introduced to a new type of hazard prediction, one can expect problems while scientists, unpracticed in communicating their predictions in ways useful to the public or public officials, learn how to do that. On the other side of the coin, journalists and public officials called on to deal with a new type of hazard prediction need to learn about the probabilities, sources of human error, and potential societal effects of both the hazard and the prediction of the hazard.

While prediction of asteroid impacts is, on its surface, more scientifically straightforward than most, if not all, of the other cases presented in this book, it still raises many similar problems. A prediction itself (as in the case of 1997 XF11) may dominantly affect public perception of the entire issue and may prove to be more significant than the predicted event. Policy responses must negotiate several obstacles, which include: competing scientific approaches to prediction; difficulties in assessing uncertainties and reporting them to the public; differing societal and international contexts for interpreting predictions; and potential conflicts of interest among both scientists and decision makers. Scientists should continue to try to educate the public about the hazards, but their lack of knowledge of the societal and political context often hampers their efforts. The ever more rapid and superficial exchange of information fostered by popular culture, educational institutions, and the Internet is inimical to the kind of in-depth understanding of complex technical issues on which society's future may depend.

Some technical experts in NEOs, as well as many of the best science writers in the world, largely failed to understand the essential realities of how the false prediction of XF11 was made and retracted. A year and a half after the XF11 frenzy, the predominant view remained that the matter was resolved by finding prediscovery observations of the asteroid on existing photographic plates. One might conclude from this interpretation that more attention should be given to increasing the efficiency of finding NEO images in archives. Actually, the predominant failings were (a) failure of the NEO community to analyze publicly available data, which would have demonstrated that the asteroid presented no hazard; and (b) failure of individuals and institutions to adopt and adhere to procedures that would increase the reliability of public announcements. We hope that these failures will foster greater understanding between those who make predictions and those who use them.

Acknowledgments

I thank David Morrison for his direct and indirect assistance. Discussions with Alan Harris, Ted Bowell, Rick Binzel, Paul Chodas, Don Yeomans, Hal Levison, and numerous other colleagues in astronomy, as well as participants in the two prediction workshops, have helped me to formulate the issues that face us.

Notes

1. Technical background on the impact hazard is found in two recent books: Gehrels (1994) and Remo (1997).
2. PHAs are asteroids for which MOID <0.05 astronomical units, or 7.5 million km. One AU = 149,600,000 km, the mean distance between the earth and the sun.
3. In 1908, near Tunguska, Siberia, a fifteen megaton-yield asteroid explosion flattened trees for miles.

References

Binzel, R.P. 1997. A near-earth object hazard index. In *Near-earth objects: The United Nations International Conference*, J. Remo, ed. *Annals of the N.Y. Academy of Science*, vol. 822, pp. 545–551.

Bowell, E., and K. Muinonen. 1992. The end of the world: An orbital uncertainty analysis of a close asteroid encounter. *Bulletin of the American Astronomical Society* 24:965.

Browne, M. 1998. Old photos helped refine progress of asteroid. *New York Times* (March 14).

Chapman, C.R. 1998. The threat of impact by near-earth asteroids (www. boulder.swri.ed u/clark/hr.html) and action plan statement to House Subcommittee on Space & Aeronautics (www.boulder.swri.edu/clark/ actnea.html).

Chapman, C.R., and D. Morrison. 1994. Impacts on the earth by asteroids and comets: Assessing the hazard. *Nature* 367:33–40.

Chodas, P.W. 1999. Presentation at AAS Division on Dynamical Astronomy meeting, Estes Park, Colorado, April.

Chodas, P.W., and D. Yeomans. 1997. Impact warning times for earth crossing asteroids. *Bulletin of the American Astronomical Society* 29:960.

Chodas, P.W., and D. Yeomans. 1998. Could asteroid 1997 XF11 collide with earth? *Bulletin of the American Astronomical Society* 30:1029–1030.

Chodas, P.W., and D.K. Yeomans. 1999. *Orbit determination and estimation of impact probability for near-earth objects*. Presented at the 21st Annual American Astronautical Society Guidance and Control Conference, Breckenridge, Colorado, February 3–7.

Chyba, C.F., G.E. van der Wink, and C.B. Hennet. 1998. Monitoring the Comprehensive Test Ban Treaty: Possible ambiguities due to meteorite impacts. *Geophysical Research Letters* 25:191–194.

COMPLEX. 1998. *Exploration of near earth objects*. Washington, DC: Committee on Planetary and Lunar Exploration, National Research Council, National Academy Press.

Gardner, M. 1998. Near-earth objects: Monsters of doom? *Skeptical Inquirer* (July–August).

Gehrels, T., ed. 1994. *Hazards due to comets and asteroids*. Tucson: University of Arizona Press.

Gerrard, M.B., and A.W. Barber. 1997. Asteroids and comets: U.S. and international law and the lowest-probability, highest-consequence risk. N.Y.U. *Environmental Law Journal* 6(1):4–49.

Gladstone, B. 1998. Report on *Morning Edition*, National Public Radio, June 4, 1998.

Goldman, S.J. 1998. The most dangerous rocks in space. http://impact. skypub.com/rocks.html (from *Sky & Tel.*, June 1998, 33).

Gordon, B.B. 1998. The asteroid caper (editorial). *Astronomy* (July):6.

Harris, A.W. 1999. Near-earth asteroid surveys. In *Collisional processes in the solar system,* H. Rickman and M. Marov, eds. Dordrecht: Kluwer ASSL series.

Harris, A.W., G.H. Canavan, C. Sagan, and S. Ostro. 1994. The deflection dilemma: Use versus misuse of technologies for avoiding interplanetary collision hazards. In *Hazards due to comets and asteroids,* T. Gehrels, ed. Tucson: University of Arizona Press, pp. 1145–1156.

Hartz, J., and R. Chappell. 1998. *Worlds apart: How the distance between science and journalism threatens America's future.* Nashville, TN: First Amendment Center, Freedom Forum.

Kunich, J.C. 1998. Posted on World Wide Web, August 1998, sac.saic.com/space_warfare/planlrev.htm.

Marsden, B.G. 1999. Presentation at AAS Division on Dynamical Astronomy meeting, Estes Park, Colorado, April.

McCord, T.B., J. Morriss, and R. Schmidt. 1995. Detection of a meteoroid entry into the earth's atmosphere on February 1, 1994. *Journal of Geophysical Research* 100:3245–3249.

Milani, A., S.R. Chesley, and G.B. Valsecchi. 1999. Close approaches of asteroid 1999 AN10: Resonant and non-resonant returns. *Astronomy Astrophysics* 346:L65–L68.

Milani, A., S.R. Chesley, A. Boattini, and G.B. Valsecchi. 1999. Virtual impactors: Search and destroy. Posted on World Wide Web, copernico.dm.unipi.it/~milani/virimp

Morrison, D. (chair). 1992. The Spaceguard Survey: Report of the NASA International Near-Earth-Object Detection Workshop. Pasadena, Calif.: Jet Propulsion Laboratory/NASA, January 25.

Morrison, D., C.R. Chapman, and P. Slovic. 1994. The impact hazard. In *Hazards due to comets and asteroids,* T. Gehrels, ed. Tucson: University of Arizona Press, pp. 59–91.

Morrison, D., and E. Teller. 1994. The impact hazard: Issues for the future. In *Hazards due to comets and asteroids,* T. Gehrels, ed. Tucson: University of Arizona Press, pp. 1135–1144.

Muinonen, K. 1999. Asteroid and comet encounters with the earth: Impact hazard and collision probability. In *Proceedings of The Dynamics of Small Bodies in the Solar System: A Major Key to Solar System Studies,* A.E. Roy and B.A. Steves, eds. Maratea, Italy: NATO Advanced Study Institute series 1999, volume 522. Boston, Mass: Kluwer Publishing, pp. 127–158.

Okrent, D. 1987. The safety goals of the U.S. Nuclear Regulatory Commission. *Science* 236:296–300.

Park, R.L. 1992. Star warriors on sky patrol. *New York Times* (March 25), A19.

Rather, J.D.G., J. H. Rahe, and G. Canavan (organizers). 1992. *Summary Report of the Near-Earth-Object Interception Workshop.* JPL/NASA, August 31.

Remo, J.L., ed. 1997. *Near-earth objects: The United Nations International Conference. Annals of the N.Y. Academy of Science,* vol. 822.

Sagan, C., and S. Ostro. 1994. Dangers of asteroid deflection. *Nature* 369:501.

Steel, D. 1995. *Rogue asteroids and doomsday comets.* New York: John Wiley & Sons.

Verschuur, G.L. 1998. Impact hazards: Truth and consequences. http://impact.skypub.com/page1.html (from *Sky & Telescope*, June 1998, pp. 26–34).

Predicting Earthquakes:
Science, Pseudoscience, and
Public Policy Paradox

Joanne M. Nigg

Damaging earthquakes, while not frequent occurrences in the United States, cause great social disruption and economic loss. The Northridge, California, earthquake of January 1994, for example, was the most costly, damaging single disaster in the history of the United States. Although relatively few people were killed in the event (57), over 11,800 people received hospital treatment for earthquake-related injuries, while tens of thousands more went unattended. Segments of two major interstate freeways collapsed, resulting in several months of traffic nightmares for commuters and commercial trucking operations. An estimated 114,039 residential structures were damaged by the quake, 14,500 of which were unsafe for temporary or permanent occupancy. These damaged structures contained over 100,000 housing units; 30,000 were vacated or had significant structural damage, and another 30,000 were at risk of being removed from the building stock because of the expense of repairs (Comerio 1995). According to estimates from the California Governor's Office of Emergency Services (OES) in January 1996, approximately $25 billion in losses due solely to damaged structures and their contents had occurred. As of December 1995, 681,710 applications for state and federal disaster assistance had been received, more than double the amount in any previous single U.S. disaster. (The previous record for applications was 304,369 following Hurricane Hugo, which struck the Carolinas, Puerto Rico, and the Virgin Islands in 1989.) We have only to look at the Great Hanshin earthquake in Japan—one year later to the day—to see how an earthquake of only a slightly larger magnitude resulted in much greater losses (estimated at almost $100 billion) and caused much more serious social disruption, killing over 5,000 people, rendering almost 300,000 people homeless (over 10,000 of whom were still in temporary housing three years later), and creating industrial losses from which the region may take another decade to recover. And the August 1999 earthquake in western Turkey, which killed more than 17,000 people, demonstrates how utterly catastrophic a huge earthquake can be, especially in a region that is not adequately prepared.

Because many regions of the United States are exposed to earthquake hazards, and because few of the jurisdictions in those regions have incorporated seismic design into their building codes or have developed land-use plans that take seismic hazards into account, substantial national concern has developed about the vulnerability of our built environment to future seismic events, especially in major metropolitan areas. Because of its frequent experience with destructive earthquake events during the past century, the state of California has mandated certain types of earthquake preparedness and mitigation by its local governments, but it is clear that much more could be done. Without "reminders" like the Northridge quake that a region is at risk from a potentially dangerous threat, little policy activity is likely to take place, even though policy makers are aware of the hazard itself. There are always other, more commanding local problems—crime prevention, educational improvements, economic development, and job creation—that take precedence over seismic matters.

With valuable infrastructure and human lives at risk, scientists stepped forward in the early 1970s and offered to provide a short-term warning—of hours to days—that a destructive event was about to occur. While predictions could not save the built environment, the hope was that they could make it possible to evacuate buildings, shut down transportation systems, lower water levels in reservoirs, enable families to reassemble or to remain together, shut down production facilities safely, ready medical facilities, and put emergency response units on alert. The purpose of such warnings would be to lessen life loss, reduce secondary economic losses, and keep social disruption to a minimum. In areas outside of California where little seismic mitigation had taken place, prediction was often seen as a cost-effective response to a low-probability, high-consequence event. In California, prediction was seen as providing additional protection in a state that was actively trying to reduce its vulnerability.

The effectiveness of earthquake prediction as a tool for reducing earthquake impacts depends, in part, on developing community response plans that can be implemented when predictions are issued. The overarching policy issue—how to lessen earthquake losses to the built environment and social systems by disseminating forewarnings of future damaging earthquake events—has continued to be the focus of governmental efforts to deal with scientific forecasts, but the specific strategies considered have varied, often due to changes in scientific approaches to prediction.

This case study will trace the interwoven strands of scientific approaches and policy responses to earthquake predictions since the mid 1970s. The state of California has been the focus for concentrated research—in both earth science and social science—on earthquake prediction. Federal policy has played an important role in identifying priorities for both scientists and state and local government officials with respect to the impact of earthquake predictions on society.

Scientific Promise and Policy Response

Terminological Specification, Continuing Confusion

What is meant by the term *earthquake prediction?* Scientists have used a variety of words to describe their efforts to foretell earthquakes: a potential precursor (e.g., the Southern California uplift, popularly known as the Palmdale bulge), a hypothesis test, an experiment (e.g., the Parkfield segment of the San Andreas Fault throughout the 1980s), an earthquake watch (e.g., Parkfield), an earthquake advisory (e.g., the San Diego earthquake swarm), a forecast, a forewarning, and a heightened probability (e.g., San Francisco Bay Area and Southern California). These terms have usually been associated with legitimate scientific projections of future earthquake events; however, pseudoscientific predictions by amateur earthquake scientists (Henry Minturn for Los Angeles in 1976 and Iben Browning for the central United States in 1989) have also been issued using these same terms. Whether or not the source of or the basis for the announcement was scientifically "legitimate," policy makers and the general public have often responded similarly to these prediction events, regardless of what they were called.

In an attempt to clarify what would be desirable in a "credible" prediction from geoscientists, the Panel on Earthquake Prediction (1976) recommended that predictions from the scientific community should include six elements:

- *lead time*—a statement about how far in the future the earthquake will occur;

- *time window*—a statement about the time period, the dates between which the earthquake will occur;

- *magnitude*—a statement about the strength (measured on the Richter or similar scale) of the predicted earthquake;

- *location*—a statement of the geographical area in which the earthquake epicenter will occur;

- *impact*—a statement about the amount of damage that is likely to occur; and

- *probability*—a statement about the likelihood or confidence that the first five parameters will occur as specified.

However, by the early 1980s, members of the scientific community were already trying to retreat from this type of specificity, saying that predictions would be "messy" for years to come and that it was highly doubtful scientists would be able to meet these criteria. While it has been the intention of the earth science community to develop announcements that characterize a future event in these ways, it has been unable to do so. However, policy makers cannot wait until the ideal prediction can be formulated because they are faced with a variety of both legitimate and pseudoscientific forecasts of future (usually damaging) earthquake events for which they must prepare.

Advances in Prediction Science

Southern California residents were introduced to scientific accomplishments in the earthquake prediction field sporadically beginning in the early 1970s. Two examples illustrate the types of scientific advances that were being made and how the science of earthquake prediction was being portrayed to the lay public (including policy makers).

In November 1973, James Whitcomb, a geophysicist then at the California Institute of Technology, predicted a small quake in Southern California based on measured change in sound wave velocity as it traveled through subterranean rock layers. The prediction was considered semi-successful by scientific criteria because Whitcomb correctly identified the time period and area (San Bernardino—Riverside) of impact but not the magnitude (the actual event was smaller). One year later, two scientists from the United States Geological Survey (USGS), Malcolm Johnston and John Healy, were credited with accurately predicting that an earthquake of up to 5.0 magnitude would hit the Hollister area within a few days based on their review of magnetic field and tilt data. The next afternoon, a 5.2 earthquake occurred in Hollister. This accurate prediction was heralded in California newspapers as proof that scientists were on the verge of accurate earthquake predictions, including being able to predict time, location, and magnitude of coming quakes.

In the midst of this optimism, the USGS announced in 1974 that a substantial "uplift" had been discovered on a section of the San Andreas Fault in the Palmdale region of California's Mojave Desert. Similar phenomena had been observed prior to earthquakes in other parts of the world; and, given the large land area that the Southern California uplift covered, some scientists believed that it could be a precursor to a very large earthquake. Other scientists, however, called for intensified research in the area, because uplifts had also been discovered in other places that weren't accompanied by seismic activity. But the recognition of the Palmdale bulge focused public attention on the scientific efforts being made to forecast an earthquake in the not-too-distant future and on what could be done to lessen its impacts.

Throughout 1974 and 1975, the public was made aware of the experimental techniques and theories on earthquake prediction being used in Japan, the USSR, and China. The "successful" Haicheng prediction—made by the Chinese in February 1975 and credited with saving the lives of tens of thousands of people through widespread evacuation and precautionary measures—was touted as a significant contribution to prediction approaches (despite widespread scientific skepticism about this claim in the United States). Similarly, the Japanese were developing an earthquake prediction system, based on monitoring anomalies in Shizuoka prefecture, that included an automated warning system.

Then in April 1975, James Whitcomb, in a paper given at a scientific meeting, hypothesized that an earthquake in the 5.5–6.5 magnitude range could occur anytime within the next twelve months somewhere near the epicenter of the 1971 San Fernando earthquake. In the media, this hypothesis was referred to as a "prediction," a "forecast," and a "warning." A flurry of public attention followed this announcement, with both the public and state and local government officials trying to decide how to respond, especially since Whitcomb didn't want this situation portrayed as a "real" prediction but as the scientific experiment he believed it to be. This distinction was lost on a lay audience, including local policy makers trying to deal with public inquiries about what they would do about this prediction.

In early 1974, the National Academy of Sciences acknowledged the development of earthquake prediction studies and techniques and commissioned two panels to assess the scientific state-of-the-art (Panel on Earthquake Prediction 1976) and to provide an overview of the societal issues raised by earthquake predictions and of the potential social, economic, and political consequences that predictions might create (Panel on Public Policy Implications of Earthquake Prediction 1975). Both studies recognized that the science was still in its infancy and likely to be inaccurate for some time.

These efforts reflect the optimism of the early and mid-1970s and the expectation that scientific breakthroughs would allow predictions, at least in California, to be made within ten years.

The Formulation of a National Earthquake Policy

The scientific developments in California were taking place against a backdrop of destructive earthquake events that had substantial impacts on the national psyche. The 1971 San Fernando earthquake has been referred to as a "watershed" event in California. It was the strongest earthquake to strike an urban area in the United States in almost forty years; it damaged or destroyed structures that were thought to have been "earthquake-resistant" (i.e., constructed to withstand earthquakes); and it almost caused the collapse of the Van Norman Dam—the threat of which led to the evacuation of over 80,000 people for the several days it took to drain the reservoir. Had the dam failed during the earthquake, the resulting flood would have created the largest loss of life ever in the United States from a "natural" disaster.

During this period, earthquakes outside the United States also highlighted the damages and casualties that could result when a large-magnitude event hit a populated area—the earthquake in Guatemala City in February 1975, in which 23,000 people were killed and over a million left homeless; the earthquake in the Friuli region of Italy in May 1975, in which over fifty small towns were partially destroyed; and the two July 1975 earthquakes in the Tangshan region of China, which were estimated to have killed as many as 750,000 people and brought the heavily industrialized region to a complete standstill.

The California congressional delegation had introduced federal legislation aimed at providing funding for earthquake research in the mid-1970s, and the National Earthquake Hazards Reduction Act (P.L. 95-124) was finally passed in 1977 (and amended substantially in 1980). This act established the National Earthquake Hazards Reduction Program (NEHRP), making earthquakes the only natural hazard with a federal mandate to lessen their impacts.[1] One of the initial objectives of the act was:

> the implementation in all areas of high or moderate seismic risk, of a system (including personnel, technology, and procedures) for predicting damaging earthquakes and for identifying, evaluating, and accurately characterizing seismic hazards.

One of the specific research elements of the act required the "development of methods to predict the time, place, and magnitude of future earthquakes." The initial implementation of the act included the development of a plan for the "evaluation of prediction techniques and actual predictions of earthquakes" as well as "warning the residents of an area that an earthquake may occur." The 1980 amendments to the act referred to the importance of NEHRP's timely evaluation of predictions and the need to coordinate with state evaluation mechanisms (discussed below) in order to minimize confusion. From the passage of the act through the early 1980s, earthquake prediction was a major focus of NEHRP, and the USGS, which carried out the bulk of the prediction research, received by far the largest appropriation among the four federal NEHRP agencies.[2]

However, by the early 1980s, the USGS and the earth scientists involved in NEHRP were revising their optimism about their ability to make short-term earthquake predictions in the near future. In 1976, the Panel on Earthquake Prediction had written:

> The Panel unanimously believes that reliable earthquake prediction is an achievable goal. We will probably predict an earthquake of at least magnitude 5 in California within the next five years in a scientifically sound way and with a sufficiently small space and time uncertainty to allow public acceptance and effective response. A program for routine announcement of reliable predictions may be 10 or more years away. (p. 31)

But by 1983, the annual NEHRP report to Congress portrayed a different reality:

> Although significant progress has been made toward earthquake prediction, in general, a simple, highly reliable and universal formula for short-term prediction of earthquakes has not been found. (FEMA 1984, p. 35)

California Responds

As public and governmental interest in scientific information on earthquake prediction intensified in early 1975 and 1976, the Southern California media began to give much greater attention to prediction. State and local officials began to worry about what effects such announcements would have on the public and about how they would actually respond to these announcements. One well-known local politician in Los Angeles made the statement that a scientist shouldn't issue a prediction without being "100 percent certain" the earthquake would occur.

Other decision makers raised the concern that a prediction might actually cause greater harm to the local area than an earthquake, by causing people to "panic," property values to decline, and businesses and families to leave the area, some temporarily but others permanently.

Some social science research suggested the possibility of modest economic and social disruption (Haas and Mileti 1976, 1977). Other studies, however, concluded that this level of public disruption was unlikely (e.g., Turner et al. 1980; Kiecolt and Nigg 1982) and suggested that the public would more likely take a "wait and see" attitude and try to gather additional information to determine how to respond to such announcements.

Also, in response to the concerns about social disruption from a prediction announcement, state and local politicians and agencies began asking questions about what appropriate response should be made to prediction announcements. Part of their answer depended on how confident scientists were about the prediction (that is, on the level of probability attached to some future event), both in terms of theoretical development and the adequacy of data.

In order to provide this type of evaluation for public decision makers, the state of California established the California Earthquake Prediction Evaluation Council (CEPEC), chaired by the state geologist and consisting of government and university earth scientists. Based on CEPEC's assessment of a prediction, the governor would decide whether to mobilize state agencies and inform local municipalities. CEPEC was to function like a "science court," holding a hearing to review the scientific evidence on which a prediction was based and to determine whether there was an increased likelihood that an earthquake would occur during the predicted time window. CEPEC was to meet twice yearly to be updated by scientists about new techniques and theories and gain additional background information about forthcoming predictions. The establishment of CEPEC created a direct linkage between the earth scientists making and assessing predictions and decision makers responsible for public safety.

CEPEC was established early in 1975, principally because of the discovery of the Southern California uplift, and in mid-April of that year the council held its first hearing to review relevant scientific evidence. As a result of the hearing, CEPEC reported to the governor that a major earthquake could occur along the southern portion of the San Andreas Fault anytime within a decade, thus supporting the research of the USGS in this region. Although the CEPEC evalution was covered by the major media in Southern California, it seemed to provoke little public reaction. However, it did cause California's Seismic Safety Commission to issue an "advisory" to cities and counties in Southern California, encouraging them to enhance their earthquake preparedness and response planning.

Shortly after CEPEC's evaluation of the uplift, Caltech's James Whitcomb issued his "hypothesis test." Although the presentation of Whitcomb's prediction experiment was made to colleagues at a scientific meeting, it created much greater popular concern than the uplift, perhaps because it "fit" the more specific public expectation of what a prediction should be. For example, Whitcomb specified a time window of one year, as well as a magnitude range, for the event in a specific area of the Los Angeles basin. These factors made his prediction sound less ambiguous than the scientific information on the uplift, which came from a scientific organization (rather than a person), had a wide time window with an uncertain magnitude, and referred to a location in the Mojave Desert rather than an urban center (Turner, Nigg, and Paz 1986).

Public information-seeking was voracious after stories on Whitcomb were carried by the media. Both Caltech and USGS's Menlo Park office (in addition to local emergency management agencies) were inundated with requests about the meaning of the prediction and what people should be doing to protect themselves. Rumors also began to circulate during this period about actions that the government was supposedly taking (some of which were, incidentally, true) and about the likelihood that the earthquake would be larger than predicted.

In late April, CEPEC reconvened to consider Whitcomb's scientific evidence. On the basis of its review, the council concluded that the probability of an earthquake in the area Whitcomb identified was not significantly different than the average for similar geologic areas of California, although the council noted that the data were sufficiently suggestive to warrant Whitcomb's continued testing of his hypothesis.

The implications of this "cautious" evaluation for the general public were apparently ambiguous, with almost half of those who remembered hearing about the prediction still taking it "seriously" or "very seriously" a year later (Turner et al. 1986). For the state's Office of Emergency Services (OES) and the city of Los Angeles, however, there was a clear message: earthquake predictions are a fact of life and must be integrated into emergency planning processes. Later that year, the Southern California Earthquake Prediction Program (SCEPP) was established with funding from the Federal Emergency Management Agency (FEMA) to work with city and county governments to include prediction response in their earthquake preparedness planning.[3] The mayor of Los Angeles also established a blue ribbon Task Force on Earthquake Prediction in 1978 that met over the next five years, developing a draft of an earthquake prediction response plan in 1983 (Mattingly 1986). When the city of Los Angeles adopted a formal plan in 1989, it was the first local government in the country to have done so.

At the national level, federal legislation gave the director of the USGS the authority to issue an earthquake advisory or prediction to appropriate federal, state, and local government officials.[4] To support this authority, the USGS established the National Earthquake Prediction

Evaluation Council (NEPEC) in mid-1976. It was intended to provide the same function as CEPEC for regions outside of California and to coordinate with CEPEC when predictions were made for California.

Scientific and Policy Outcomes

The Parkfield "Experiment"—An Attempt
at an Integrated Warning System

In the early 1980s, as earth scientists began to acknowledge that break-throughs in earthquake prediction research were unlikely, they were also being pushed by Congress to develop a "prototype prediction network in Southern California" (FEMA 1983). Among the accomplishments identified in the 1982 NEHRP report to Congress was the recognition that earthquakes occur in "seismic cycles" on the southern San Andreas and that a "time-predictable" model was being developed to describe earthquake recurrence on a portion of the San Andreas near Parkfield, California.

In 1984, USGS scientists predicted, with 95 percent certainty, that a moderate earthquake (magnitude 5.0–6.0) would occur along the Parkfield segment of the San Andreas Fault between 1985 and 1993 (Bakun and Lindh 1985). This prediction was based on research that showed a twenty-two-year recurrence interval for moderate events along the Parkfield segment since the nineteenth century. The Parkfield prediction met all of the Panel on Earthquake Prediction's earlier criteria for a "true" prediction.

The area around Parkfield is quite rural, with sparse population and no engineered structures or multistoried buildings. This was a much different, simpler situation than a prediction targeted at a major metropolitan area. However, further analysis of the geologic data raised the concern that, on the basis of past events, an earthquake along the Parkfield segment could "trigger" a larger earthquake (of 6.5–7.0 magnitude) to the southeast along the San Andreas Fault, with potential impacts in seven highly populated counties covering a wide area of central California.

Because of this expanded concern about potential societal impacts, NEPEC evaluated the prediction in November 1984 and endorsed it as a "long-term" prediction; that is, the council concurred that there was sufficient scientific evidence that an earthquake in that area was likely to occur within the parameters specified. CEPEC, which was convened to review the prediction for the governor of California, issued a similar endorsement in February 1985. In concurrence with OES and the California Office of Mines and Geology, USGS issued the first public announcement of the prediction in April 1985 (Goltz 1985).

As this was the first officially endorsed earthquake prediction ever issued in the United States, OES contacted the four counties that could be most affected by a "triggered" earthquake to inform them of the prediction and to assure them that the state would be working with them on their response and warning plans.

OES initiated two policy activities related to the Parkfield prediction. First, in conjunction with the USGS, an imminent alert system was established, based on monitored geophysical activity—strain, tilt, fault creep, and foreshocks.[5] Based on predetermined levels of change in those measurements, alerts were to be sent to OES. The alert system provided for four levels of notification corresponding to increasing levels of probability of a shock within seventy-two hours. OES established different response criteria for each alert level and developed its own response plans (Office of Emergency Services 1990a, 1990b).

Second, OES began working with local jurisdictions in four of the at-risk counties, to prepare them to respond to the prediction. Two workshops were held for local officials in June and July of 1985 to inform them of the basis for the prediction and how the alert system would function. OES also hired a consultant to help the local governments draw up earthquake prediction response plans. These plans were to be developed for both a 6.0 and a 7.0 earthquake and for three time frames—long-term (which applied to the Parkfield prediction), short-term (days to hours), and very short-term (15–30 minutes) (e.g., Reitherman 1986a, 1986b).

From the system's inception in October 1985 through May 1987, forty level C and level D alerts (the two lowest levels) were issued to OES. Later, level A alerts, indicating a significant increase in the probability that a damaging earthquake was imminent, were issued on two occasions. The first, on October 19, 1992, expired three days later with no event occurring (Mileti and Fitzpatrick 1993). After the second level A alert in 1993 (also a false alarm), OES met with county emergency preparedness representatives to determine whether they wanted to continue with the experiment. They did. Procedures were then revised by OES to institutionalize the monitoring and alert processes.[6]

The original prediction period ended in 1993, but the interest of the scientific community is still focused on the Parkfield segment. Although the emergency management community saw this experiment as a way to establish a warning system for an imminent earthquake, the scientific community had always viewed it primarily as an effort "to trap a moderate earthquake within a densely instrumented network" (NEPEC Working Group 1994, p. 9), in order to better understand the last stages of the process that leads to an earthquake. For the scientific community, the experimental surveillance of the fault segment continues, although at a somewhat reduced level, in an effort to identify the types of measurable premonitory signals that indicate that an earthquake is about to occur.

The Parkfield experiment gave both the scientific and policy communities an opportunity to consider what the components of a warning system should consist of and how response systems should be developed. Sociological studies conducted during this period evaluated how successful the state had been in communicating earthquake risk and response information to the residents of the four-county area (Mileti, Farhar, and Fitzpatrick 1990; Mileti and Fitzpatrick 1993). These studies yielded advice on how to improve public information campaigns about earthquake hazards, predictions, and protective actions.

The scientific prediction experiment was a "failure" because an earthquake did not occur within the time window specified. But years of cooperation between scientists and the emergency management community forged significant links, at least at the state and federal levels. Unfortunately, as recent interviews with the chief county emergency managers show, the extent to which earthquake prediction response plans were actually incorporated into formal emergency response plans varied greatly among the four at-risk counties. One county had no special plan for responding to an earthquake prediction; another county had a plan, developed in 1988, but did not incorporate it into the county's formal emergency planning document; a third county added a section in its regular disaster response plans on responding to an earthquake prediction, but that was in 1993—the last year of the prediction window. Only one county took the Parkfield prediction seriously enough to conduct a prediction-response exercise and modify its response plan based on the results of the exercise.

The Browning Prediction
In the midst of the Parkfield prediction experiment, another earthquake prediction was made, this time for the New Madrid Fault in the central United States. This prediction differed from the Parkfield experiment in many ways, not the least of which was its repudiation by the scientific community. Nevertheless, the New Madrid Fault had been the site of huge earthquakes in 1811 and 1812, and the prediction of a new event attracted national attention in late 1989 and 1990. This was the first U.S. earthquake prediction outside of California for which NEPEC made a scientific assessment.

Iben Browning, a self-taught climatologist with a Ph.D. in zoology, ran a consulting business that predicted long-term climate trends and their impact on agricultural futures. In October 1989, Browning told attendees of an equipment manufacturing conference that a major earthquake could occur around December 2 or 3, 1990. This prediction attracted the attention of the Missouri Governor's Conference on Agriculture in early December 1989. Two of the region's largest newspapers ran stories on the prediction and also reported (incorrectly) that Browning had successfully predicted the October 1989 Loma Prieta, California, earthquake (Tierney 1994).

Although media attention was intermittent during the early part of 1990, it became increasingly more intense as December approached. By late summer and early fall, major national media stories about the prediction began to appear. While the news stories often pointed out that seismologists still considered it impossible to predict earthquakes, they also continued to give credit to Browning for earlier predictions. At the same time, the director of an earthquake information center at a university in the New Madrid region—a geologist—was quoted frequently in the local and national media as saying that he considered Browning to be highly credible, which created legitimacy for the prediction (Tierney 1994). Subsequent analyses of newspapers in the region found that throughout this period, stories were either neutral about Browning's credibility (Shipman, Fowler, and Shain 1991) or were supportive of his theory (Dearing and Kazmierczak 1991). This "sensational but cautious" approach to the prediction by the media gave it an aura of credibility.

Fueled by increasing media attention, public discussion and concern about the prediction also increased during the late fall. Intense public information-seeking about the meaning of the prediction *and* how to prepare for an earthquake overwhelmed emergency management agencies, the Red Cross, and universities in the region. Prior to this time, very low levels of earthquake preparedness had existed in the New Madrid region. Given this "window of opportunity," state and local emergency management agencies undertook a variety of activities to get information out to the public. Also, to illustrate the extent to which they were prepared to respond to an earthquake, some state emergency management agencies and local jurisdictions conducted earthquake response exercises during this period, some even in the week leading up to December 2. Unfortunately, these activities gave many people (including some local government and school officials) the mistaken impression that the prediction was being taken seriously by scientists, which further stoked public concern.

Fifteen years earlier, when a similar type of pseudoscientific prediction was announced in Southern California by Henry Minturn, the legitimate scientific community had been extremely reluctant to become involved in the issue (Turner et al. 1986), and this same reluctance was on display for the Browning prediction. As early as spring 1990, some of the state geologists in the region—who considered the prediction to be completely without scientific merit—attempted to have NEPEC undertake an evaluation in order to lessen public concern. NEPEC, however, resisted conducting a formal hearing on the prediction because its members mistakenly believed that by evaluating the prediction they would give it more credibility. Informally, the legitimate earth science community had already determined that the theory upon which the prediction was based was not valid, that the evidence Browning used was flimsy, and that Browning himself had no scientific credentials in the earth sciences arena; but it had issued no formal, countervailing statement. Much of the public, meanwhile, was apparently taking the prediction seriously, as indicated by increased information-seeking about protective measures, inquiries about and purchases of earthquake insurance, and the common decision to keep children out of school during the first week of December.[7] Finally, under congressional pressure, NEPEC convened in late fall 1990 and publicly issued a statement to the effect that no evidence existed that an earthquake was any more likely to occur in the New Madrid region in early December than at any other time (USGS 1990a).

This prediction "event" illustrates the complex relationship between a developing science, public understanding, and governmental policy responses. Perhaps most important, the mechanism established to provide legitimate assessments of scientific predictions, NEPEC, did not provide sufficient—or sufficiently timely—guidance to citizens who lived in an area of the country that was not earthquake prone, and who were not familiar with the developing science of earthquake prediction. NEPEC attempted to deal with this prediction in much the same way it dealt with approximately three hundred other less public predictions that had come to its attention since its inception—by not evaluating it because it was deemed unscientific (USGS 1993). Eventually, NEPEC recognized that it needed to respond to the *social*—rather than the scientific—significance of the Browning prediction by conducting a very public evaluation. But by then, much of the damage had already been done.[8]

Changes in Scientific Approaches to Forecasting Earthquakes

The failure of the Parkfield experiment ended optimism about scientific capabilities to provide short-term predictions. In fact, optimism had been declining for several years, as the earth science community investigated other ways to characterize the earthquake hazard for earthquake-threatened areas of the country. With the exception of the Parkfield experiment, by the mid-1980s scientific efforts had begun to shift to long-term "forecasts" (Ellsworth 1986) and "early-warning" systems (FEMA 1986).

Long-Term Forecasts

In the 1985 NEHRP report to Congress, the development of a short-term predictive capability was said to be "more difficult" than anticipated. As an alternative, a "second generation" approach was being taken toward earthquake prediction by closely monitoring localized segments of both the northern and southern San Andreas Fault where "heightened risks" were identified. Over the next few years, the reports to Congress rarely mentioned the term "prediction," except in reference to the Parkfield experiment, but substituted the terms "high, long-term seismic potential" or "heightened probabilities."

These efforts resulted in characterizing various segments of the San Andreas and other fault systems in Southern California and the San Francisco Bay area in terms of their likelihood of generating a certain magnitude earthquake within a given period of time. These forecasts (as they were routinely called) were long-term predictions; that is, they had no lead times but were instead characterized by time windows that started in the present and continued from one to several decades into the future. They were primarily based on calculations of recurrence intervals from historical records or field analyses. The results of these studies were displayed on maps in order to provide visual representations of high-hazard areas with respect to earthquake potential.

Interestingly, in 1990 the USGS took the lead in developing a lengthy supplement for the Sunday newspapers in the San Francisco Bay area to present to the public information on the revised probabilities for local faults (USGS 1990b). The USGS eventually enlisted the assistance of state emergency management officials as well as social scientists who had studied risk and warning communication. The supplement included a lay explanation of the earthquake hazard and the associated probabilities, as well as preparedness guidelines and references to local agencies for additional information. A similar brochure was released in Southern California in 1997.

This approach to forecasting earthquakes has diminished the concerns of local policy makers about the impact that short- and intermediate-term predictions may have on local citizens and economies. Predictions had the cognitive benefit of focusing a lay audience's attention on a specified threat—a place, a time, a magnitude—for which people could imagine likely consequences as well as protective actions. The longer-term forecasts—stating, for example, that there is a 50 percent chance of an earthquake sometime in the next twenty years along a particular fault segment—lack the drama necessary to capture public (and many local politicians') attention. These extremely wide time windows don't convey the same sense of urgency as does a "prediction." While such forecasts can lead to substantial localized efforts to mitigate earthquake risks (Bakun 1995), recent evidence from research conducted by Mileti and his colleagues in the San Francisco Bay Area and by the Disaster Research Center, both in the Bay Area and in Southern California (Nigg et al. 1996), indicates that such broad forecasts are not being used to make substantial improvements in mitigation. The general nature of the forecasts may be insufficiently threatening to mobilize local policy makers to take other than the most rudimentary steps (and usually those required by law) to lessen the vulnerability of their communities (until after an earthquake has occurred).

Early Warning Systems

A second approach currently being pursued by the earth science community is the development of earthquake "early warning systems" (Holden, Lee, and Reichle 1989). Contrary to what the name may imply, an early warning system does *not* function to predict a seismic event. Rather, it functions *after* an earthquake has begun, to warn distant communities that significant ground shaking will begin within seconds. Such a system takes advantage of the fact that an electronic signal transmitted by the warning system travels much faster than the seismic waves that propagate away from the epicenter of the quake. Thus, the farther away the community is from the epicenter, the longer the time between receipt of the transmitted warning and the arrival of the first ground motion.

Mexico developed and implemented a successful early warning system that in 1995 provided seventy-five seconds of notice to residents of Mexico City from an earthquake off the Pacific Coast, over three hundred miles away. Japan and Taiwan also have preliminary systems under development, and the Taiwan system was partially tested in the mid-

1990s. Although such a system was first mentioned in the NEHRP report to Congress in 1985 (FEMA 1986), the United States did not devote extensive funding to early warning systems until after the Northridge earthquake. Currently, great efforts are being made to design telemetry systems and upgrade and integrate seismograph networks in Southern California, in order to provide early warning to the Los Angeles basin area of a major earthquake on the San Andreas Fault. To date, however, most effort has focused on the collection, interpretation, and transmission of earth science data—the technical assessment stage of a warning system. Social science research on the dissemination component for such a warning system has only recently begun. Caltech, with funding from FEMA, has commissioned a study to determine what can be done in less than sixty seconds that could save lives and lessen social disruption. Some major policy questions that have not yet been answered are: Who are the likely users of this information? Do they have the capability to use the information effectively? Will the information be transmitted to the general public? If so, for what purpose (what are people expected to do in these seconds)? Given the uncertainty associated with such a warning system, what are some of the problems associated with false alarms?[9] A key policy question is whether spending millions of dollars on the technical development of the warning system is justified by the social benefits expected from such an extremely short "warning period."

Prediction as a Basis for Policy—A Paradox

Scientific approaches to predicting damaging earthquakes in California have contributed to earthquake preparedness despite the failure to achieve a short- and intermediate-term predictive capability. This failure has also led earth scientists to develop alternative ways of providing information that can meet the basic NEHRP mission of reducing earthquake consequences. For example, the earth science community has enhanced the national seismic mapping program to provide each state with earthquake fault maps and some indication of the hazard associated with them. This was not one of the early goals of NEHRP but developed over time as the promise of prediction waned and federal legislators began to expect policy-relevant results from earth science (as opposed to the basic research studies that were funded during the early years of the program).

Public-policy makers have reacted to the promise of prediction capabilities by trying to develop response plans and communicate earthquake threat information more effectively to the general public. However, as the scientific strategies have changed to emphasize both long-term forecasts and instantaneous warnings, it seems that the concerns of public-policy makers have either waned (in the former case) or have yet to develop (in the latter). While we should not expect science to develop in an orderly or linear fashion, rapidly evolving scientific approaches to characterizing and responding to future earthquakes have resulted in reactive, and often short-lived, public policies that do not seem to have had any long-term programmatic effects. In this respect, earthquake prediction has been a disappointment in both science and policy.

On the other hand, the focus on earthquake prediction has had a positive, if indirect, impact on earthquake loss reduction policies, not just in California but in other earthquake-threatened areas across the country. Awareness of the earthquake threat among both the public and policy makers—kept alive in part by the promise of prediction—stimulated progress in reducing social and structural vulnerability to earthquakes. In fact, it was the prospect of predicting earthquakes that resulted in congressional action in 1977 to establish NEHRP, a program that has since contributed to significant reduction of earthquake risk and vulnerability throughout the nation. Another equally important benefit of the early emphasis on prediction was the opportunity for local, regional, and state governments to engage scientists in a discussion of probabilities, uncertainties, and the vagaries of geologic processes, while, at the same time, working with them to develop procedures that provided for quick consultation on potential precursory events and rapid warning notification systems. This close relation between scientists and decision makers has been critically important to the continued development of earthquake hazard mitigation policies, especially in California.

In this context, the Browning prediction is troublesome. In the mid-1980s, social science studies indicated low levels of public awareness of, and governmental preparedness for, an earthquake in the New Madrid Fault zone (Mushkatel and Nigg 1987). Today, however, there is a much wider appreciation of the seismic threat among the general public, and several midwestern states now have seismic elements in their building codes. The pseudoscientific Browning prediction unquestionably contributed to these desirable results.

We are left with a paradox. While earthquake prediction itself was not successful, the policies that resulted from its early, unfulfilled promise were. In considering linkages between scientific prediction and policy initiatives, such nonlinear, serendipitous outcomes should not be ignored. One could argue that because no one can accurately predict when and where an earthquake will occur, policy makers have had to consider and implement policies to actually reduce the vulnerability of the built environment. Yet one would hardly want to base earthquake hazard policies on the expectation that failed science—not to mention pseudoscience—will lead to beneficial societal outcomes. One can easily imagine a scenario in which failed predictions would lead to a backlash against both science and preparedness. Perhaps the wiser course would be for scientists not to predict success at predicting, and for policy makers to respond accordingly.

Notes

1. The "wind" community—those atmospheric scientists, engineers, and social scientists who are concerned with the effects of tornadoes, severe storms, and hurricanes—have been trying to get a similar federal program developed since the early 1980s, but they remain unsuccessful.
2. The four federal agencies identified to implement NEHRP, as specified in the 1980 amendments to the act, were the newly created Federal Emergency Management Agency (as lead agency among the four), the National Science Foundation (NSF), the U.S. Geologic Survey, and the National Bureau of Standards (now the National Institute of Standards and Technology).
3. Although SCEPP originally emphasized prediction, about a year after its founding the penultimate word of its name was changed to "Preparedness."
4. Federal legislation included the Disaster Relief Act of 1974 and the Earthquake Hazards Reduction Act of 1977 and its 1980 amendments.
5. Both the USGS and the State of California invested substantial funding—approximately $1 million each—in instrumenting and monitoring the Parkfield segment of the fault beginning in 1985. NSF sponsored additional, academic research projects in that area.
6. I would like to thank an anonymous reviewer of this chapter for this insight on the process.
7. For a discussion of how Southern California responded to the pseudoscientific prediction by Henry Minturn of an earthquake for December 1975, see Nigg 1982.
8. For an example of a NEPEC evaluation of a scientific prediction involving U.S. scientists in another country, see Olson, Podesta, and Nigg 1989.
9. For example, some electrical and gas utility companies believe that this type of information could allow them to automatically shift loads and switch off service to potentially affected areas; however, any assessment of the usefulness of this very short-term alert system will also have to investigate the

time needed to bring these systems back on line, and the effect of such delay, especially in areas that are not damaged by the event.

References

Bakun, William H. 1995. "Reducing Earthquake Losses Throughout the United States—Pay a Little Now, or a Lot Later." USGS Fact Sheet 169-95. Washington, DC: U.S. Department of the Interior, U.S. Geological Survey.

Bakun, William H., and Allan G. Lindh. 1985. The Parkfield, California earthquake prediction experiment. *Science* 229:619–624.

Comerio, Mary. 1995. *Northridge housing losses*. Berkeley: Center for Environmental Design Research, University of California.

Dearing, J. W., and J. Kazmierczak. 1991. *Making iconoclasts credible: The Iben Browning earthquake prediction*. Paper presented at the Southern Illinois University at Edwardsville Research Conference on Public and Media Response to Earthquake Forecasts, Edwardsville, IL, May 16–18.

Ellsworth, William. 1986. Progress toward reliable earthquake prediction. In *Future directions in evaluating earthquake hazards in Southern California*, William M. Brown III, William J. Kockelman, and Joseph I. Ziony, eds., pp. 249–251. USGS Open File Report 86-401. Menlo Park, CA: USGS.

FEMA (Federal Emergency Management Agency). 1983. *The National Earthquake Hazard Reduction Program: A report to Congress, fiscal year 1982*. Washington, DC: FEMA.

FEMA (Federal Emergency Management Agency). 1984. *National Earthquake Hazards Reduction Program: Fiscal year 1983 activities. Report to the United States Congress*. Washington, DC: FEMA.

FEMA (Federal Emergency Management Agency). 1986. *National Earthquake Hazards Reduction Program: Fiscal year 1985 activities. Report to the United States Congress*. Washington, DC: U.S. Government Printing Office.

Goltz, James D. 1985. *The Parkfield and San Diego earthquake predictions: A chronology*. Los Angeles: Southern California Earthquake Preparedness Project, California Office of Emergency Services.

Haas, J. Eugene, and Dennis S. Mileti. 1976. *Socioeconomic impact of earthquake prediction on government, business, and community*. Boulder: University of Colorado, Institute of Behavioral Science.

Hass, J. Eugene, and Dennis S. Mileti. 1977. Socioeconomic impact of earthquake prediction on government, business, and community. *California Geology* 30:147–157.

Holden, Richard, Richard Lee, and Michael Reichle. 1989. *Technical and economic feasibility of an earthquake warning system in California*. Special Publication 100. Sacramento: California Division of Mines and Geology.

Kiecolt, K. Jill, and Joanne M. Nigg. 1982. Mobility decisions based on perceptions of a hazardous environment. *Environment and Behavior* 14:131–154.

Mattingly, Shirley. 1986. Responding to earthquake forecasts: A local government view. In *Future directions in evaluating earthquake hazards in*

Southern California, William M. Brown III, William J. Kockelman, and Joseph I. Ziony, eds., pp. 270–274. USGS Open File Report 86-401. Menlo Park, CA: USGS.

Mileti, Dennis S., Barbara C. Farhar, and Colleen Fitzpatrick. 1990. *How to issue and manage public earthquake risk information: Lessons from the Parkfield earthquake prediction experiment.* Fort Collins: Hazards Assessment Laboratory, Colorado State University.

Mileti, Dennis S., and Colleen Fitzpatrick. 1993. *The great earthquake experiment: Risk communication and public action.* Boulder, CO: Westview Press.

Mushkatel, Alvin H., and Joanne M. Nigg. 1987. Opinion congruence and the formulation of seismic safety policies. *Policy Studies Review* 6:645–656.

NEPEC Working Group. 1994. *Earthquake research at Parkfield, California, for 1993 and beyond.* USGS Circular 1116. Washington, DC: U.S. Government Printing Office.

Nigg, Joanne M. 1982. Communication under conditions of uncertainty: Understanding earthquake forewarnings. *Journal of Communication* 32:27–36.

Nigg, Joanne M., and Kathleen J. Tierney, with James M. Dahlhamer, Melvin J. D'Souza, Lisa Reshaur, and Gary Webb. 1996. *Utilization and impact of earth science information among local governments and businesses in Southern California.* Final Report no. 39. Newark: Disaster Research Center, University of Delaware.

Office of Emergency Services. 1990a. *California short-term earthquake prediction response plan.* Sacramento: State of California, OES.

Office of Emergency Services. 1990b. *Parkfield earthquake prediction response plan.* Sacramento: State of California, OES.

Olson, Richard Stuart, Bruno Podesta, and Joanne M. Nigg. 1989. *The politics of earthquake prediction.* Princeton, NJ: Princeton University Press.

Panel on Earthquake Prediction. 1976. *Predicting earthquakes: A scientific and technical evaluation with implications for society.* Washington, DC: National Research Council, National Academy of Sciences.

Panel on Public Policy Implications of Earthquake Prediction. 1975. *Earthquake prediction and public policy.* Washington, DC: National Academy of Sciences.

Reitherman, Robert. 1986a. *County-specific emergency planning information relative to the Parkfield earthquake prediction.* Redwood City, CA: Scientific Service, Inc.

Reitherman, Robert. 1986b. *State earthquake prediction response planning report.* Redwood City, CA: Scientific Service, Inc.

Shipman, J.M., G.L. Fowler, and R.E. Shain. 1991. *Iben Browning and the fault: Newspaper coverage of an earthquake prediction.* Paper presented at the Southern Illinois University at Edwardsville Research Conference on Public and Media Response to Earthquake Forecasts, Edwardsville, IL, May 16–18.

Tierney, Kathleen J. 1994. Making sense of collective preoccupations: From research on the Iben Browning earthquake prediction. In *Self, collective action and society: Essays honoring the contributions of Ralph H. Turner*, Gerald Platt and Chad Gordon, eds. Greenwich, CT: JAI Press.

Turner, Ralph H., Joanne M. Nigg, and Denise Heller Paz. 1986. *Waiting for disaster.* Los Angeles: University of California Press.

Turner, Ralph H., Joanne M. Nigg, Barbara Shaw Young, and Denise Heller Paz. 1980. *Community response to earthquake threat in Southern California, part ten: Conclusions, problems, and recommendations.* Los Angeles: Institute for Social Science Research, University of California.

USGS (U.S. Geological Survey). 1990a. *Evaluation of the December 2–3, 1990, New Madrid Seismic Zone Prediction.* Washington, DC: U.S. Department of the Interior.

USGS (U.S. Geological Survey). 1990b. "The Next Big Earthquake in the Bay Area May Come Sooner Than You Think. Are You Prepared?" 23-page insert distributed in San Francisco Bay Area newspapers, September 9.

USGS (U.S. Geological survey). 1993. *Responses to Iben Browning's Prediction of a 1990 New Madrid, Missouri Earthquake.* USGS Circular 1083. Washington, DC: U.S. Government Printing Office.

PART THREE
Foreordained Conclusions: Prediction and Politics

Natural hazards happen to us. But sometimes we happen to nature. Beaches are eroding, in part because of our efforts to protect beachfront property. Hard-rock mines release toxic chemicals into groundwater. Nuclear weapons and energy facilities generate long-lived radioactive waste, including some radionuclides that don't exist in nature. When human action is implicated in adverse environmental impacts, it is easy to start pointing fingers. Decisions have to be made. Even a decision to do nothing—to let a beach erode, to let radioactive waste sit around in corroding metal drums—is a decision to do something, a decision that will create winners and losers. On the other hand, a decision to take action, to intervene, is a conscious decision to shape the future. Shaping the future of a beach or a nuclear waste repository is not a simple engineering problem. It is a prediction of how the beach will behave for the next ten years, and how the repository will behave for the next ten millennia.

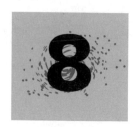

What You Know Can Hurt You: Predicting the Behavior of Nourished Beaches

Orrin H. Pilkey

Over the past century, population pressures, general affluence, the attraction of beautiful coastal beaches, and demands for increased recreation have accelerated the exploitation of our beaches. Meanwhile, diminishing beach sand supplies and rising sea level lead to shoreline retreat. Because beachfront property owners are rarely willing to move their structures farther from the water, human development of shorelines comes into sharp conflict with natural coastal processes. Beach nourishment—adding sand to beaches—has become the favored solution to this conflict, a means of "saving" both beaches and buildings.

As much as 80 percent of the U.S. shoreline is eroding. (The term *erosion*, although a poor descriptor of shoreline retreat, is deeply ingrained in real-world usage and will be used throughout this chapter.) Erosion is caused by many factors, including sea level rise. The *erosion problem*, however, is caused by people who have purchased beachfront property adjacent to a retreating shoreline (see figure 8.1), a relatively small number of people compared to beach users.

The long-term prognosis for the erosion problem is that it will increase in severity because: (1) the number of buildings at risk is growing; (2) beach sand supply—nature's way of nourishing beaches—is being reduced due to channel dredging, damming of rivers, and other engineering activities; and (3) sea level is rising, which increases erosion rates. In other words, the natural sources of beach sand are decreasing, and the natural causes of sand removal are increasing, leading to more net erosion of beaches. Although a number of states have instituted secondary controls on beachfront development (e.g., construction setback lines), no state government has instituted a long-term solution to the erosion problem. Setback lines only postpone the problem to the next generation.

Fig. 8.1   Erosion scarp in a dune on a North Carolina beach. In the foreground are pipes and pavement fragments from a house that was once prime beachfront property.

Responses to shoreline erosion face two societal values that often conflict. The first is the preservation of property adjacent to the shoreline by any means. Beachfront property owners tend to be influential people with the ability and will to defend this value. The second value is preservation of the public recreational beach. The beach is utilized and valued by numbers of people much larger than the numbers of property owners. Conflict arises when, for example, seawalls that destroy beaches are used to protect property.

Our society can "solve" the erosion problem on a developed beach in three ways:

- *hard stabilization,* which is any way of holding the shoreline in place using hard, immovable objects, usually seawalls, groins, or offshore breakwaters;

- *soft stabilization,* or emplacement of new sand, called beach nourishment or beach replenishment; or

- *relocation,* or abandonment of beachfront buildings.

Hard stabilization is the best approach if the dominant societal goal is protection of property adjacent to the shoreline; that is, if preservation of the beach is not considered a priority. Most often hard structures start small and grow in size and length with time. This has been the experience in New Jersey, where shoreline armoring has proceeded for more than a century, leading to a process of total beach degradation now called *New Jerseyization*. Beach loss occurs as the retreating beach backs up against the seawall (or any other fixed object at the back of a beach), gradually narrows, and eventually disappears (see figure 8.2).

If preservation of the recreational beach is the highest priority, relocation is the best approach. Relocation, on a small scale, has been carried out along North American shorelines for decades. This has involved letting an occasional building fall into the ocean, or moving it, as its time comes. Nags Head, North Carolina, loses buildings every year. One cottage, the Outlaw family home, has been moved back five times, a total of six hundred feet, in one hundred years and is currently at the ocean's edge. In recent years, many Nags Head cottages have been replaced by high-rise buildings, a common occurrence on the world's more developed shorelines. The high costs and technical difficulties of moving high-rise buildings reduce the options for a community's erosion response.

Howard and colleagues (1985) argue that retreat, ultimately, is the only option. They conclude: "[Sea] level is rising and the American shoreline is retreating. We face economic and environmental realities that leave us two choices: (1) plan a strategic retreat now or (2) undertake a vastly expensive program of armoring the coast line and, as required, retreating through a series of unpredictable disasters."

Beach nourishment, however, is increasingly the chosen erosion-response alternative, especially on the U.S. East and Gulf Coasts. Some states (South Carolina, North Carolina, Rhode Island, Maine, Texas, and Oregon) have outlawed hard stabilization, making soft stabilization particularly attractive. The United States spends approximately $100 million annually on nourishment (not including the Pacific Coast). Once started, nourishment is a never-ending process. A beach must be repeatedly renourished at intervals—usually between one and ten years—that depend on local wave energy and storm frequency. The new beaches usually lead to intensification of development, providing an ever larger political base for additional funding of future nourishments. Continued nourishment allows the affected communities to ignore the realities of sea level rise and increasing erosion rates as they progressively increase building density and size, which decreases their flexibility to respond to the inexorable dynamics of beach systems.

Most beach nourishment in the United States is carried out by the U.S. Army Corps of Engineers (COE), a federal agency that Congress funds on a project-by-project basis. In order for a federally supported nourishment to be approved, a benefit–cost ratio greater than 1.0 and an

Fig. 8.2   South Myrtle Beach, South Carolina, after the passage of Hurricane Hugo in 1989. The buildings to the left, behind the rock seawall, were partially protected from the storm, but the beach is gone. To the right, the beach retreated, the buildings were flooded by the storm surge, but the beach is wide and healthy. This scene illustrates why society must ultimately choose between beaches and buildings.

environmental impact statement are required (Pilkey and Dixon 1996, p. 10). Both of these requirements are met by predicting the rate of loss of the artificial beach and then predicting the volumes of sand that will be required to maintain a beach of certain dimensions in place for a specified length of time.

The U.S. Army Corps of Engineers Coastal Engineering Research Center (CERC) is the principal organization involved in beach nourishment research. Most of the mathematical models and other concepts used in support of beach nourishment in this country were developed there. The basic procedures used by the Corps are outlined in its *Shore Protection Manual* (USACE 1984).

Predicting the rate of loss of a nourished or natural beach requires an understanding of some very complex surf zone processes. The pace of beach change is extremely rapid, at rates of feet per year, and we have a great number of measurements of eroding beaches. Our direct experience with the phenomenon of shoreline retreat provides at least a coarse check on predictions.

Implicit in the prediction of future beach behavior is the expectation that the sea level will continue to rise. But the mechanics or underlying causes of the current rise in sea level (one foot per century along the U.S. East Coast) are uncertain, as are the mechanics of how a gradual sea level rise causes shoreline erosion. Sea level rise is a much more gradual event than beach changes, which can occur overnight in a big storm. We have not had enough experience with long-term sea level rise to do more than assume it will continue as it has in the past. Thus, in the following discussion of beach behavior, it is assumed that whatever impact sea level rise is having on beach behavior, it will continue at the same rate and in the same fashion for the foreseeable future.

Beaches and Models

Coastal engineers, rather than coastal scientists, produce most of the models used to predict beach behavior. The engineering profession has successfully used mathematical models for many years in the design of engineered structures. Probably in much of engineering, design without modeling would now be considered unsophisticated, even unacceptable. But it is difficult to transfer the mathematical modeling approach from predicting the behavior of steel and concrete structures to predicting the course of earth surface processes.

Models of Beach Behavior

A model represents a system and typically emphasizes those aspects of the system most useful for achieving the purposes for which the model is built. A mathematical model uses equations to represent the system. For example, an engineer could use a mathematical model of a steel bridge to simulate the deflection of a bridge as a truck drives across it: The equations calculate the reactions of the structure as the load is applied. Because the bridge beams and joints are both well characterized and well understood, such a model can be used to design a safe bridge. Modeling the dynamics of a beach to predict its erosion is different, because the dynamics of a beach are chaotic, being subject to random external influences, and because the components of the beach system change as they interact with each other.

The coastal zone is an immensely complex environment, linked to other earth environments in ways that scientists are only beginning to understand. Movement of beach sand and changes in beach shape are initiated by large- and small-scale atmospheric changes (climate and weather) that drive ocean water movement (waves and currents). Organisms constantly modulate the interactions among moving atmosphere, moving water, and moving sediment. They burrow, sort, compact, cement, and extrude sediment, while leaving organic slimes and mats covering the sea floor. These individual complex systems that together constitute the shore—the atmosphere, the bottom sediment, the water column, and the biota—interact in ways that are nonlinear and unpredictable, given our current level of data and understanding. Therefore, any model of a beach must be carefully constructed and tested, and its users must recognize its limitations.

At this point, it is important to emphasize that this discussion is concerned with mathematical models used for applied or engineering purposes as opposed to models used for basic scientific purposes. The distinction is important. The questions that are of direct interest to society when planning beach nourishments involve *where*, *when*, and *how much*—i.e., predicting where and when we need to put how much sand to nourish a beach. Scientific models may be mostly concerned with *how*, *why*, and *what if*—i.e., predicting how a beach changes, and why and what if such changes occur. In chapters 1 and 2, Sarewitz and Pielke and Oreskes discussed the distinction between science aimed at predictive precision and science that seeks to deepen insight. The case of beach behavior starkly illustrates the real-world dangers of failing to understand this distinction.

The difference can be illustrated by considering two examples. A mathematical modeling effort concerned with the *how* of beaches is the study of beach cusps by Werner and Fink (1993). In this type of modeling, the number of parameters is held to a minimum: the fewer the better, the simpler the better. Process variables suspected to be key elements of the natural phenomenon are then examined individually. If, under these simple and rudimentary conditions, variation of the parameter produces in the model what has been observed in nature, the parameter is considered to be an important element of the process. On the other hand, engineering models used to plan a beach nourishment must be concerned with the precise future behavior of the beach. To achieve such precision, these models must attempt to include all known significant elements of a process such as the erosion behavior of a particular beach, thus making them much more complex and inclusive.

An early approach to predicting future beach behavior made use of simple, analytical models. Two important examples of such analytical models are the "overfill factor" (USACE 1984) and the "length equation" (Dean 1983).

The *overfill factor* is used to determine the amount of extra sand needed because of the difference in the grain size of the native (original) sand and the fill (nourishment) sand. This model assumes that if the fill sand is finer than the native sand, waves will winnow away the fine fraction of the fill until the native grain size has been achieved. The overfill factor is a simple multiple that determines the amount of extra sand needed to allow for this loss and still have the desired beach. This calculation is implicitly based on the assumption that an equilibrium grain size exists that is determined by the wave "climate," i.e., the average wave conditions over a period of years, of a particular beach. But no such correspondence between beach wave climate and grain size has been recognized. Around the world, beaches with similar wave climate have very different grain sizes depending upon the sources of the sand. Beaches adjust for different grain sizes by changing their shapes and slopes, not only by winnowing away the finegrains.

The *length equation* is used to determine how a nourished beach's length—its longshore dimension—affects its durability. It is based on the assumption that beaches erode at the ends only, and thus the longer a beach, the longer the time required to remove the sand. But years of experience with nourished beaches clearly shows that they do not erode only at their ends. For example, during storms, sand loss often occurs in a direction perpendicular to the shore, and erosion "hot spots" are often not at the ends of nourished beaches.

These analytical models give incorrect results, and users could do simple studies to show they are wrong. Such studies would investigate, respectively, whether fine grains are winnowed away from nourished beaches, and whether nourished beaches of great length systematically last longer than short ones. The observed behavior of beaches shows that these models do not provide accurate nourishment volumes.

Numerical models seek to capture more of the complexity of beach behavior than can be included in an analytical model. Perhaps the two numerical models that have been most commonly used in the United States are GENESIS (see Hanson and Kraus 1989) and SBEACH (see Larson and Kraus 1989), which were developed primarily at the COE Coastal Engineering Research Center by Nicholas Kraus. Both models are user friendly and are available for free on floppy disks from CERC. Both have been used in beach nourishment design. In the following two sections, I describe a key assumption and a key parameter that underlie numerical models to illustrate the empirical challenge of predicting beach behavior. I then review some attributes of the GENESIS model.

The Shoreface Profile of Equilibrium

The most fundamental assumption behind virtually all models of beach behavior is the "shoreface profile of equilibrium." It has been defined as "a long-term profile of the ocean bed produced by a particular wave climate and type of coastal sediment" (Schwartz 1982). The shoreface is usually considered to be the portion of the inner continental shelf across which beach sand is readily exchanged in offshore and onshore directions (figure 8.3). For practical purposes it is the concave upward portion of the inner shelf. On the East Coast, the base of the shoreface, where the much flatter continental shelf begins, is typically ten to twenty meters water depth, but that depth can vary widely. For example, off some barrier islands in the Arctic it is two meters deep, but it extends to a depth of seventy meters on the extreme-wave-energy coast of southeast Iceland.

An important aspect of the concept of equilibrium profile is the closure depth. As used in the mathematical models, it is the depth beyond which seaward sediment transport is insignificant. Closure depth is thus a fence that prevents sediment from moving farther away from shore.

Nourished beach design assumes that the introduced nourishment sand will achieve the "equilibrium profile" once it is sorted out by the

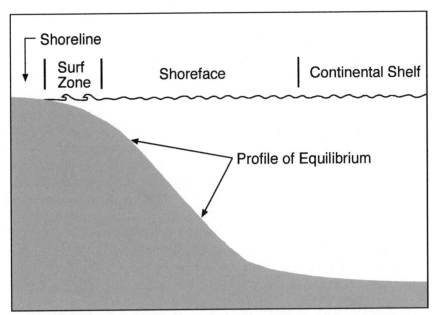

Fig. 8.3   A schematic cross-section of a sandy beach to show some of the terminology used in this chapter. The slope of the shoreface is greatly exaggerated.

waves. Since it is assumed that no sediment is transported beyond the closure depth, the profile ends at that point. In other words, once the equilibrium profile has been achieved, the beach is assumed to be stable. The equilibrium profile is the principal basis for estimating needed sand volumes for nourished beaches. The shallower the closure depth, the less sand required to achieve the correct profile because, as figure 8.4 shows, the beach nourishment sand is assumed to build up the entire shoreface profile to its equilibrium state, and a shallower depth demands less sand. Originally, closure depth was assumed to be at eighteen to twenty meters water depth off east Florida (Bruun 1962). Politics have intervened and closure depth on the east coast of Florida, for purposes of beach design, is now assumed to be four meters, in order to lower the amount of sand putatively needed and thus lower the estimated cost of the beach nourishment, making it politically more feasible.

The shoreface profile is assumed to be described by the following equation relating water depth to distance offshore:

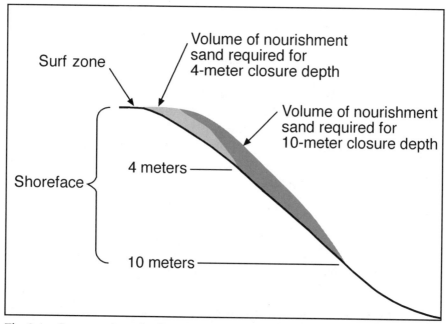

Fig. 8.4   Cross-section of a shoreface showing why the choice of a shallow closure depth results in a smaller volume of sand for a nourished beach. Ordinarily, sand is not pumped onto the shoreface when a beach is nourished. The assumption is made that after the sand is emplaced on the beach, it will be redistributed by wave action, and the beach will eventually achieve the profile of equilibrium. The slope of the shoreface here is greatly exaggerated.

$$h = Ay^n,$$

where h = water depth, y = distance offshore, n = 0.66, and A is a factor related to grain size. The equation was derived by measuring and averaging over a large number of real shorefaces (Dean 1977). In the equation, the only parameter representing the character of a particular beach is A, which is assumed to depend only on the grain size of the beach's sand, so that if the grain size of any shoreface anywhere in the world is known, its shape is known. *The problem is that no relationship between grain size and A has been shown to exist* (Pilkey et al. 1993). Furthermore, it is highly likely that wave climate and underlying geology, among other factors, also control shoreface slope.

The concept (as opposed to the equation) of the profile of equilibrium is based on the following oceanographic assumptions:

- ASSUMPTION: ALL SEDIMENT MOVEMENT ON THE SHOREFACE IS DRIVEN BY WAVE ORBITAL ACTION ACTING ON A SANDY SURFACE. Contrary to this assumption, Wright (1995) has shown that on some shores, bottom currents (as distinct from waves) transport large amounts of sediment, yet this transport is not considered in the profile of equilibrium model.

- ASSUMPTION: THERE IS A SEDIMENT-FENCE CLOSURE DEPTH. This assumption is an outgrowth of the idea that all sediment is transported by wave orbital interaction with the sediment. In deeper waters such interaction ceases because the action of waves does not reach bottom, hence the significance of the closure depth. However, there is extensive evidence both in known processes (e.g., bottom currents, the depth of wave action during storms) and observed sediment distribution paths (Thieler et al., 1992) that such a sediment fence cannot exist.

- ASSUMPTION: THE SHOREFACE IS COVERED BY A THICK LAYER OF UNCONSOLIDATED SAND. This assumption means that shoreface heterogeneity plays no role in the model because it is assumed away. However, there is contrary evidence in the literature, i.e., evidence that the geology can play a major role in the evolution of certain beaches. For example, recent studies (Riggs, Cleary, and Snyder 1995) have indicated that in many instances, underlying geology (e.g., outcropping rocks and mud layers) plays a paramount role in shoreface and shoreline evolution.

- ASSUMPTION: OFFSHORE BARS AND OTHER BATHYMETRIC FEATURES DO NOT AFFECT SEDIMENT TRANSPORT OR SHOREFACE BEHAVIOR. This assumption is an outgrowth of the fact that the shoreface profile equation produces a smooth curve without sand bars. But offshore bars in the surf zone are real and are critical factors in surf zone dynamics and sediment transport.

For all of the above reasons, the equation and the concept of the shoreface profile of equilibrium, as used in mathematical models of beach behavior, are wrong. Therefore, any model based on this concept—and most are—has a profound and fundamental weakness.

Modeling Wave Height and Sand Movement

Wave height and direction are the single most important parameters in modeling beach sand transport, because waves are the strongest determinants of beach erosion. Indeed, wave height has been termed the "Rosetta stone" of beach modeling. Wave action at a beach is complex and dynamic. For example, several sets of waves typically impact the beach simultaneously, yet in a model one height and one direction must represent this complicated situation. It is instructive to review how wave height is obtained and used in most models of beach behavior. It is a five-step process, each with significant potential for error.

- STEP 1: PREDICT THE DEEP-WATER WAVE CLIMATE. This is based on wave "hindcasting," i.e., determining a distribution of past wave heights and directions, using published local weather information. Since waves reaching the shoreface can be generated anywhere in an ocean basin and not just locally, the random occurrence of large distant storms is a source of error.

- STEP 2: BRING THE WAVES ASHORE FROM DEEP WATER INTO SHALLOWER WATER. As waves encounter shallow water, their direction changes; a wave refraction model is used to bring waves to the surf zone. Since waves can and often do come from several directions, choosing a single direction is a source of error. Knowing how waves will come ashore requires good knowledge of nearshore bathymetry, but most models mathematically calculate bathymetry using the "profile of equilibrium" equation, which is a source of error. Friction between waves and the sea floor, which varies according to sand grain size and the presence of rock outcrops, is not considered (another source of error).

- STEP 3: CHOOSE A WAVE HEIGHT. Models use a single wave height, and different models make different choices. The model (GENESIS) discussed in the next section uses an average of the highest one-third of waves over a given time interval, while another commonly used model (SBEACH) uses the highest annual wave height averaged over a twelve-hour interval. The former approach considers no storms, and the latter considers only storms. All the waves coming ashore are assumed to be of the same height and frequency and to come from the same direction, completely unlike a natural surf zone, where waves of different sizes may come from several directions at once, an obvious source of error.

- STEP 4: BREAK THE WAVES. The shape of a beach determines how waves will interact with the bottom—how they will break and how they will affect sand transport. Therefore, it is important to know the actual shape of the bottom. Most models assume and use a calculated design beach (from the profile of equilibrium equation) whose shape is determined by sand grain size. This offshore profile is assumed to be constant, while real beaches are dynamic. Sand bars are usually not considered (GENESIS), but when they are (SBEACH), these dynamic features are assumed to be static and unchanging. These assumptions can introduce significant errors.

- STEP 5: MOVE SAND. In most models this is done using parameters that relate waves to sand transport and are determined in laboratory "wave tanks" and "sediment flumes" (which are themselves physical models of beaches). GENESIS adjusts the amount of sand calculated to be moved by a factor, K, discussed below. Actual measurements of sand transport in the surf zone are difficult and error-prone (Bodge and Kraus 1991). No comprehensive field information exists to determine the relationship between wave height and transport on a particular beach, or on beaches in general. The lack of information on, and understanding of, real beach sand transport is a source of error.

GENESIS
The Generalized Model for Simulating Shoreline Change (GENESIS) is intended to be used on beaches influenced by engineering structures. As it is now the most widely used model and illustrates well the typical problems with beach behavior modeling, it is emphasized here. GENESIS is designed to simulate the long-term coastal changes resulting from spatial and temporal differences in longshore sediment transport (i.e., movement of sand parallel to the beach) at sites with engineered structures (Hanson and Kraus 1989). The model is used by engineers and planners to predict how a shoreline will move in response to beach nourishment and shoreline armoring or other activities that could alter longshore transport. It is the state-of-the-art model for prediction of shoreline changes (Komar 1998a).

Young et al. (1995) critically review GENESIS and conclude that the model cannot predict beach behavior. Assumptions used in GENESIS, starting with the shoreface profile of equilibrium, either cannot be met or are such simplifications that the model's effectiveness as a predictive tool is limited at best. Frequently, averaged values of parameters are used, smoothing over great potential variability in data sets (waves, grain size, shoreface profile). GENESIS does not consider storms, normally the driving force of beach change. Modelers justify this by saying that using the average of the highest third of waves is adequate to cover the impact of storms, but a single storm (see figure 8.5) can do more than a decade of "normal" weather (Carter and Woodroffe 1994). Moreover, when predictions are made, it is not possible to quantify the error in the predicted results because there is inadequate data to allow useful statistical analysis; in particular, there is typically little preproject monitoring. GENESIS does not provide the modeler with statistical answers.

Fig. 8.5   Storm waves striking the shoreline at South Nags Head, North Carolina. Ordinarily, the greatest changes in beach size and shape occur during storms. In this photo, overwash by waves can be seen entering the town. Onshore loss of sand by this method is not considered in the GENESIS model.

The model consists of two components: a longshore transport equation and a shoreline change equation. The basic physical data required to run the program include:

- current shoreline position (i.e., location on a map);

- wave characteristics (height, period, and direction);

- engineering structures and activities (e.g., the location of groins, jetties, and beach nourishment);

- measured or calculated beach or shoreface profiles oriented perpendicular to the shoreline (i.e., beach geometry); and

- boundary conditions, which can be anything that defines the limits of the beach to be modeled (e.g., a groin or inlet), and estimates of the amount of sand that flows into or out of the beach across this boundary (e.g., due to the permeability of a groin).

All other factors are either physical constants (e.g., the density of water), user-specified constants (e.g., shoreface slope), or calibration parameters (otherwise known as fudge factors).

As the technical manual for GENESIS emphasizes, the user must constantly rely on his or her own technical expertise (Hanson and Kraus 1989). All of the uncertainty discussed above makes GENESIS at best a qualitative rather than a quantitative model and at worst a model that, after a certain amount of adjusting of input parameters, produces a result that the coastal expert expected—a way of backing up one's judgment with what appear to be objective numbers (Young et al. 1995). In spite of these serious shortcomings, many of which are explicitly noted by the inventors of GENESIS (Hanson and Kraus 1989), the model has found wide application.

Have Models Worked?

The long-term costs of beach nourishment have been consistently underestimated, in part because models have not produced reliable predictions of beach life spans and nutrition costs (figures 8.6 and 8.7; Leonard, Clayton, and Pilkey 1990). This underestimation of costs, by itself, is not proof of the failure of mathematical models, however, because politics are involved in cost estimates. That is, the Corps of Engineers, the agency that designs and funds most nourished beaches, is typically under huge political pressure from the Congress to approve proposed projects. Approval is easier to get if the costs are expected to

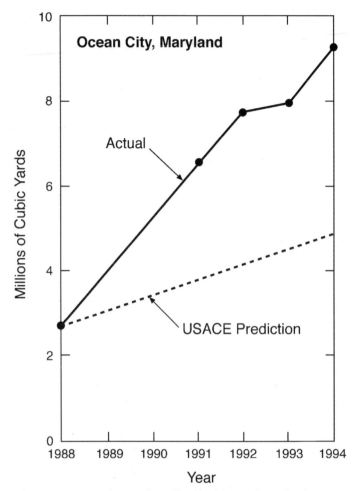

Fig. 8.6 Comparisons of predicted and actual sand volumes required for the Ocean City, Maryland, beach nourishment project. Predicted volume of sand was significantly less than what was actually needed.

be low. Nevertheless, even leaving out politics, models have failed to make good predictions.

One obvious reason for predictive failure is that the design life spans, costs, and sand volumes are given in nonprobabilistic fashion, i.e., as single numbers without error bars. Since nourished beaches are strongly affected by randomly occurring storms, i.e., storms whose magnitude, duration, direction of approach, and frequency cannot be predicted years in advance, large error bars are needed on these numbers. The Corps, however, argues that the Congress and other policy makers cannot understand probabilities and error bars. This argument leads Corps

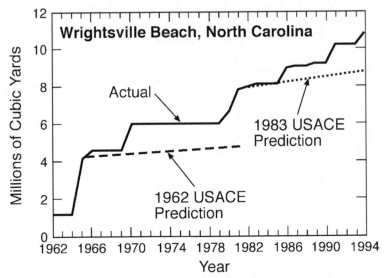

Fig. 8.7 Comparison of predicted and actual sand volumes for Wrightsville Beach, North Carolina, project. Despite fourteen years of experience nourishing this beach, the second sand volume prediction (1982) was still very low in comparison with reality.

scientists and engineers to produce numbers they cannot rigorously defend (Pilkey and Dixon 1996).

Although the failure of beach designers to successfully predict beach life spans seems clear to many, the Corps of Engineers conducted an analysis indicating that their predictive design effort succeeded quite well (USACE 1994; Hillyer and Stakhiv 1997). In short, the Corps argued that predicted costs of nourishment projects were quite similar to the actual costs. The problem with this conclusion, as I have pointed out in the past (Pilkey 1995, 1997), is that the agency did not consider whether a beach stayed in place between renourishments. Often the beach disappeared. One important lesson here is the need for completely independent evaluation of the predictive success of beach nourishment design.

Why Failure Is Not Recognized

For a number of reasons, the lack of success of modeling is apparent neither to the public nor to some scientists and engineers. From the public's standpoint, a time delay of as little as a year seems to be sufficient to let the dust settle from the societal debate over nourishment and for the details of the issues to fade from memory. The design predictions based on mathematical models become forgotten details in a sea of claims and counterclaims. When a nourished beach disappears too quickly, a number of statements are commonly made in defense of the original prediction. These include:

- THE STORM WAS UNUSUAL AND UNEXPECTED. Our society would not routinely accept the "unusual storm" excuse for a failed highway bridge, but it is routinely accepted for the storm-driven loss of artificial beaches.

- THE LOST SAND (FROM THE NOURISHED BEACH) IS JUST OFFSHORE CONTINUING IN ITS FUNCTION OF STORM PROTECTION. But in East and Gulf Coast beaches, where storm surges may be very large, the principal design element for storm protection (as stated in Corps of Engineer design memoranda for each project) is the berm or artificial dune on the upper beach. Storm protection is the principal justification required for federal funding. Needless to say, the argument that "the sand is just offshore" is very unimpressive to local politicians and chambers of commerce whose principal goal in beach nourishment is to gain a recreational beach and bring back the tourists.

From a technical viewpoint, the problems with model-designed nourished beaches are not recognized by coastal engineers for several reasons.

- *Lack of monitoring.* No nourished beach has been monitored for its design life. Part of the problem is that federal agencies tend to fund monitoring for two to three years, after which it becomes a lower priority and is not funded. Another part of the problem might be a political unwillingness to generate facts that might reveal a low success rate.

- *Lack of monitoring criteria.* Because nourished beaches disappear unevenly in both time and space, declaration of success or failure is complex. Nourished beaches always have erosion hot spots, so it is possible to stand at one location and declare it to be a success. If one kilometer of a three-kilometer beach is lost in a storm, is that a failure? Success or failure should be based on how well the beach is solving the initial perceived problems that led to beach nourishment. The monitoring programs that do exist vary in what is actually measured, so there is no body of agreed-upon standard data or common criteria to compare nourishment efforts.

The Problems with Models

As noted above, the beach is an immensely complex environment linked to other environments, for example, the atmosphere. The beach itself consists of linked components—for example, wind, waves, sediment, and biota—whose behavior and interactions are still not fully understood by the scientific community. This makes modeling challenging but not impossible, depending on one's goals. Models might be used to elucidate principles governing beach behavior; however, *prediction* of specific beach behavior remains very uncertain.

A common modeling objective is prediction of the behavior of a nourished beach for a fifty-year life span. But we are no closer to predicting the fifty-year behavior of a beach than we are to predicting fifty years of weather.

Haff (1996) critically reviewed the sources of uncertainty or error in predictive models of earth surface processes. These include: model imperfections; omission of important processes; lack of knowledge of, and sensitivity to, initial conditions; and external forcing. External forcing—the influence of phenomena outside the beach system—is critically important; for example, randomly occurring storms are not modeled.

Table 8.1 lists some model uncertainties that fall in Haff's categories. *Model imperfections* refers to errors in the characterization of processes inherent in the model itself. One such error is the assumption that parameters in the model are universally applicable to all beaches. A good example of this is the use in GENESIS of sediment transport coefficients, designated by the letter K. The model's longshore transport equation calculates the volume of sand expected to be moved, using two coefficients (or Ks) to obtain an amount of sand transport thought to be "reasonable." This "fudge factor" arose because observed and predicted longshore transport volumes on two California beaches did not match (Komar and Inman 1970). The COE *Shore Protection Manual* (USACE 1984) suggests for general usage K = 0.77, and, as a result, many of the longshore transport volume figures used in the United States are based on this multiple. But why should a single multiple be valid everywhere? Is the longshore transport equation wrong? Or perhaps field measurements of longshore transport are fraught with errors, as demonstrated by Bodge and Kraus (1991). Other field studies have been carried out to determine a better value for K. By comparing observed and predicted longshore transport volumes, values of K ranging from 0.04 to 1.6 have been obtained, representing a span of nearly

TABLE 8.1

Four sources of model uncertainty identified by Haff (1996), as applied to models of beach behavior. Modified from Thieler et al. (2000).

Model Imperfections	Unknown Initial Conditions and Sensitivity to Initial Conditions	External Forcing	Omission of Significant Processes
Assumption of an equilibrium shoreface profile	No (or poor) wave data	Areal and temporal variations in sediment supply	Sediment transport beyond closure depth
Scaling up short-term relationships to long-term (minutes to decades)	No (or poor) historical shoreline retreat data	Multiple, randomly occurring storm events	
Assumption of universal applicability of parameters	Degree of instability of nourished beaches	Storm surge	
		Tidal currents	Water table and/or pore pressure effects on sediment erodibility
Use of adjustable constants as fudge factors	Geology underlying the shoreface		Liquefaction or bed ventilation of surf zone sediments by breaking waves
Use of wave tank data for modeling the prototype			Wave refraction/diffraction effects
	Offshore bars and bedform configuration		
			Storm surge ebb currents
			Wind-driven up/downwelling currents
			Wind-driven longshore currents
			Wave-current interactions

two orders of magnitude in potential longshore transport volumes (Wang, Kraus, and Davis 1998; Stutz and Pilkey 1999). This raises the question: Do we know longshore transport volumes on any beaches?

Perhaps the most important *process omission* in the models is the exclusion of wave-current interactions. The importance of the latter process is that as long as breaking waves suspend sand into the water column in the surf zone, very small currents can transport the sand laterally for long distances. Offshore bars are typically an *unknown initial condition*. Beach behavior may be highly sensitive to such initial conditions, as illustrated by coarse shell lags, the shell pavements found on many beaches and shorefaces that are concentrated when finer materials are winnowed away. If shell or gravel lags exist on a beach, they can delay or alter the response of a beach to storms, resulting in a significantly different storm response than would have occurred in a beach of uniform grain size. Perhaps the most important omission of all in models applied to beach nourishment is some measure of apparent instability of nourished beaches, as indicated by the very high loss rates of artificial beaches relative to their natural predecessors. Rates of loss of artificial beaches can be 1.5 to 12 times those of natural ones, according to Leonard et al. (1990).

Response to Criticism

Criticism of models is usually not well received by modelers; beach model developers and other scientists and users do not engage in a constructive dialogue. Objective oversight is resisted, perhaps because models are involved in determining the allocation of significant funding. It is a sensitive bread-and-butter issue. Critical reviews of models of earth surface processes can be difficult to publish (Haff, personal communication). A common reviewer response is "We are already aware of these problems."

Published criticism of mathematical models of beach behavior has rarely been met with straightforward and balanced critical response in the literature. The bullets below represent typical comments made, both in print and informally, in response to my criticism of mathematical models and their use:

- Models are valuable "as a learning structure for making predictions prior to construction, subsequently monitoring the project, then later comparing predictions with monitoring results . . ." (Houston and Dean 1997).

- We use models primarily to fine-tune results as a check on the various other approaches and tools that we use to design nourished beaches.

- This is the best model we have at our current state of knowledge and until we find something better.

- Simply criticizing the assumptions behind the models is insufficient. It is necessary to run the model and observe its veracity on the basis of actual beach behavior.

- We're learning from our mistakes.

- Model skeptics are "neo-luddites" (Komar 1996).

- Don't throw the baby out with the bathwater.

Impact of Models on Coastal Science

One interesting impact of beach behavior modeling has been the "contamination" of coastal science with concepts originally devised as simplified assumptions for models. Two examples of this are the profile of equilibrium concept and its corollary, closure depth as a sediment fence. Their widespread use in models led to their acceptance, at least in some circles, as valid scientific principles. At present, both concepts, especially closure depth, are widely criticized (e.g., Wright 1995).

In the view of some, the maturity of an earth science specialty can be measured by its success in the use of numerical models for description and prediction of geological processes (Komar 1998b). Such a view holds that the more mathematical a science, the better. The strength of geological sciences, in the view of others, has always been its observational foundation (Baker 1994; chapter 2, this volume).

Ironically, Komar's assertion that coastal geology is a mature science is based on the use of the models critically described in this paper. This view that modeling for prediction is the ultimate in sophistication lends unjustified authority to the modeling approach. Sophistication in predictive modeling is considered so desirable by many scientists and engineers that shortcuts are taken to construct models before the system to be modeled is understood, even though this is actually the opposite of sophistication. Thus, modeling has fostered a climate of carelessness. Assumptions that would never be accepted if examined critically by themselves are tucked into models and never revisited.

Alternatives to Models in Nourished Beach Design

If model-based predictions are unreliable, what alternatives exist? There are three approaches to the design of beaches that do not involve the use of mathematical models (Pilkey et al. 1994):

- *Imitate nature.* This involves studying the beach prior to the nourishment and assuming that the nourished beach will behave like its natural predecessor. Such studies take years. Two examples are the Dutch approach and the Gold Coast, Australia, experience. According to Verhagen (1992), the following process is used by the Dutch to design nourished beaches:

  1. Take beach profile measurements for at least ten years.

  2. Calculate the loss of sand in cubic meters per year.

  3. Multiply this by the desired life span of the nourished beach.

  4. Add 40 percent volume as an "unexpected loss" factor.

  5. Put the sand on the beach.

  Smith and Jackson (1992) discussed the Gold Coast, Australia, approach used on a beach that is very sand rich, relative to those on the U.S. East Coast. More than twenty years of combined wave and profile observations there demonstrated that typhoons cause the formation of large offshore bars that gradually move ashore as part of the storm recovery process. While the bar exists, beach erosion is nonexistent. In two recent nourishment projects, the Australians have constructed an artificial storm bar with nourishment sand (a politically difficult process, since citizens disapprove of pumping sand into the sea rather than onto the beach), and the approach seems to be working by reducing costs.

- *Employ the kamikaze beach approach.* This approach involves placing the sand on the beach with no design effort beyond plans for sand placement and distribution on the beach. This approach to nourishment is funded in the navigation category by the U.S. Army Corps of Engineers. Such nourishment activities are carried out by the COE as an adjunct to navigation projects and do not require their own separate and favorable cost-benefit ratio because they are justified by the improvement in navigation. Beach nourishment in this case is actually a way to dispose of dredge spoil. (The COE must, however, prove that nourishing a beach is the cheapest way to dispose of dredge spoils.) Some of these beaches are very large, including one on Atlantic Beach, North Carolina, of 5 million cubic yards—equivalent to 500,000 dump truck loads.

- *Learn from nearby nourished beaches.* This approach assumes that the nourished beach will behave similarly to other nourished beaches nearby. There are strong regional differences in the durability of nourished beaches on U.S. East Coast barrier islands (Pilkey and Clayton 1989). For example, nourished beaches on the east coast of Florida, south of Cape Canaveral, have typical life spans of seven to nine years. In New Jersey, nourished beach life spans are almost always less than three years. Of course, there are always beaches that don't follow regional trends. That is, despite these regional life-span averages, there are significant intraregional variations. Some are related to occurrences of storms, but, in most cases, the reasons for the variations are not apparent.

This approach (Pilkey 1988) is essentially a kind of adaptive management—in which decisions are based on recent experience rather than mathematical models—but it remains to be seen whether it will work in the administrative and regulatory framework that governs beach nourishment projects (or whether that framework can be adjusted to be more adaptive).

Conclusions

Mathematical modeling of the behavior of beaches for applied purposes has not been successful in providing reliable predictions to support beach nourishment. Among a number of other things, dependence of beach behavior on randomly occurring storms and a strong sensitivity to initial conditions may doom future efforts to answer *when, where,* and *how much* with sufficient certainty to be an effective decision-making tool. Models currently used to predict beach behavior suffer from the following shortcomings:

- The assumptions behind models are wrong. Assumptions with recognized significant weaknesses (see table 8.1) are routinely used.

- Assumptions aside, adequate data are rarely available for the purposes of setting model initial conditions, determining model parameters, and calibrating the model (including determining uncertainties). (See the list of model limitations from Hanson and Kraus 1989, from the GENESIS manual.)

- The models are gross oversimplifications of complex systems that scientists do not fully understand. Models have not improved apace with our understanding of beach processes (e.g., bottom current studies and recognition of geologic control of erosion), and the assumptions used in the models have thus become increasingly untenable.

- There is a general lack of monitoring, hindsighting, and independent evaluation of model predictions. Not the least of the problems is the widespread acceptance of the "unusual storm" excuse for too rapid beach loss, which precludes objective evaluation of the use of models.

- Adjustment of the model to give generally "reasonable" results or even to match observations in one set of conditions does not imply that the model will give accurate results in other conditions.

- Most models are assumed to have universal applicability on all types of beaches in any wave climate. Clearly, the natural system is too complex for this to be true.

In short, currently used predictive models of beach behavior are inadequate for the engineering tasks for which they were designed. All the same, models remain an important political tool in the ongoing battle over management of our coasts. Despite their proven inadequacies, these models bring scientific credibility—and an aura of certainty—that helps justify the decision to nourish beaches and to continue development of the shore. Thus, an engineering tool has been applied to a scientific problem in order to justify policy decisions that are often controversial. Nature, however, pays no attention, and shorelines continue to retreat.

References

Baker, V.R. 1994. Geomorphological understanding of floods. *Geomorphology* 10: 139–156.

Bodge, K.R., and N.C. Kraus. 1991. Critical examination of longshore transport rate amplitude. *Proceedings of Coastal Sediments '91*, pp. 139–155. New York: American Society of Civil Engineers.

Bruun, P. 1962. Sea level rise as a cause of storm erosion. Proceedings of the American Society of Civil Engineers. *Journal of the Waterways and Harbors Division* 88WWI: 117–130.

Carter, R.W.G., and C.D. Woodroffe. 1994. Coastal evolution: An introduction. In *Coastal evolution: Late quaternary shoreline morphodynamics*, R.W.G. Carter and C.D. Woodroffe, eds. Cambridge: Cambridge University Press.

Dean, R.G. 1977. *Equilibrium beach profiles: U.S. Atlantic and Gulf coasts.* Technical Report no. 12. Newark: Department of Civil Engineering, University of Delaware.

Dean, R.G. 1983. Principles of beach nourishment. In *CRC handbook of coastal processes and erosion*, P. Komar, ed., pp. 217–231. Boca Raton, FL: CRC Press.

Haff, P. 1996. Limitations on predictive modeling in geomorphology. In *The scientific nature of geomorphology: Proceedings of the 27th Binghamton symposium in geomorphology, New York*, B.L. Rhoads and C.E. Thorn, eds., pp. 337–358. New York: John Wiley & Sons.

Hanson, H., and N.C. Kraus. 1989. *GENESIS: Generalized model for simulating shoreline change.* Technical Report CERC 89-19. Vicksburg, MS: USACE Waterways Experiment Station.

Hillyer, T.M., and E.Z. Stakhiv. 1997. Discussion of O.H. Pilkey, The fox guarding the hen house (editorial). *Journal of Coastal Research* 13: 259–264.

Houston, J.R., and R. Dean. 1997. Review of the book *The Corps and the Shore*, O.H. Pilkey and K. Dixon. *Shore and Beach* 65: 27–31.

Howard, J.D., W. Kaufman, and O.H. Pilkey. 1985. National strategy for beach preservation. *Proceedings of Conference on America's Eroding Shoreline.* Savannah: Skidaway Institute of Oceanography.

Komar, P. 1996. A maverick on the beach. *New Scientist* 151: 41.

Komar, P.D. 1998a. *Beach processes and sedimentation* (2nd edition). New York: Prentice-Hall.

Komar, P.D. 1998b. The modeling of processes and morphology in the coastal zone—Reflections of the maturity of our science. *Shore and Beach* 66: 10–22.

Komar, P.D., and D.L. Inman. 1970. Longshore transport on beaches. *Journal of Geophysical Research* 75 (30): 5514–5527.

Larson, M., and N.C. Kraus. 1989. *SBEACH: Numerical model for simulating storm induced change.* Technical Report no. 89-9. Vicksburg, MS: USACE Waterways Experiment Station.

Leonard, L., T. Clayton, and O.H. Pilkey. 1990. An analysis of replenished beach design parameters on US East Coast barrier islands. *Journal of Coastal Research* 6: 15–36.

Pilkey, O.H. 1988. A thumbnail method for beach communities: Estimation of long-term beach replenishment requirements. *Shore and Beach* 56: 23–31.

Pilkey, O.H. 1995. The fox guarding the hen house. *Journal of Coastal Research* 11: iii–v.

Pilkey, O.H. 1997. Reply to Hillyer and Stakhiv. *Journal of Coastal Research* 13: 265–267.

Pilkey, O.H., and T.D. Clayton. 1989. Summary of beach replenishment experience on U.S. East Coast barrier islands. *Journal of Coastal Research* 5: 147–159.

Pilkey, O.H., and K.L. Dixon. 1996. *The Corps and the Shore.* Washington, DC: Island Press.

Pilkey, O.H., R.S. Young, D.M. Bush, and E.R. Thieler. 1994. Predicting the behavior of beaches: Alternatives to models. *Littoral* 94: 53–60.

Pilkey, O.H., R.S. Young, S.R. Riggs, A.W.S. Smith, H. Wu, and W.D. Pilkey. 1993. The concept of shoreface profile of equilibrium: A critical review. *Journal of Coastal Research* 9: 255–278.

Riggs, S.R., W.J. Cleary, and S.W. Snyder. 1995. Influence of inherited geologic framework on barrier shoreface morphology and dynamics. *Marine Geology* 126: 213–234.

Schwartz, M.L., ed. 1982. *The encyclopedia of beaches and coastal environments*. Stroudsburg, PA: Hutchinson Ross.

Smith, A.W.S., and A. Jackson. 1992. The variability in width of the visible beach. *Shore and Beach* 60: 7–14.

Stutz, M.L., and O.H. Pilkey. 1999. Discussion of Wang, P., N.C. Kraus, and R.A. Davis: Total longshore sediment transport rate in the surf zone. *Journal of Coastal Research* 14: 269–282.

Thieler, E.R., A.L. Brill, C.H. Hobbs, and R. Gammisch. 1992. Geology of the Wrightsville Beach, NC shoreface: Implications for the concept of shoreface profile of equilibrium. *Marine Geology* 126: 217–287.

Thieler, E.R., O.H. Pilkey, R.S. Young, D.M. Bush, and F. Chai. 2000. The use of mathematical models to predict beach behavior for coast engineering: A critical review. *Journal of Coastal Research* 16.

USACE (U.S. Army Corps of Engineers). 1984. *Shore Protection Manual*. Vicksburg, MS: U.S. Army Corps of Engineers, Coastal Engineering Research Center, Waterways Experiment Station.

USACE (U.S. Army Corps of Engineers). 1994. *Shoreline erosion and beach erosion control study, phase 1: Cost comparisons of shoreline protection projects of the U.S. Army Corps of Engineers*. IWR Report no. 94 PS-1, Alexandria, VA: U.S. Army Corps of Engineers.

Verhagen, H. 1992. Method for artificial beach renourishment, *Proceedings of the 23rd International Conference on Coastal Engineering*, Venice, Italy, pp. 2474–2485.

Wang, P., N.C. Kraus, and R.A. Davis. 1998. Total longshore sediment transport rate in the surf zone: Field measurements and empirical predictions. *Journal of Coastal Research* 14: 269–282.

Werner, B.T., and T.M. Fink. 1993. Beach cusps as self-organized patterns. *Science* 260: 968–971.

Wright, L.D. 1995. *Morphodynamics of inner continental shelves*. Boca Raton, FL: CRC Press.

Young, R.S., O.H. Pilkey, D.M. Bush, and E.R. Thieler. 1995. A discussion of the Generalized Model for Simulating Shoreline Change (GENESIS). *Journal of Coastal Research* 11: 875–886.

Is This Number to Your Liking? Water Quality Predictions in Mining Impact Studies

Robert E. Moran

If one is to mine gold commercially in the western United States, the operation is likely to be at least partly on federally managed land—most often managed by the U.S. Bureau of Land Management (BLM) or the U.S. Forest Service. Such federal lands constitute about 50 percent of the eleven western states and 90 percent of Alaska. Portions of mines may also be on private lands. Most such operations are huge open-pit mines, sometimes more than a thousand feet deep, nearly a mile wide, and more than a mile in length. The land management agencies oversee the issuing of permits as well as subsequent operations with the intent of minimizing future impacts to the site and its resources. However, the construction of such huge pits inevitably involves moving and exposing massive volumes of rock, often hundreds of feet below the water table. Once mining ceases and the dewatering pumps are shut off, a lake will form within the excavated pit up to the level of the water table. Because pits of this scale were first constructed at gold sites only in the late 1980s, we have no appropriately long-term information on the chemistry of the resultant pit waters; many of these pits are still being excavated, and the pit lakes have yet to form.

Mining regulatory agencies like the BLM face a dilemma when issuing permits for such mining activities. On the one hand, they are required to prepare environmental documents that should disclose any anticipated significant damages to the resources of the site and describe appropriate mitigation procedures. On the other hand, they are mandated by agency policy, and apparently by legislation as well, to promote mining activities on federal lands. My discussions with numerous BLM staff in Idaho, Nevada, Utah, Montana, and Colorado have confirmed that their operations are strongly guided by the mandate to promote mining. Hence, it is not surprising that these same BLM employees could not name a site where a formal request to conduct large-scale hard-rock mining on BLM land had ever been denied for environmental reasons—an observation corroborated by Roger Flynn, an attorney for the Western Mining Action Project, an environmental organization in Boulder, Colorado. Thus, because the mining will go forward, environmental documents describing proposed mining activities carry a large burden in assuring the public that impacts to water resources, wildlife, and other natural assets will be acceptable.

Can the BLM assure the general public that site surface water and groundwater quality will not be degraded as a result of these activities? Traditionally, it has not been acceptable for the BLM to tell citizens that it is uncertain about future impacts. Thus, the BLM usually requires the mining company to present *predictions* of future water quality in the environmental documents prepared for public review. Until about ten years ago, such documents often simply presented a qualitative opinion about the likelihood that future water quality problems would develop. Based on my review of dozens of such mining environmental documents, as well as discussions with others involved in this process, significant, long-term water quality problems were not typically anticipated or disclosed. But public skepticism about such simple, rosy characterizations is justified, especially as actual field observations have shown that unforeseen problems have frequently developed, at such mine sites as Summitville, Colorado; Zortman-Landusky, Montana: Gilt Edge, South Dakota; Thompson Creek, Idaho; and Sleeper, Nevada.

In an attempt to make the predictions appear more scientific and trustworthy, regulators in recent years have usually required that some form of predictive testing or geochemical computer modeling be included in the environmental studies. Tests and models lend an apparent sense of certainty. However, these predictions and studies are prepared by consultants who are chosen and paid by the mining companies being regulated. The majority of the dozens of such studies I have reviewed continue to anticipate few, if any, significant, long-term water quality problems and tend to predict overly optimistic scenarios.

Since most of the computer simulations of open-pit mining water quality were performed in recent years, insufficient time has elapsed to reasonably judge the success of these predictions. To some extent the water quality predictive science is still in its infancy. Nevertheless, it is surprising that no sudy has ever been done to compare predicted versus observed water quality at general metal mining sites.

In this chapter, I argue that the cause of overly optimistic scenarios lies not in the science, but rather in the economic and political pressures placed on the technical consultants and the government managers, which lead them to use inadequate models and to misuse their predictive results. This chapter focuses on one example of mining on BLM land, but the conclusions pertain more generally to mining on federally managed lands. Also, it comments specifically on predictions about pit lake water quality, but the same general issues are relevant to predictions about the quality of water from similar disturbances: mine workings, waste rock piles, tailings, and heap leach piles.

To summarize, the appropriate use of water quality models is important because predictions generated by the models are being used to justify federal and state approval of massive mining projects, implying that we truly know what the future water quality impacts will be. The models give the public a false sense of certainty. Where unforeseen problems develop many years after mine closure, the taxpayers may have to bear the remediation costs and the environmental consequences, as they have had to do with the majority of abandoned metal mines in the West.

Nevada Case History

Description of an actual example of hard-rock mining on federal land will illustrate how predictions have typically been used. The example is an open-pit gold site in north-central Nevada. This site will be referred to as the Aguirre mine.[1] The technical details come from draft and final environmental impact statement (EIS) documents released in 1994 and 1996, and supporting documents. I participated in the preparation of the final EIS.

A large Canadian company had proposed to construct, mostly on BLM land, the open-pit Aguirre gold mine to a depth of nearly 1,000 feet, which would take the pit approximately 800 feet below the local water table. While most of this part of Nevada is harsh desert, it is frequently underlain by highly permeable alluvial and carbonate aquifers that yield tremendous amounts of groundwater. As that was the case here, an extensive system of extraction wells and pumps would have to be constructed around the perimeter of the proposed pit to dewater the rock, so that it could be mined. Once all of the economically suitable ore had been removed—in an estimated twelve years—the pumps would be shut off, and water would begin to flood the pit, eventually forming a lake between 700 and 800 feet deep. While Nevada already has more than thirty open-pit gold mines, most have resulted in relatively shallow lakes. None of the few other deep pits had been flooded at the time the Aguirre EIS was being prepared. Thus, no examples of comparable gold pit water quality were available in the literature. (At least one compara- ble, moderately deep pit, the Sleeper mine, has filled and begun to react chemically.) An operation involving the pumping of tens of thousands of gallons of groundwater per minute generates considerable concern— always among neighboring landowners and frequently with local and national environmental groups. The proposed Aguirre mine also aroused the concerns of several Native American groups. Some worried that the mine dewatering might dry up existing springs or wells that are used for livestock watering and domestic purposes. Others worried that the water in the pit lake would become contaminated and in turn cont- aminate surrounding groundwater and springs. Hence, various federal and state regulatory agencies became involved in the permitting process.

In Nevada, the actual enforcement of water yield and water quality regulations at mining sites falls to the state agencies. However, since most of the operations are on federal lands, the federal agencies (the BLM and the Forest Service) make the decisions about appropriate land use. Theoretically, they can approve, deny, or modify any proposed activity such as mining, logging, or grazing. However, many federal land management staff attest that they do not have the legal authority to oppose a mining operation, that they can only attempt to minimize the negative impacts (personal communications). Many resource experts disagree with this view of the land managers' authority, feeling that it is simply a comfortable political position the agencies have chosen to take (Wilkinson 1992). I will not pursue the legal aspects of this issue beyond noting that the controlling legislation, the Mining Law of 1872, clearly instructs the agencies to promote mining on federal lands (Leshy 1987).

This leaves the regulators in a delicate position. They feel obligated to allow mining to occur, even to encourage it, but they must ensure the public that negative impacts will be minimal. Hence, they grasp predic- tions of future water quality conditions and impacts, and of other impacts, as tools to help them negotiate the situation. Obviously, there is a conflict of interest here and pressure to predict a largely benign future.

The situation at the Aguirre mine was typical in that the company hired its own consultants to collect data, perform studies, and predict future conditions. In essence, these studies stated that there would be no negative consequences to any nearby water resources. That is, nearby wells and springs would not have their yields reduced, and the chemical quality of these waters would not be degraded.

The post-mining pit water quality was predicted by the company's consultant by coupling two computer models known as PHREEQE and MINTEQA2. Both are well known within the hydrological modeling community and are often used to gain a better understanding of which chemical species may be stable and what reactions may be occurring in a specific environment. However, only the consulting community routinely uses these models to predict specific concentrations of minor and trace constituents far into the future.

The originally predicted pit water quality concentrations, as shown in the Aguirre mine consultant's report and the draft EIS, are presented in table 9.1. Using these modeled results, the mining company's consultant

TABLE 9.1

Predicted post-mining pit water quality—draft EIS. All concentrations in mg/L, unless noted.

Element	Valley Center	Valley Margin	Mixture	At equilibrium w/ Calcite, Quartz & Illite	At equilibrium w/ 0.002 moles/l (250 mg/l) Pyrite
Calcium	72.45	57.01	62.18	54.83	increase
Magnesium	12.07	17.03	15.57	15.58	no change
Sodium	69.17	97.25	87.95	88.00	no change
Postassium	9.06	14.92	12.98	12.98	no change
Iron	0.561	0.085	0.244	0.244	increase
Manganese	0.028	0.039	0.035	0.035	no change
Aluminum	<0.05	0.091	0.061	0.061	no change
Barium	0.053	0.045	0.048	0.048	no change
Strontium	0.361	0.847	0.686	0.686	no change
Silicon as SiO$_2$	14.40	12.17	13.28	5.99	no change
Chloride	57.25	26.57	35.14	35.16	no change
Carbon			167.4	160.75	increase
Sulfur as SO$_4$	88.48	128.84	115.50	115.52	increase
Nitrogen as NO$_3$	1.525	1.19	1.302	1.303	no change
Fluoride	0.639	2.36	1.788	1.788	no change
Lithium	0.016	0.260	0.179	0.179	no change
pH	7.78	8.01	7.95	7.48	7.19
pE	−0.80	−0.70	−0.9256	0.4801	−3.6926
Temp	16.15	27.2	23.52	23.52	23.52
Total Alk	167.33	263.32	232.21	209.79	404.45

• pH in standard units.   • pE in millivolts.   • Temperature in degrees centigrade.

stated that the pit water was expected to have near-neutral pH, a total dissolved solids (TDS) concentration of less than 500 mg/L, and low dissolved metals concentrations. This description is essentially what one would expect for water suitable for human consumption. In fact, while speaking at several public meetings with local citizens during and after preparation of the draft EIS, mining company representatives, their consultants, and BLM staff stated that the post-mining pit water would be of drinking-water quality.

Under the National Environmental Policy Act (NEPA 1969), which became effective in 1970, all federal agencies must prepare a "detailed statement" for all "major federal actions significantly affecting the quality of the human environment." Thus the need for EISs, or similar reports, when large-scale mining is proposed on federal lands. Given the cutbacks in federal budgets and staffs, the land management staff are seldom able to prepare such an EIS, especially not on the schedule the mining company desires. Also, agencies like the BLM often do not have adequate technical staff, such as geochemists, qualified to perform the more technical analyses. As a result, another consultant is usually hired to advise and assist the BLM—the third-party consultant. One would assume that the third-party consultant is hired to give an independent perspective to the management agency, to balance the biases inherent in the industry consultant's viewpoints. But here the process becomes even more convoluted and conflicted. The mining company generally has considerable influence on the BLM staff's decision as to which third-party consultant is selected. More important, it is the mining company that ultimately pays all the invoices of the third-party consultant.

The third-party consultant has a conflict of interest, as he or she works for both the federal management agency and the company being regulated. Both "masters" may review and approve estimated costs, cost modifications, and schedules. The company also supplies critical environmental baseline data, project design information, and results of *its* alternative analyses. In principle, the lead federal agency directs the technical effort of the third-party consultant and decides on the final language of the EIS. In practice, depending on the individual personalities involved, their technical background, and the workload of the BLM staff, much of the EIS preparation may actually be directed by the mining company.

The draft EIS prepared for the Aguirre mine contained general water quality predictions and impact assessments based on the industry consultant's modeling. The actual predicted pit water quality data (table 9.1) were never shown in the draft EIS; only the consultant's report was cited. Since the industry consultant's conclusions were quoted in the draft EIS with no substantive changes or additions, it is obvious that the third-party consultant accepted those model results without significant independent scrutiny. When the Aguirre draft EIS was released in the summer of 1994, it received much criticism. Most centered on the unreasonableness of the pit chemistry predictions. How was it possible that a lake formed in the scorching Nevada desert sun would still have dilute waters suitable for drinking decades or even hundreds of years after mining ceased? Why didn't the EIS discuss the future concentrations of many toxic metals not shown in table 9.1?

In this instance the modeling had been too simplistic. It considered only conditions of chemical equilibrium and did not allow the pH to vary as theoretical chemical conditions changed. It did not account for evapoconcentration through time, or the differing speeds of chemical reactions. Although the deep site water was of geothermal origin, it made no allowance for changes in reaction rates or solubilities due to water temperatures and pressures above standard conditions (i.e., 25 degrees C, and 1 atmosphere pressure). And it did not realistically deal with the fate of metals once they became trapped on solid particles of clays or iron hydroxide. For example, this model assumed that once a copper ion attached to iron hydroxide particles, the suspended iron-copper mass would settle to the bottom of the pit lake, and that there were no conditions under which the copper, or other such sorbed metals, might be released back into the lake waters. The pit lake was simulated as a one-way sink for metals; hence, the concentrations were guaranteed to decline.

Further, the model made no allowance for the roles of microorganisms in the chemical reactions. This is a common shortcoming of geochemical models and can render such simulations all but useless, since microorganisms can drastically change the rates at which many reactions occur (Chapelle 1993). The model did not consider variations in chemistry with depth in a deep lake. Also, the model assumed that the quality of waters entering the pit from the weathering and oxidation of the pit walls could be represented by data from extremely simplistic, short-term leaching tests (meteoric water mobility tests), which cannot possibly reproduce the very complex, long-term chemical reactions that actually take place. A related shortcoming is the assumption that only the materials exposed on the two-dimensional face of the pit walls are oxidized when a pit is dewatered. This seems unreasonable since the dewatering wells would lower the water table to below the level of the pit bottom, allowing much of the three-dimensional rock mass within this dewatered zone to become oxidized during the many years of mining. Most important from a practical point of view, the baseline water quality and rock geochemistry data needed for input to the model were inadequate, especially with regard to the deeper zones.

Nevertheless, the computer-generated results could appear to be quite formidable to a lay audience—and to the local BLM staff. The data were calculated to several significant figures; for example, predicted calcium was 54.83 mg/L. There was no discussion of the uncertainty in these predictions. The implication was, "This is *truth.*" Such predictions generate a sense of confidence in the minds of the audience, which allows the regulators to move the regulatory decision-making process forward.

The Aguirre mine site is located on federal land involving ownership disputes between the BLM and Native Americans of the Western Shoshone tribe. These disputes tended to polarize the discussions regarding potential environmental impacts, making the ordinarily routine review process contentious and visible. Public criticism, together with significant newspaper and TV coverage, forced the BLM to enlist more experienced third-party consultants and to reexamine the predictions seriously. As the new water quality and geochemistry expert for the third-party team, I attempted to make obvious the model shortcomings, but, most important, I tried to shift the focus toward highlighting the uncertainty of such predictions—even when performed at the state-of-the-art. Given such uncertainty, the only way to move forward is to attempt to minimize or spread the risk, as insurance companies do.

Uncertainty is often a threatening concept to regulators and public officials in general, who in many cases either do not understand it or are wary of presenting it to the public. The public, in turn, may be understandably reluctant to accept and fund a project for which the outcomes are not well understood. However, there is a way to bring some comfort to the situation. Often, similar projects have been completed numerous times before, and the regulators may have a database compiled from the population of such projects from which expectations based on statistics can be drawn. It is common for a technical manager to consult historical data on, for example, predicted maximum flood heights or average dam construction cost overruns, in order to anticipate the range of expected outcomes. These data are routinely presented in statistical terms, such as means, medians, and error bars or confidence intervals.

Because deep pit lakes are a relatively new phenomenon, data on such sites are quite limited. But one might expect that the BLM would have databases showing actual and predicted water quality data from, for example, a population of waste rock piles at hard-rock gold mining sites throughout the West where predominantly oxidized or reduced ores were mined. Although it would be necessary to consider other relevant factors such as climate, a land manager could use such comparisons to get a statistical sense of the risk of future water quality problems based on observed developments in these analogous settings. However, such data compilations do not exist within the BLM. Instead, consultants are asked to generate site-specific, seemingly accurate predictions of future water quality for such sites as the Aguirre pit lake.

Neither the BLM nor the Aguirre company representatives wanted the revised EIS to discuss the uncertainty of predictions, probably for all the reasons already mentioned. Also, if the uncertainty was obvious, the BLM might be obligated to increase the dollar amounts of financial bonding required from the mining company. Such a situation might even cause the land managers to consider the need for environmental liability insurance. Bonding and environmental liability insurance for long-term water quality problems are very sensitive subjects in the mining business—especially since they have so seldom been required. Better, from the industry and BLM points of view, to attempt to refine the predictive models and maintain the facade of certainty.

Given this political reality, the new third-party consultants suggested that whatever predictions were made, even if simplistic, ought to reflect common sense. Anyone could go to the desert geochemical literature and find frequent references to the fact that most natural desert lakes evolve toward high alkalinity. Such a condition would increase the dissolved concentrations of many metals and metalloids (e.g., arsenic, mercury, selenium, molybdenum, uranium, and nickel) in the right circumstances. The new consultants suggested that the revised impacts analysis ought to indicate that the pit lake would likely be strongly alkaline after many decades, and that predictions of metal concentrations would be subject to a wide margin of error.

Most of a year passed while the company hired an additional geochemical consultant, and all parties gave direction to the original company consultant on how to improve the predictions. Many different assumptions and new input data were incorporated into a revised model, which, not surprisingly, yielded totally different results. After many runs of the new model and much tinkering to try to correct for internal inconsistencies, the predictions shown in table 9.2 were generated and ultimately appeared in the final EIS. Unlike the tapwater-quality liquid of the draft document, this pit fluid would be highly alkaline, with high dissolved solids content and significantly elevated concentrations of several metals and other constituents. Such water might prove toxic to fish, birds, and livestock and by a large margin would not meet Nevada drinking water standards.

TABLE 9.2
Predicted post-mining pit water quality—final EIS. All concentrations in mg/L, unless noted.

Constituents	5 yrs	50 yrs	100 yrs	150 yrs	200 yrs	250 yrs	Nevada Drinking Water Standards
pH	8.57	8.81	9.05	9.24	9.37	9.46	6.8–8.5
Alkalinity	198	351	648	1,083	1,580	2,060	
Chloride	30	40	52	66	80	95	250
Fluoride	1.4	1.9	2.5	3.0	3.5	3.9	2.0
Nitrate	0.8	1.1	1.4	1.7	2.1	2.5	10
Sulfate	100	133	171	214	258	301	250
Arsenic	0.02	0.02	0.03	0.04	0.04	0.04	0.05
Barium	0.06	0.09	0.12	0.13	0.15	0.16	2.0
Cadmium	0.003	0.004	0.005	0.006	0.007	0.008	0.005
Calcium	7.2	2.7	1.1	0.6	0.4	0.4	
Chromium	0.005	0.007	0.009	0.011	0.013	0.015	0.1
Copper	0.004	0.006	0.007	0.008	0.009	0.01	1.3
Iron	0.0006	0.0009	0.0015	0.0023	0.0031	0.0038	0.3
Lead	0.001	0.001	0.002	0.002	0.003	0.004	0.015
Magnesium	12	16	21	26	32	38	150
Manganese	0.03	0.03	0.04	0.05	0.06	0.06	0.05
Mercury	0.001	0.001	0.002	0.002	0.003	0.003	0.002
Potassium	10	14	18	22	26	30	
Selenium	0.006	0.008	0.01	0.012	0.014	0.016	0.05
Silver	0.3	0.3	0.4	0.6	0.7	0.8	0.1
Sodium	84	110	142	182	223	264	
Zinc	0.09	0.12	0.16	0.22	0.21	0.21	5.0
TDS	444	671	1,058	1,610	2,206	2,795	500

• pH in standards units. • Alkalinity as $CaCO_3$. • Charge imbalance increases from 2%, 8%, 19%, 29%, 37%, to 41% from year 5 through year 250.

The new model results were more realistic but still quite simplistic, retaining most of the shortcomings previously mentioned. One of the more severe weaknesses was presenting a uniform water chemistry for the entire lake for each time period. As most competent limnologists and oceanographers would attest, it is extremely unlikely that the chemistry of such a deep lake would be uniform throughout its depth, at all seasons of the year (Kuhn, Johnson, and Sigg 1994; Miller, Lyons, and Davis 1996).[2]

Again, the important point is not the technical details but how the predictive modeling was used. In the last revision of the final EIS for review by the BLM, text was added warning readers that the latest predicted pit concentrations (see table 9.2) should not be taken as "gospel." One important EIS paragraph read:

> It should be noted that hydrogeochemical models are most useful as tools to better understand, qualitatively, how a complex interactive system will behave. Such models are less successful at making accurate or precise quantitative predictions of future metals concentrations. . . . Therefore, the predicted pit metals concentrations should be interpreted as general approximations having considerable potential for error, both positive and negative. Only through future monitoring will the actual concentrations be known.

This one paragraph was the only portion of the revised EIS noted for revision by the BLM water quality specialist. He wanted it removed! To discuss uncertainty in modeling was heresy. Fortunately, senior staff of the BLM in Nevada agreed to leave the offending paragraph in the final EIS.

Final Observations and Recommendations

The Aguirre mine example elucidates several problems inherent in the use of predictions to demonstrate the environmental impacts of mines:

- The regulatory agency encourages the use of numerical models to lend a sense of certainty to the predictions and to advance the agency's perceived mission.

- The BLM staff feel obligated to promote mining on public lands. Hence, political pressures can cause the impacts analysis to be conducted so that it indicates a generally favorable outcome. In the example of the Aguirre mine, the NEPA review process did succeed in revising the predictions such that they were less optimistic and more reasonable, but this result seems to be an exception. In any case, the Aguirre mine was ultimately permitted.

• For the most part, consultants in the present system are neither financially nor politically independent; in particular, they often depend on mining companies. Overly optimistic predictions are the result. Some means must be found to allow both the regulators and their consultants to feel free to give more independent evaluations.

• The recent cutbacks in federal staffs and budgets make it unlikely that the BLM (or similar management agencies) can adequately oversee the complex permitting process for mine projects. The BLM modeling expert involved with the Aguirre EIS stated that he had been advising on approximately sixteen to eighteen different projects at the same time.

• The reliance on modeled predictions lends a false sense of certainty about the future. That, coupled with the tendency to report optimistic predictions, means that the agencies will often underestimate the value of bonds needed to cover future cleanup operations. If an unforeseen environmental problem surfaces years after mining has ceased and the bond has been released, the taxpayers may be stuck with the bill—or the impacts. This is especially true where small, foreign-owned companies are involved. There may be no practical means for attaching their overseas assets to pay for later cleanup.

In light of such realities and constraints, a different view of science and risk management needs to be applied to the process of permitting mines in a broader context of environmental protection. The following recommendations could be central to implementing such a view.

• Predictive models should be used to improve the *conceptual* understanding of how rock-water-chemistry systems work, not to generate apparently precise predictions of, for example, the arsenic concentration, to the nearest 5 micrograms/L, in the surface layer of a particular pit lake one hundred years in the future. As an alternative to such false precision, most scientists knowledgeable about the overall uncertainty would likely state that the metal concentration was predicted to be between 5 and 50, or 5 and 500 micrograms/L, for example. Studies by the U.S. Geological Survey (Plumlee et al. 1993) report techniques that "predict" only broad ranges of expected metal concentrations using graphical techniques. Such a low level of certainty seems more appropriate for use in preparing mining environmental reports than the approaches used in preparation of the Aguirre EIS.

- Fundamental research needs to be conducted on the adequacy of data from short-term leaching or kinetic tests as input to hydrogeochemical models. Existing tests have been developed largely within the mining industry. There is considerable reason to believe such short-term tests do not give reasonable predictions of future leachate chemistry (Li 1997; Lawrence and Wang 1997; Robertson and Ferguson 1997). Independent, long-term testing is warranted.

- Existing regulations, which are complicated and duplicative, encourage a "command and control" style of oversight by both the state and the federal governments. Much paper is generated, but detailed oversight is often meager. Much of the regulatory nitpicking could be relaxed if a simple, comprehensive liability bonding program could be developed and implemented. This approach does not depend on predictions. If, for example, a $50 million bond was held by the BLM specifically to cover potential long-term water quality problems, the mining company would willingly do whatever was necessary to get back its money. Another alternative might be to require the company to purchase some form of environmental liability insurance, adequate to cover unforeseen water quality problems. A branch of the World Bank Group—MIGA, the Multilateral Investment Guarantee Agency—has for years sold currency and political risk insurance to mining and other companies overseas, so it is obvious that the international lending agencies believe the concept has some merit. This approach, however, is subsidized by the taxpayers of the cooperating countries.

Notes

1. As the announcer on 1950s radio and television mysteries would say, "The names have been changed to protect the innocent."
2. Almost a year after publication of the final EIS, I examined data for the Sleeper pit, corroborating that pit lake stratification can occur. A lake approximately 300 feet deep had begun to form at this now inoperative mine site. Pit lake waters at a depth of 120 feet were highly acidic, with a pH of 3.75 (Water Management Consultants 1996). While comparable pit lakes in this setting are expected to be alkaline, the acid conditions at depth within the Sleeper pit lake may reflect the high sulfide content of the host rock or waste rock drainage.

References

Chapelle, F. H. 1993. *Ground-water microbiology and geochemistry*. New York: John Wiley & Sons.

Kuhn, A., C. A. Johnson, and L. Sigg. 1994. Cycles of trace elements in a lake with a seasonally anoxic hypolimnion. In *Environmental chemistry of lakes and reservoirs*, L.A. Baker, ed., pp. 473–493. Washington, DC: American Chemical Society.

Lawrence, R.W., and Y. Wang. 1997. Determination of neutralization potential in the prediction of acid rock drainage. In *Proceedings of the Fourth International Conference on Acid Rock Drainage*, pp. 449–464. Vancouver: British Columbia Ministry of Environment, Lands, and Parks.

Leshy, J.D. 1987. *The mining law: A study in perpetual motion*. Washington, DC: Resources for the Future.

Li, M.G. 1997. Neutralization potential versus observed mineral dissolution in humidity cell tests for Louvicourt tailings. In *Proceedings of the Fourth International Conference on Acid Rock Drainage*, pp. 149–164. Vancouver: British Columbia Ministry of Environment, Lands, and Parks.

Miller, G.C., W.B. Lyons, and A. Davis. 1996. Understanding the water quality of pit lakes. *Environmental Science and Technology* 30(3):118A–123A.

NEPA (National Environmental Policy Act). 1969. *National Environmental Policy Act of 1969*. 42 U.S. Code, section 4321–4347 (1988).

Plumlee, G.S., K.S. Smith, W.H. Ficklin, P.H. Briggs, and J.B. McHugh. 1993. Empirical studies of diverse mine drainages in Colorado: Implications for the prediction of mine-drainage chemistry. In *Proceedings, Sixth Billings Symposium on Planning, Rehabilitation and Treatment of Disturbed Lands*, pp. 176–186. Billings, MT: Reclamation Research Unit publication no. 9301, March 21–27.

Robertson, J.D., and K.D. Ferguson. 1997. Predicting acid rock drainage. *Mining and environmental management* 3(4):4–8.

Water Management Consultants. 1996. Sleeper mine pit closure. In *Third quarter 1996 status report, prepared for Amax Gold Inc.*, October.

Wilkinson, C.F. 1992. *Crossing the next meridian: Land, water, and the future of the West*. Washington, DC: Island Press.

From Tin Roof to Torn Wet Blanket: Predicting and Observing Groundwater Movement at a Proposed Nuclear Waste Site

Daniel Metlay[1]

In 1944, while producing the plutonium that eventually would devastate Nagasaki, the United States also generated in Washington State the first batch of high-level radioactive waste.[2] In the decades that followed, additional defense production facilities were built, and more than one hundred commercial nuclear power reactors began to dot the landscape. The waste from both military and civilian uses of nuclear energy now lies scattered across forty-three states in huge metal tanks, cylindrical glass logs, basins filled with water, and dry storage casks. The amount of waste likely to be produced in the next thirty years will be nearly twice what exists today. Finding a more permanent means of isolating all that radioactive material from people and the environment represents a formidable scientific, engineering, and political challenge.

A strong consensus prevails within the technical community that the most effective method for long-term isolation of radioactive waste is burial in a deeply mined geologic repository. A committee appointed by the National Academy of Sciences (NAS) advocated this approach as early as 1957 (NAS 1957). Twenty-four years later, the United States officially selected geologic disposal as the preferred means of isolating radioactive wastes for thousands of years (DOE 1981). Subsequently, many other countries—including Sweden, Finland, Belgium, England, Canada, France, Germany, Spain, Russia, and China—have adopted the same course.

What no country has done, however, is to demonstrate that the technical and political hurdles to building a high-level radioactive waste repository can be overcome at a specific site. For the United States, until that task can be accomplished, the policy objectives of safeguarding human health, protecting the environment, fulfilling the federal government's moral and legal responsibility for stewardship of the waste, and ensuring that significant risks are not exported to future generations will not be fully achieved.

This chapter traces how the U.S. Department of Energy (DOE), through its Office of Civilian Radioactive Waste Management (OCRWM), is developing intricate predictions of how a complex repository system sited at Yucca Mountain, Nevada, might perform for as long as a million years.[3] I begin with a background discussion of why Yucca Mountain might be an appropriate disposal site, then narrow the focus to examine attempts at predicting one key parameter affecting repository performance: percolation flux, or the volume of water flowing through a unit area of rock per unit of time at the proposed underground repository horizon. I consider how those predictions have evolved over the last fifteen years and try to explain those changes. Finally, I explore how the predictions had important public policy implications. More generally, however, I examine how scientists tried to understand and deal with predictive uncertainties in a hotly contested and politically charged domain.

Predicting Percolation Flux in the Unsaturated Zone at Yucca Mountain

In a geologic repository, groundwater is the chief mechanism for transporting nuclear waste to the accessible environment. In light of this, the 1957 NAS report argued that "abandoned salt mines or cavities especially mined to hold waste are, in essence, long-enduring tanks" (p. 5). Two factors make salt formations an appropriate and, in some sense, elegant location for a repository. First, because salt is highly soluble in water, the existence of a salt formation suggests the absence of water. Second, should any fractures arise in the salt to compromise the isolation potential of the formation, they would soon be self-sealed because salt flows plastically under the high pressures found at typical repository depths. Intermittently over the next two decades, efforts were made to explore specific salt sites for a high-level waste repository; they all proved futile, stymied either by technical missteps or by political opposition.[4]

In 1979, the salt-centric strategy was overturned, and a new paradigm took its place. The Interagency Review Group on Nuclear Waste Management (IRG), established by President Jimmy Carter, relying heavily on recent technical analyses (APS 1978; USGS 1978), concluded that the behavior of the host rock is *only one of many factors* that affect repository performance. Other important influences are:

- the waste form—either the unaltered spent fuel rods or the material, such as glass, within which the reprocessed waste is embedded;

- the waste package, which holds the waste form—this can be either simple, such as a thin-walled, stainless-steel canister, or complex, such as a thick-walled canister composed of several layers of material;

- the design of the repository structure—for example, the surface temperatures of the waste packages can be kept below or above 100° C for thousands of years; and

- the hydrogeological environment that surrounds the repository facility—i.e., the factors that control the behavior of the groundwater in and around the repository (IRG 1978).

On the basis of this conclusion, the IRG cautioned repository developers that what matters is *the behavior of the disposal system as a whole*, not the behavior of any particular component in isolation.

This new, but hardly radical, paradigm provided the conceptual foundation for the 1982 Nuclear Waste Policy Act (NWPA). That law established a site-selection process in which at least five different hydrogeologic environments would be compared. The NWPA requires the DOE to set forth criteria for comparing sites. The three sites that best satisfied those preestablished criteria would be selected as the prime candidates for repository development (NWPA Section 112). Nine environments were investigated early on, and that number was quickly reduced to the requisite five. The DOE employed sophisticated decision-aiding methodologies to reduce further the number of sites, and by 1986, only three locations—the Deaf Smith salt site in Texas; a basalt site at Hanford, Washington; and Yucca Mountain, which features volcanic tuff—remained in the so-called horse race.

Although political considerations likely played some role in reducing the number of sites, it appears that a desire to examine locations having diverse geologies was more important. But with political opposition mounting in the states still in contention, the cost of investigating each environment increasing dramatically, and widespread anger at the DOE for unilaterally "postponing indefinitely" efforts to find a site in the eastern United States for a second repository, Congress in 1987 passed the Nuclear Waste Policy Amendments Act (NWPAA). Among other things, the NWPAA limited site investigations to Yucca Mountain unless that site was found to be unsuitable, and Congress authorized the characterization of another site [NWPAA, Section 160(g)(3)].[5]

The selection of Yucca Mountain appears to have been driven by a variety of technical and political factors (Colglazier and Langum 1988). The state of Nevada, for example, has claimed that the choice was dictated almost entirely by a hostile constellation of political forces that overwhelmed Nevada's three-person congressional delegation. This perspective notes that among those political forces in 1987 were the vice president and the speaker of the House, who were from Texas, where the Deaf Smith site is located, and the majority leader of the House, who was from Washington, where the Hanford site is located.

Although it would be naive to discount political considerations in the passage of the NWPA amendments, Yucca Mountain did appear attractive *technically* to many policy makers, especially when compared to Deaf Smith, which overlay the huge Ogallala aquifer, and Hanford, which is highly fractured and close to the Columbia River. Thus, while the scientific arguments that pointed to the suitability of the Nevada site likely were insufficient for its selection in 1987, those arguments were probably necessary.

Waste Isolation at Yucca Mountain

Several features of Yucca Mountain make it a potentially attractive location for a repository. Situated within a larger federal reservation, the site is isolated far from major population centers. It is also located within a closed drainage basin, so surface waters do not flow into major river systems. Yucca Mountain is in an arid environment, receiving less than two hundred millimeters of precipitation a year, and there are no perennial streams nearby.

Initially, scientists envisioned constructing a facility more than six hundred meters below Yucca Mountain's crest, deep within the saturated zone (SZ)—that is, below the water table, where groundwater moves continuously through pore spaces and fractures in the rock. On the basis of suggestions from the U.S. Geological Survey (USGS) (Robertson, Dixon, and Wilson 1982), however, the DOE decided to explore the possibility of placing a disposal facility at a lesser depth, within the thick unsaturated zone (UZ), above the water table. In the UZ, water and gases are present in the pore spaces in varying proportions, but always at less than saturation, and the water either does not move at all through the pore spaces, or it moves extremely slowly. This new approach rested on the argument that the amount of water reaching the repository would be very small and that a repository could be designed so that any such water would pass into the permeable rocks below, with minimal contact with the canisters of waste.

An important study by Eugene Roseboom (1983) of the USGS more fully explained the advantages of storing the waste in the UZ:

- The UZ forms a natural barrier that can promote waste isolation. It thereby contributes to "defense-in-depth."

- The environment is relatively dry and should remain so for many thousands of years.

- The behavior of any water in the UZ is far more predictable than the behavior of water in the SZ.

- Predicting the effect of heat on water in the UZ is easier than predicting its effect on water in the SZ.

- Relatively simple engineering and design features can be developed to divert water from waste packages and to drain the water into faults and fractures.

- Emplacement of waste in the UZ makes access, monitoring, and retrieval much simpler.

At the same time, Roseboom noted two disadvantages. If climate change caused the water table to rise to the level of the repository, radionuclides would no longer be isolated. Moreover, if spent nuclear fuel was not reprocessed, gaseous radioactive isotopes of iodine and carbon could be released. Roseboom discounted both the likelihood and the seriousness of the two drawbacks.

During the following year, the DOE formally compared the advantages and drawbacks of locating a repository in either UZ or the SZ (Johnstone, Peters, and Gnirk 1984). In its official discussion of the decision, the DOE observed that none of the horizons were unsuitable as a repository (DOE 1986, pp. 2-45–2-47). However, rock units in the SZ were effectively eliminated because they had either poor mechanical strength or high groundwater temperatures (Carter 1987, p. 172). In the end, the DOE selected the Topopah Spring rock unit, a thick layer of volcanic ash and larger fragments that was "welded" into a strong, relatively impermeable rock—or welded tuff—because of the high temperature conditions under which it was deposited. The choice of the Topopah Spring unit departed from convention because it occurred at depths considerably less than the 700–1,000-meter depths usually envisioned for deep geologic repositories.

How Dry Is Dry?

Percolation flux is one of the key parameters affecting the performance of a repository at Yucca Mountain. Water can percolate through rocks by moving through interconnected pore spaces and along cracks and fractures. The higher the percolation flux, the more water will seep into the tunnels—or drifts—where the radioactive material is emplaced. Water in the repository has two potentially problematic effects. First, it can accelerate the corrosion of the waste packages. Second, the more water contacting the waste, the more likely that radionuclides will be mobilized and released from the repository.

Unfortunately, percolation flux cannot be measured directly because of the large scale and inaccessibility of the rock units involved, as well as the slow movement of water through those units. Thus, percolation flux is predicted from indirect measurements on rock samples in the laboratory, as well as from mathematical models (NWTRB 1991, pp. 18–21). The remainder of this section describes how predictions of percolation flux at Yucca Mountain have evolved over time. In following the narrative, the reader may find it helpful to refer to figure 10.1.

Roseboom's assessment of the UZ as a potential host for a repository benefited from the work of another USGS scientist, Isaac Winograd, who had spent many years at the adjacent Nevada Test Site investigating both the regional and the local hydrology to assist the DOE in constructing tunnels within which nuclear weapons tests were conducted. Although Roseboom never developed quantitative models to estimate the amount of water that might reach the repository horizon, his informal analysis relying on "expert opinion," suggested that a percolation flux of roughly four millimeters per year (mm/yr) was in the right ballpark.[6] In particular, Roseboom was reluctant to characterize the UZ as "dry." He noted that water would flow along fractures, that water might very well drip onto waste packages continually, and that water could even pond within the facility, thereby coming in contact with a substantial fraction of a waste package's surface.

At about the same time that the UZ was selected as the reference repository site at Yucca Mountain and as the DOE was still investigating alternative sites in a half dozen locations around the country, a study conducted by two other USGS scientists (Montazer and Wilson 1984) reached the following conclusions:

- Average precipitation was somewhat higher than the 125 mm/yr suggested by Roseboom. It was probably closer to 150 mm/yr.

- Based on studies of other arid environments, net infiltration of rainwater into the shallowest rocks above the repository site—the Tiva Canyon unit—probably averaged 0.4 to 4.5 mm/yr.

- The Paintbrush unit, which lies directly above the proposed repository, could act like an umbrella—or what investigators would later call a tin roof—and laterally divert up to 100 mm/yr of water away from the repository—water that would otherwise travel down toward the stored waste. The actual amount of diversion was unknown.

- A maximum of approximately 0.2 mm/yr of water could be flowing through the pore spaces—or matrix—of the Topopah Spring unit, where the proposed repository would be located. The flux in fractures in the rock, however, was not known.

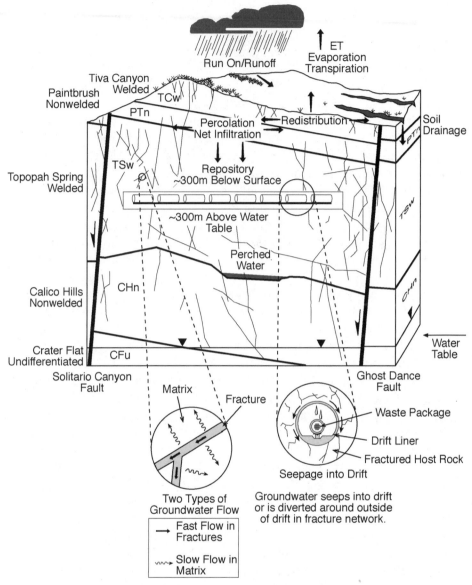

Fig. 10.1   Flow in the unsaturated zone.

Nonetheless, on the basis of the physical properties of the rocks, Montazer and Wilson (1984) believed that "of the conservatively estimated 4.5 mm/yr net infiltration, probably only a *maximum of approximately 1 mm/yr* is transmitted through the Topopah Spring unit [emphasis added]" to the repository horizon.

Thus, Roseboom's 1983 prediction of a 4-mm/yr percolation flux was reduced the next year to 1 mm/year by Montazer and Wilson's research.

How important is this 3-mm/yr difference? The performance of a repository built at Yucca Mountain will depend on how much of the percolation flux seeps into the emplacement drifts and comes into contact with the waste. When the percolation flux increases, it "could cause *more than a proportional increase* in the seepage [flux] in the drifts" (NWTRB 1998, p. 38, emphasis added). This nonlinearity arises because pores in the rock will locally fill with water until some threshold is reached, at which point the water will be able to move from the pores into fractures that lead into the repository. In other words, changes up or down in predictions of the percolation flux could produce *even larger changes* in predictions about seepage flux and, therefore, repository performance. The evolution of estimates of percolation flux needs to be viewed and understood from this perspective.

As the investigations by Montazer and Wilson were being completed, the DOE office in Nevada finished preparing a draft Environmental Assessment (EA), a key document in the site-selection process mandated by the NWPA (DOE 1984). When final, the EA would become the foundation for choosing which three of the five sites still in contention would become the prime candidates for development into a repository. Citing the work by Montazer and Wilson, the DOE observed: "Despite the uncertainty about the exact conditions and processes of the hydrologic system at Yucca Mountain, especially in the unsaturated zone, the conservatism of the assumptions and the analyses allows confidence in the general conclusions about the hydrologic system [i.e., low percolation flux]" (DOE 1984, p. 6–120). On the basis of those conclusions, the DOE projected that there would be *zero release* of any radionuclide to the accessible environment for the first 10,000 years after closure of a repository at Yucca Mountain (DOE 1984, p. 6–247). The choice of a 10,000-year prediction reflected the expectation that the Environmental Protection Agency (EPA) would establish a 10,000-year period for regulatory compliance. This regulatory time frame is, of course, arbitrary and represents a policy judgment related loosely to the heat decay of the waste as well as to the toxicity of the waste in comparison to the original uranium from which the waste derives. Other regulatory periods have been suggested, such as 1,000 years or the time at which an exposed population would receive the peak dose (NAS 1995).

By the time the EA became final in 1986, the advantages of developing a repository at Yucca Mountain seemed only to have increased. Although the DOE scientists realized that significant uncertainties remained, they began to highlight the possibility of lateral diversion of water by the Paintbrush unit, that is, the "tin roof" that overlay the potential repository site (DOE 1986, p. 6–137). In fact, lateral diversion coupled with an estimate of lower rainfall hitting the surface directly above the proposed repository led the DOE to conclude that Montazer and Wilson's original prediction of 1 mm/yr may have been too high. "Although no firm value for moisture flux in the Topopah Spring unit has yet been established, all preliminary field and laboratory estimates *are less than 0.5 mm/yr*" (DOE 1986, p. 6–151). Performance over 10,000 years, to be sure, cannot be improved over zero release. But the DOE was sufficiently encouraged by its preliminary assessments to claim the following (DOE 1986, p. 6–295):

> The [Nuclear Regulatory Commission] limits for the . . . release rate from the engineered barrier system can be met *without any engineered barriers other than the waste form* because the amount of water likely to be in contact with the waste is insufficient to cause higher rates of waste dissolution [emphasis added].

Refining the Predictions, 1988–95

By 1988, a year after Congress had prevented the DOE from characterizing any site other than Yucca Mountain, the DOE's assessment of percolation flux dropped even further. Citing later work by Montazer et al. (1985) and Wilson (1985), the DOE scientists in Nevada maintained that "the percolation flux through the [Topopah Spring] unit . . . may well be about or much less than the 0.5 mm/yr predicted by Wilson" (DOE 1988, p. 3–208). Thus, over a period of about five years, predictions of percolation flux fell by a factor of as much as forty—from Roseboom and Winograd's 4 mm/yr, to a DOE estimate of *as low as 0.1 mm/yr.*

Over the next seven years, the generally accepted mean value for percolation flux ranged from 0.02 to 1.0 mm/yr. In 1991, for example, DOE sponsored two performance assessments (PA). The PAs use complex mathematical models to predict the range of potential radionuclide releases from a repository. And of course the PAs are themselves dependent on the prediction of future percolation flux.[7] The first assessment used the values 0, 0.01, 0.05, 0.1, and 0.5 mm/yr for percolation flux. This analysis concluded that a repository would meet the 10,000-year standard set by the Environmental Protection Agency in 1985 (but remanded by federal appeals court order in 1987),[8] as well as the regulations developed by the Nuclear Regulatory Commission (PNL 1992). The second assessment used a distribution of values for percolation flux—ranging from 0 to 39 mm/yr with a mean of 1.0 mm/yr—and a "weeps" model that allowed for the independent flow of water through random and discrete fractures, rather than simply through pore spaces. This analysis concluded that, under some scenarios, the remanded EPA standard probably could not be met at Yucca Mountain (Sandia National Laboratory 1992).

Two years later, the DOE sponsored another pair of PAs. In the assessment performed by Sandia National Laboratory (1994), percolation flux ranged from 0 to 7 mm/yr. In the assessment performed by TRW (1993)—the management and operating contractor for Yucca Mountain—percolation flux ranged from 0 to 3 mm/yr (Van Luik 1994). Those ranges were based on informal expert judgment that, in turn, rested on the limited experimental data that were available at the time. The mean value for percolation flux presumed in both studies was relatively low: 0.5 mm/yr, or eight times less than the Roseboom estimate. Once again, both sets of predictions showed that release of radionuclides to the accessible environment was strongly dependent on percolation flux. Perhaps more surprising was the conclusion of one assessment that if percolation flux were much above 0.1 mm/yr, long-term releases (post-100,000 years) would exceed the remanded EPA standard by as much as an order of magnitude (TRW 1993).

Similar findings were obtained from a third set of performance assessments two years later (TRW 1995). Yet by then, thinking about percolation flux had almost achieved the status of conventional wisdom. For example, although cautioning against accepting the DOE's estimates of percolation flux uncritically, the independent U.S. Nuclear Waste Technical Review Board (NWTRB, the board) observed: "If it can be shown definitively that the percolation flux at the repository horizon is primarily matrix flow, less than 0.1 mm/yr, then it probably will be difficult not to deem the site 'suitable' on hydrologic grounds" (NWTRB 1996, p. 28).

Observing Percolation Flux, 1996–99

Until 1995, predictions of percolation flux were derived from indirect measures using data obtained from corings and supplemented by the results of computer simulations. In generating these predictions, scientists had to make assumptions, many of which seemed plausible at the time. One key assumption was that water flowed mainly in the matrix, or pore spaces, of the rock; another was that lateral diversion by the tin roof formed by the Paintbrush tuff formation was generally effective. For the most part, neither assumption came under direct challenge, although Flint's (1995) work on infiltration suggested that lateral diversion would have to be *extremely* effective to limit percolation flux to approximately 0.2 mm/yr.

By 1996, however, the Exploratory Studies Facility (ESF), a five-mile tunnel that parallels the eastern boundary of the potential repository emplacement area, offered a new opportunity to obtain rock samples from deep within the UZ. What was discovered took the Yucca Mountain project personnel by surprise.

Scientists at Los Alamos National Laboratory began systematic sampling every 200 meters within the ESF, as well as sampling in or adjacent to faults and concentrations of fractures. Analyzing these specimens in June 1996, they discovered that some contained significantly elevated levels of chlorine 36 (^{36}Cl). This radioisotope can occur naturally, but most ^{36}Cl was created and entered the atmosphere as a result of the above-ground nuclear weapons tests that took place before 1963—the so-called "bomb pulse." Thus, the presence of ^{36}Cl strongly suggests that water can move nearly 300 meters from the top of Yucca Mountain to the nominal repository horizon in less than fifty years. The presumption that water might flow so far down via *fast paths* was strengthened because most of the samples containing elevated ^{36}Cl levels were collected near faults and fractures that appeared to extend from the ESF tunnel to the surface of Yucca Mountain.

Strictly speaking, the ^{36}Cl discovery provided direct evidence only that the groundwater travel time, i.e., the time it took water to move down approximately 300 meters from the surface of Yucca Mountain to the repository horizon, was much faster than previously anticipated. It did not say anything about the volume of water moving through the fractures. Nonetheless, the findings by Los Alamos led the DOE to revisit its fundamental assumptions about how much water flows in fractures versus how much flows through the rock matrix in the UZ. By October 1996, the DOE reviewed saturation and moisture tension data, pneumatic data, fracture coating data, temperature data, and perched water data through a distinctly different lens. Out of that review came a strikingly different conceptual model of flow in the UZ (Williams and Bodvarsson 1996).

The NWTRB summarized the changed thinking stimulated by the discovery of ^{36}Cl at the repository horizon. The board first addressed the question of water distribution. The new data strongly suggested that two distinct flow systems coexist at Yucca Mountain. In one flow system, the water travels rapidly through interconnected fractures. Moreover, the amount of water present must be sufficient to allow gravity—which pulls the water downward toward the repository—to overcome the capillary action that tends to "suck" water out of the fractures and into the porous matrix of the rock. These conditions could be met—and the "fast paths" activated—episodically, by infrequent, intense precipitation events. In the other flow system, the water travels continuously through interconnected pores in the rock and may take as long as thirty-thousand years to reach the repository level, compared to the fifty years or less indicated by the ^{36}Cl. The board then considered the question of lateral diversion and concluded that the tin roof of the Paintbrush was, at best, leaky (NWTRB 1997, pp. 13–14). Nonetheless, the critical question of *how much water* flowed in each system remained unanswered.

Over the next year, the DOE tried to integrate the implications of the ^{36}Cl findings into its thinking. Unable to mount experiments to improve its predictions about percolation flux, the project formally elicited the views of seven experts. After reviewing the available data, an independent facilitator intensively probed the opinions of each of these scientists. Based on that interrogation, each expert developed his own distribution of what the average percolation flux at the repository horizon might be. The distributions of these seven opinions were combined to produce an aggregate distribution that had a mean of approximately 10 mm/yr and a fifth to ninety-fifth percentile range of 1 to 30 mm/yr (UZFM 1997). In short, rather than being protected by a tin roof, a repository at Yucca Mountain would likely sit under something more akin to a torn wet blanket.

Why Did Predictions of Percolation Flux Change?
In 1983, Roseboom, with the help of Winograd, suggested that percolation flux at Yucca Mountain was roughly 4 mm/yr. Fourteen years later, a panel of experts convened by the DOE came up with an estimate that was about a factor of two higher. Yet, for most of the period stretching from 1985 to 1996, scientists involved in the repository project held that percolation flux was approximately one hundred times smaller. These oscillations are illustrated in figure 10.2. In this section, I propose two

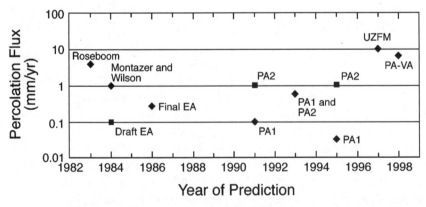

Fig. 10.2   Changing predictions of present-day average percolation flux at Yucca Mountain. Points on the graph refer to predictions mentioned in the text. Note that the vertical scale is logarithmic. (Abbreviations: EA = environmental assessment; PA = performance assessment; UZFM = unsaturated zone flow model; VA = viability assessment.)

complementary explanations of why estimates of percolation flux remained so low for so long.

I should note at the start that although "explaining" organizational behavior is more an art than a science, the effort is informed by persuasive theoretical constructs and an expanding volume of empirical research. Nonetheless, definitive answers are almost impossible to establish; generally the best that one can do is to suggest explanations that are plausible and consistent with a large number of known facts and actions.

Given this caveat, the first, and the most straightforward, possible explanation is that geologists associated with the project did the best they could in developing the predictions, given the limits on obtaining relevant data and the state of the science. The calculations, completed during the mid-1980s, relied on indirect methods (Montazer and Wilson 1984; Wilson 1985), limited data, or computer simulations (Sinnock, Lin, and Tierney 1986). Only after direct access was gained to the Topopah Spring unit through the ESF tunnel could scientists easily obtain samples that would speak more directly to the question of percolation flux. Without that data, the prediction of low percolation flux did not seem unreasonable. Nevertheless, it appears as if the predictions did underestimate the value of the percolation flux.

A second possible explanation recognizes that a variety of psychological, bureaucratic, political, economic, and regulatory influences, at various times, could have affected predictions made by project scientists and managers. By suggesting that these institutional forces also might have been at work, I am not claiming that the technical analysts were biased or unprofessional. My position is much more nuanced: I believe that those institutional forces all moved in the same direction. Consequently, when faced with the need to resolve uncertainty about percolation flux, the scientists had little organizational incentive to settle on a higher value or, more important, to question whether a lower value was correct. This approach to addressing uncertainty need not have been adopted consciously; in fact, it probably was not. More likely, it arose simply because organizational norms and culture have a well-documented and pervasive effect on individuals' actions and judgments (Steinbrunner 1974).

Managing uncertainty by reducing the size of the estimated percolation flux was not scientifically unreasonable and probably offered the path of least resistance to bureaucratic momentum, especially early on, when it appeared that performance was not very sensitive to the magnitude of the flux (Thompson, Dove, and Krupa 1984; Sinnock, Brannen, and Lin 1984). Later, however, institutional pressures probably became less subtle. For that reason, considering separately what took place before 1988 and what happened afterward is useful.

Running a Horse Race.

Before 1988, site-characterization work was parceled throughout the DOE complex. Three contractor organizations had the responsibility for conducting the technical analyses of the nine sites initially selected by the DOE. Rockwell International studied basalt at Hanford, Washington, and reported to the DOE's nearby Richland Operations Office. Battelle Memorial Institute investigated salt at seven locations stretching from Utah to Mississippi and reported to the DOE's Chicago Operations Office. SAIC was a major contributor to the work at Yucca Mountain and reported to the DOE's Nevada Operations Office.

These contractors offered their services to the government for profit; whoever survived the winnowing process from nine to five to three candidate repository sites would receive a steady stream of funds for perhaps as long as ten years. By the same token, the scientists working for each contractor understood that their future jobs depended on their making a persuasive case that "their" site could perform well. Finally, because the contractor associated with the site that was selected for repository development would secure a federal commitment for many more years, each DOE operations office also had a programmatic and bureaucratic stake in the outcome of the horse race, albeit to a somewhat lesser degree than the contractor firms and their employees. Beyond these rather pragmatic considerations, scientists and managers at each site realized that helping to develop the country's first high-level radioactive waste repository would be a significant professional and personal accomplishment.

Jockeying for position began in 1983 as the DOE's headquarters staff started to put together drafts of the site-suitability guidelines (10 CFR 960), the criteria by which the three finalists and the winner of the horse race would be judged. Each DOE operations office, aided by its associated contractor corps, filed comments, many of which were designed either to remove language that placed its site at a disadvantage or to include language that gave its site a boost. These efforts intensified in 1984 as the guidelines were close to being published. For example, Jeffrey Neff, head of the Salt Repository Project Office in Chicago and responsible for overseeing characterization of a salt site in rural Utah, sought to remove language limiting the transportation of waste through a national forest (Neff 1984). Stephan Whitfield of the Richland Operations Office expressed concern about a disqualifier having to do with groundwater travel time, which, if accepted as written, would have hurt

Hanford's chances (Whitfield 1984). On the other hand, Donald Vieth, director of the Waste Management Project Office in Nevada, strongly supported the wording of the criteria for groundwater travel time, believing that water moved very slowly in the unsaturated zone at Yucca Mountain (Vieth 1984).

Finalizing the guidelines did not end the competition but only intensified it. The approach of looking at diverse geologic repository settings had been established years earlier by the Interagency Review Group on Nuclear Waste Management (IRG 1978) and was incorporated into both the NWPA and the Nuclear Regulatory Commission's licensing regulations. Given this approach, insiders generally expected that one of the salt sites, the Hanford basalt site, and the Yucca Mountain tuff site would be the three finalists. But nothing could be taken for granted. Investigators at Hanford had to overcome the seemingly persuasive technical argument that fractures in basalt would be almost impossible to model. Scientists at Yucca Mountain had to present a compelling case that disposal in the UZ made as much sense as disposal in the SZ. In addition, the DOE had an organizational and budgetary incentive to keep the competition going among the final three sites (U.S. Senate 1987, p. 167). Neff expressed a high level of confidence in the suitability of the Deaf Smith, Texas, salt site (U.S. Senate 1987, p. 166). Nonetheless, the confidence expressed by Donald Vieth about the prospects for Yucca Mountain is especially striking (U.S. Senate 1987, p. 71):

> We have looked at the site fairly thoroughly since 1977. I think we understand the nature of the forces that are acting on the site. If one takes the information we have now, and tries to project the kinds of things that are liable to be discovered in the next five or six years of site characterization, *it is not conceivable to me that we would discover something of a major nature that would cause us to change our mind about it.* . . . The processes of doing the modeling and the calculations that estimate the radioactive releases from the repository tells us that *we may be five orders of magnitude below a very conservative EPA standard.* I think that we are very confident about the potential of that piece of earth to isolate the waste if it is placed there. [emphases added]

Vieth's confidence flowed from the reasoning in the final EA, published barely a year earlier (DOE 1986). In that document, the Yucca Mountain scientists developed an implicit waste isolation strategy that appeared just as technically elegant as the one that had been advanced for salt. If very little water contacts the waste, then even relatively thin

waste packages will take a long time to corrode. If the packages are cor-
roded, there will be little water for dissolving the waste. And if the waste
is dissolved, there will be little water for transporting the waste out of
the repository. Moreover, because groundwater travel time through the
UZ below the repository horizon was thought to be very long, the waste
would take thousands of years to travel to the accessible environment.
Finally, a dry repository also might be cheaper. Robust waste packages
that resist corrosion might not have to be purchased.

To clinch the argument, the Yucca Mountain scientists reminded
everyone that other means of isolating and containing the waste were
being held in reserve, to be marshaled if needed to improve repository
performance (DOE 1986, p. 2–296). Heat from the waste packages
would dry out nearby rocks, thereby reducing even further the amount
of water seeping into the drifts. Moreover, between the Topopah Spring
unit, where the repository would be located, and the water table below,
lay a unit of absorptive volcanic sediments that would act as a sponge
and further retard the migration of radionuclides to the accessible
environment.

One anonymous reviewer of this chapter asked whether political
considerations, such as the fact that Yucca Mountain is located in a
sparsely populated and relatively politically weak state, might have
engendered these relatively optimistic conclusions. Although members
of Congress sought to enact legislation that would "disqualify" sites in
their constituencies while the horse race was unfolding, I believe that
far stronger internal bureaucratic influences sought to demonstrate the
suitability of various sites, thereby keeping them in contention. And in
this context, the key to waste isolation at Yucca Mountain remained
percolation flux. Notwithstanding analyses that suggested repository
performance was insensitive to the magnitude of the percolation flux
(Thompson et al. 1984; Sinnock et al. 1984), it would be much easier to
demonstrate that a repository would perform satisfactorily if a case
could be made that the repository horizon was dry to begin with.

Crossing the Finish Line.
The 1987 amendments to the NWPA made the Yucca Mountain team
the presumptive winner not only of the horse race to become a finalist
but also of the contest to become the repository. Yet there was the matter of crossing the finish line. The newly created independent Nuclear
Waste Technical Review Board, made up of distinguished scientists,
would have to validate the DOE's technical work, and the Nuclear Regulatory Commission would have to grant a repository construction
license using an adjudicatory and potentially very adversarial process.

Complicating matters further, the project found itself caught up in a large number of political, bureaucratic, economic, legal, and regulatory controversies and debates between 1990 and 1993. Officials from Nevada launched an intensive legal, technical, and administrative campaign against the selection of Yucca Mountain. Coordination between the DOE headquarters and the Nevada Operations Office began to break down; by 1991, the project director was operating relatively autonomously. Site-characterization costs seemed to be mounting at a rate that was not sustainable much further into the future. As that effort experienced serious delays in schedule, utilities that owned reactors began to complain that the DOE would not meet its contractual obligation to begin accepting spent nuclear fuel on January 31, 1998. Industry representatives and state public utility commissions started filing suits to protect the money collected from nuclear power consumers that was intended to pay the costs of disposal. So turbulent was the environment facing the Yucca Mountain project that Senators Pete V. Domenici and J. Bennett Johnston, two of the most influential lawmakers on energy policy, suggested, apparently out of frustration, that Congress terminate the effort altogether (U.S. Senate 1992, pp. 34–40).

By early 1995, the Yucca Mountain project was virtually under siege. Calls for its termination increased. In the view of the project's critics, too much money was being spent, too little progress had been made, and the prospects for final regulatory approval were too problematic. Although the program ultimately survived because the need for a repository remained, Congress cut the project's budget request by 50 percent, and more than one thousand contractor employees were laid off. These events led many people to believe that the very idea of developing a repository within the next fifty years was hanging in the balance. One of the most articulate proponents of that view was the director of DOE's Office of Civilian Radioactive Waste Management (OCRWM), Daniel Dreyfus. Speaking to the NWTRB in October 1995, Dreyfus observed:

> The issue confronting us is whether the program can sustain meaningful progress towards a future decision on geologic disposal with a funding level that is significantly below that which was required for our current program approach. We inside the program gave serious consideration to this question, and we believe, albeit tentatively, that it can. We must, however, convince the Congress that continued pursuit of geologic disposal is, first of all, worth at least $250 million a year, and second of all, that it will have meaningful results. To do this, we have to ensure that scientific investigation *can produce results within a reasonable time frame* . . . [emphasis added; Dreyfus 1996].

Over the next year, the urgency of producing demonstrable results dominated the public pronouncements of the OCRWM's senior and mid-level managers and, perhaps more important, the ongoing dialogue that took place inside the program. Speaking bluntly before the National Academy of Science's Board on Radioactive Waste Management, Dreyfus strongly linked his vehicle for showing progress—the so-called viability assessment (VA)—to the future of geologic disposal over the next several generations. Although this same theme had been sounded in an earlier talk to the NWTRB, it had been more muted then. The VA, which was subsequently mandated by Congress, would develop a preliminary waste package and repository design, a safety analysis, a compilation of key research issues that would be addressed prior to site licensing, and a cost estimate for operating and closing the repository. Now, Dreyfus sent an explicit signal to supporters of geologic disposal that although the technical precision of the VA might not overwhelm them, they had better be prepared to give the assessment at least qualified support. If they were not willing to do so, they might not like the consequences—elimination of the geologic disposal option—and they would have only themselves to blame.

These external threats to the Yucca Mountain project were emerging and spreading just as the DOE's performance assessments were showing problematic repository performance if percolation flux rose appreciably above 0.5 mm/yr. No wonder, then, that the first formal version of a waste containment and isolation strategy released in July 1996 relied almost entirely on the existence of a dry repository (DOE 1996). Any other view would have opened the door to a new round of intense criticism. Thus, the predictions of low percolation flux that scientists converged on between 1984 and 1995 had become almost an article of faith upon which the future of the project rested.

Outcomes and Implications

How Are Scientific Activities Carried Out in a Large Public Bureaucracy in a Turbulent Social and Political Environment?

In the so-called realism school of the philosophy and sociology of science, scholars recognize that scientific activities do not generate absolute or final "truth." Instead, those investigators understand that scientific activities can produce contingent knowledge that is always subject to recall. In their view, science is an *error-correcting* activity, in which new information continually challenges accepted wisdom; over time, knowledge accretes.[9]

What happened at Yucca Mountain is consistent with this perspective. The DOE did modify its predictions after data were obtained calling into question the low estimates of percolation flux. In fact, it even issued a press release notifying the public about the Los Alamos ^{36}Cl findings. The credit for these actions, in the first instance, belongs to the scientists and managers of the DOE. They were willing to put their "faith" to a test by sponsoring the Los Alamos research; they also resisted the urge to reinterpret or explain away the unexpected findings. This does not mean that the implications of the Los Alamos work on ^{36}Cl were universally or gracefully accepted within the project, either then or now. In fact, there was considerable internal debate over how much significance should be attached to those findings. For example, the project commissioned an external peer review to examine whether the methodology used by Los Alamos was appropriate. That review generally supported the research. Even so, in January 1999, the DOE began a validation study of bomb-pulse ^{36}Cl occurrences in the ESF test tunnel.

Nonetheless, steps were taken almost immediately to incorporate a higher percolation flux in key project activities. The formal expert elicitation was launched (UZFM 1997); a revised waste containment and isolation strategy was published (DOE 1998a); and the evaluation of repository performance being conducted as part of the VA was redirected (DOE 1998b). In the areas of repository and waste package design, however, the project did experience significant difficulties in integrating the information that Yucca Mountain might be wetter than previously thought. As a result, the designs contained in the VA had to be revised (DOE 1998c; TRW 1999).

Technical experts familiar with attempts to predict percolation flux seem convinced that the current, higher, estimates for that parameter are much more valid than predictions made at the beginning of 1996. But are these predictions "right"? Even scientific realists would be reluctant to claim more than that knowledge is accreting along a jagged path that approaches, but does not necessarily reach, a "correct" answer. Thus, it is still unclear today what the correct answer is.

To reduce further the uncertainty in percolation flux, the NWTRB urged DOE to excavate a new drift across the proposed repository block, more or less perpendicular to the ESF tunnel (NWTRB 1997, p. 26). This so-called east-west crossing would enable the DOE, among other things, to gather additional information about the distribution of fractures at the repository horizon. Chairing a review of the VA for the director of the USGS, Isaac Winograd, the father of disposal of waste in the UZ at Yucca Mountain, detailed why having this better information could be important (Winograd et al. 1998, p. 10):

It is our view that the [viability assessment] overestimates percolation [fluxes] at the repository horizon and overestimates
seepage into the emplacement drifts by an even wider margin.
Consequent to these over-estimations are various proposed
engineering measures to protect against the deleterious effects
of seepage. We believe that some of these engineering measures may be unnecessary and others counterproductive with
respect to the natural assets of the repository system.

But, in the final analysis, how much information is enough and how
good predictions need to be are policy judgments. Exercising its discretion, the DOE initially refused to commit to when it would construct the
east-west crossing; under pressure from the NWTRB, the DOE eventually excavated this drift in 1997. As of mid-1999, however, very few
experiments had been carried out in the new drift. For example,
although scientists from Los Alamos collected samples in the east-west
crossing to detect the presence of ^{36}Cl, many of those specimens have
not been analyzed. Preliminary results indicate that bomb-pulse ^{36}Cl
has been found along two known faults and at two of three previously
unidentified faults in the east-west crossing. At one of these previously
unidentified faults, the bomb-pulse signal was the strongest yet measured at the site. But samples taken between faults have not been analyzed. This work, therefore, probably will not be able to reduce the
uncertainty in predicting the contribution of fast-flow paths to the overall percolation flux. Thus, it is unclear at this time how much better
information the DOE will have about percolation flux at Yucca Mountain when it decides on the site's suitability in mid-2001.

Can Institutions Be Designed to Increase the Validity of Technical Undertakings?

Starting with the earliest efforts by the DOE's predecessor agency,
the Atomic Energy Commission, to site a high-level waste repository,
concerns have been raised about whether political considerations
would overwhelm the agency's technical assessments (Carter 1987;
SEAB 1993). Consequently, the law creating the DOE required that
an independent agency, the Nuclear Regulatory Commission, license
a high-level waste repository. The events noted above leading to the
passage of the NWPA amendments, however, reinforced the concerns of many policy makers about the program's credibility. The

Nuclear Waste Technical Review Board was established to ensure that DOE's decisions are technically valid (U.S. House of Representatives 1987).

According to some theories about organizational design, the involvement of two independent technical monitors—the NRC and the NWTRB—ought to have increased the likelihood of discovering in less than a dozen years the DOE's probable underestimation of percolation flux (Landau 1969). But reality is more complicated. None of the monitors had any data other than what was available to the DOE. For example, commenting on the 1986 EA, the NRC noted (1986, p. 46):

> The choice in the draft EA of a value of [percolation flux equal to 1 mm/yr] . . . was considered to be inadequately supported and the suggestion was made that higher values be considered [by the DOE]. In the final EA, this value has been *reduced* to a constant value of 0.5 mm/yr. . . . The NRC staff concludes that the values of flux have not been adequately considered in [the DOE's] analysis.

But the NRC had no scientific basis for going a step further and asserting that the DOE made a predictive error. Moreover, although the NWTRB urged the DOE to construct the ESF tunnel, in part so that better percolation flux data could be gathered, it was not until 1995 that the board systematically examined the DOE's low predictions of percolation flux. Perhaps one could argue that both monitors fell down on the job. But given the dearth of data and the underlying uncertainty in the models—which made it impossible to arrive at definitive conclusions—that argument strikes me as a product of 20/20 hindsight.

These institutional arrangements probably facilitated, albeit indirectly, the DOE's own acknowledgment that its earlier predictions of percolation flux might not be correct. First, all data were gathered under a quality-control regime established by the NRC. The data then were made available to the public. Second, the project's technical overseers recognized and carefully evaluated implications of the new data for repository performance. For example, the NWTRB continually pointed out to DOE that its proposed repository design might not be compatible with a "wetter" Yucca Mountain. Third, DOE officials knew they might ultimately have to defend their technical positions before an NRC licensing board in an adjudicatory process. The state of Nevada is on record that it will intervene in that process and contest the DOE's application for a permit.

This likely intervention forces the DOE into substantiating and documenting its predictions. Thus, taken together, a variety of mechanisms designed to safeguard the scientific process appear to have worked.

I would conjecture that at least some of these mechanisms were especially effective because the DOE's senior managers came to realize that the Yucca Mountain project was not immediately threatened by the finding that the percolation flux was considerably higher than predicted. To be sure, it took the organization a while to coalesce around a new vision of how a repository at a wetter site might function, but ultimately it did so. On the surface, this vision does not appear much different from the earlier ones. Its major attributes—limited water contacting waste packages, long waste-package lifetimes, slow rates of radionuclide release, and reduction in concentration of radionuclides during transport—are similar, if not identical, to the attributes of the old concept of how a repository is likely to perform.

What is strikingly different, however, is how performance is now allocated between the natural and the engineered components of the repository system. Results from the latest performance assessment sponsored by the DOE were published as part of the VA (DOE 1998b). For the first five thousand years, average percolation fluxes assumed in the assessment ranged from about 4 mm/yr to about 11 mm/yr. As a result, for the first ten thousand years, the expected value of releases to the accessible environment was projected to be at least two orders of magnitude lower than the remanded EPA standard. This low dose derives from the expected behavior of the spent fuel cladding, which holds the fuel pellets in the fuel rods, and the expected robustness of the then current reference design of the waste package, which consists of two shells. The outer shell is made of carbon steel; the inner shell is a relatively new nickel alloy, C-22, that appears to have very high corrosion resistance. Using mathematical models, the DOE has made predictions about how the cladding and waste package would degrade over time. Those models, however, critically depend on what are in essence two assumptions: the integrity of the cladding after the spent fuel has been removed from the reactor and the corrosion rate for C-22. The VA indicates that even after 100,000 years, these waste packages will maintain sufficient integrity that the predicted value of the release would be approximately 20 percent of the remanded EPA standard. Only after 150,000 years, as the waste packages begin to fail in greater numbers, would radionuclide release to the accessible environment begin to exceed that limit; it then would continue to rise for another 150,000 years, when the dose would be roughly an order of magnitude higher than permissible under the remanded EPA standard. DOE is examining various engineering enhancements to reduce these projected doses further.

How Helpful Are Geologic Predictions in Developing Public Policy?

It would seem that the original vision of a repository built at Yucca Mountain has been revised significantly: In 1986, no engineered barriers appeared to be needed because the percolation flux seemed so low. Because the flux was underestimated, both as a result of the perils of predicting with little observational data and as a result of the subtle influence of many institutional forces, engineered barriers now may be indispensable.[10]

I use the word "may" intentionally. The natural barriers associated with Yucca Mountain, such as retardation in the UZ and dilution and dispersion in the SZ, could very well be more effective than the latest performance assessment holds them to be. The dilemma for the DOE is that this effectiveness may be quite difficult to demonstrate because of the complexity of the geology at and around Yucca Mountain. For example, modeling the flow of radionuclides in the SZ has proven more difficult than earlier anticipated. So, at the moment, at least, the DOE finds itself in the awkward position of promoting a geologic repository whose performance appears to depend more on the robustness of the waste package and other engineered elements than on the attributes of the natural system such as the geology and the hydrology of the site.

There may be other outcomes and implications that have not yet fully manifested themselves; only the passage of time will clearly reveal them. Some probably will be relevant to policy makers because they arise from a vision of geologic disposal that has changed substantially over the last fifteen years (although it is consistent with the views set out by the IRG in 1978). Others probably will bear on whether the DOE will be able to meet its policy objectives, which include protecting human health and the environment and preventing the export of significant risks to future generations. At least in the view of DOE leaders, fulfilling those objectives requires that Yucca Mountain receive a construction license by the year 2010. Will DOE scientists be able to demonstrate convincingly the protective power of the engineered components in that time frame? Can a more technically defensible case be made that Yucca Mountain's natural elements of the repository system contribute significantly to waste containment? The answers to these questions, as we have seen, of course depend on science and engineering, but also on economics, project schedules, and political judgments.

Conclusions

More than a century ago, the German sociologist Max Weber suggested the value of the organizational form known as a bureaucracy. Such organizations exercise power legitimately because they can bring knowledge to bear to solve problems. Too often, however, policy makers and members of the general public hold strong negative images about bureaucracies, especially public ones. Bureaucracies are perceived to be inflexible, inefficient, and, most important, incapable of learning (Crozier 1964). This essay should reinforce other works that illuminate modern bureaucracies using a more variegated light. In particular, I believe that the following lessons can be drawn from this case.

- *Notwithstanding institutional pressures that appeared to lower estimates of percolation flux, when relevant new data were acquired, those estimates were revised.* The scientific process, as understood by the realist school, seems to be working. I suspect that the dramatic and relatively unambiguous nature of the ^{36}Cl data facilitated the process of revising predictions of percolation flux. In advancing this cautiously positive conclusion, I realize that the process has not been entirely smooth. The DOE has exercised its discretion in determining the resources and schedule for gathering new information about percolation flux. That judgment has not always been supported by those responsible for project oversight.

- *Independent technical reviewers often can help detect predictive mistakes, but their utility may be less than many organization theorists assert.* The absence of meaningful data not only made the DOE's job of predicting percolation flux more difficult, but it also hindered the efforts of the two independent technical overseers, the NRC and the NWTRB. Complicating matters further, just because attention is called to the existence of technical errors does not mean that they will be corrected. The NRC, for example, possesses the authority to license a repository, but it is often reluctant to intervene on particular technical issues long before a license application is submitted. The NWTRB, in contrast, has no regulatory authority. Its stature as an independent, presidentially appointed body may or may not be sufficient reason for the DOE to accept its recommendations.

- *A variety of institutional mechanisms can be effective in increasing the likelihood that the strengths and weaknesses of geologic predictions will be subjected to technical and public scrutiny.* Senior managers can facilitate organizational learning by creating a supportive culture and environment. In the case of federal agencies, Congress and the Office of the President can establish processes that also facilitate accountability in the predictive effort. For example, the NRC's current plan to hold an adjudicatory licensing hearing on Yucca Mountain has forced the DOE to be much more careful in documenting and establishing its scientific and technical positions.

- *Senior managers who diversify their technical approaches to a problem (multiple designs, defense-in-depth, a systems perspective rather than an individual-component perspective) may be better positioned to alter courses if required.* To expect organizational leaders to engage figuratively in bureaucratic hara-kiri may not be reasonable. The higher the potential stakes involved in modifying predictions, the more difficult the process may be. If key leaders in technically based organizations can identify alternative paths to securing desired policy objectives, the chances of acknowledging predictive mistakes increase. For example, because engineered containment was seen as a possible option, the DOE may have more readily accepted the implications of the ^{36}Cl studies.

Disposal of radioactive waste has been on the public agenda for more than forty years. Within the next decade, the Yucca Mountain project probably either will have crossed the finish line or will have faltered down the stretch. I believe, however, that the lessons of this chapter on DOE's scientific work at Yucca Mountain do offer some hope that within the next generation a start can be made in dealing with one of the most significant environmental issues confronting the United States.

Acknowledgments

I am indebted to all my colleagues, especially Victor Palciauskas and Leon Reiter, for educating me about flow in the unsaturated zone. Leon Reiter also suggested the image of the torn wet blanket. Carl Di Bella taught me much about waste canisters. I have drawn on their thoughts and writings in this discussion of predicting percolation flux. These individuals are obviously not responsible for any of my failures to listen and to learn.

Notes

1. The views expressed in this chapter do not represent the views of the U.S. Nuclear Waste Technical Review Board, a presidentially appointed independent federal agency charged by Congress with evaluating the scientific and technical validity of the U.S. Department of Energy's high-level radioactive waste disposal efforts.
2. As used in this chapter, the term *radioactive waste* means spent nuclear fuel from commercial, research, and defense production reactors, as well as the solidified products of reprocessing such spent nuclear fuel. For a more formal definition, see 10 CFR 50, appendix F.
3. Yucca Mountain is in the desert, approximately 75 miles by air northwest of Las Vegas, Nevada. Part of the land that has been set aside for the potential repository lies on the Nevada Test Site (NTS), where hundreds of nuclear weapons tests were conducted.
4. In 1998, the U.S. Environmental Protection Agency (EPA) certified that the Waste Isolation Pilot Plant (WIPP), a repository located in a salt formation in southeastern New Mexico, could begin to receive transuranic-contaminated (TRU) radioactive waste. The certification came twenty years after Congress authorized the construction of the facility. In March 1999, the first shipment of TRU waste arrived at WIPP for disposal.
5. For a discussion of this history, see my "Radioactive Waste Management Policymaking," appendix A in OTA 1985; and Carter 1987.
6. A year earlier, Robertson et al. 1982, had presented a rough estimate of 3–6 mm/yr.
7. The 1991 analysis was primarily undertaken to gain experience with the methodology and was not viewed as being especially valid. In the discussion that follows, the performance assessment results are accepted at face value, but I recognize that their conclusions are strongly influenced by model assumptions and approximations.
8. In the 1992 Energy Policy Act, Congress directed the EPA to develop new dose-based standards for Yucca Mountain by 1994. As of April 2000, the EPA still had not promulgated a new standard for high-level radioactive waste. This lack of a standard prompted the DOE to examine in its PAs several different periods of compliance.
9. For example, see Taubes 1993.

10. Once again, neither explanation is designed to cast aspersions on project scientists.

References

APS (American Physical Society). 1978. Report to the American Physical Society by the Study Group on Nuclear Fuel Cycles and Waste Management. *Report of Modern Physics* 50 (January).

Carter, L. 1987. *Nuclear imperatives and public trust: Dealing with radioactive waste.* Washington, DC: Resources for the Future.

Colglazier, E.W., and R.B. Langum. 1988. Policy conflicts in the process for siting nuclear waste repositories. *Annual Review of Energy* 13:317–357.

Crozier, M. 1964. *The bureaucratic phenomenon.* Chicago: University of Chicago Press.

DOE (Department of Energy). 1981. *Final environmental impact statement on the management of commercially generated radioactive waste.* Washington, DC.

DOE (Department of Energy). 1984. *Draft environmental assessment, Yucca Mountain site, Nevada Research and Development Area,* DOE/RW-0012. Washington, DC.

DOE (Department of Energy). 1986. *Environmental assessment, Yucca Mountain site, Nevada Research and Development Area,* DOE/RW-0073. Washington, DC.

DOE (Department of Energy). 1988. *Site characterization plan, Yucca Mountain site,* DOE/RW-0199. Washington, DC.

DOE (Department of Energy). 1996. *Repository safety strategy: U.S. Department of Energy's strategy to protect public health and safety after closure of Yucca Mountain,* Rev. 0, YMP/96-01. Washington, DC.

DOE (Department of Energy). 1998a. *Repository safety strategy: U.S. Department of Energy's strategy to protect public health and safety after closure of Yucca Mountain,* Rev. 2, YMP/96-01. Washington, DC.

DOE (Department of Energy). 1998b. *Viability assessment of a repository at Yucca Mountain: preliminary design concept for the repository and waste package,* DOE/RW-0508. Washington, DC.

DOE (Department of Energy). 1998c. *Viability assessment of a repository at Yucca Mountain: Total system performance assessment,* DOE/RW-508. Washington, DC.

Dreyfus, D. 1996. Presentation to the Nuclear Waste Technical Review Board, Arlington, VA, October 17, 1995.

Flint, A. 1995. *Shallow infiltration processes at Yucca Mountain, Nevada: Neutron logging data 1984–1993,* USGS-WRI-95-4035. Denver, Colo: U.S. Geological Survey.

IRG (Interagency Review Group). 1978. *Subgroup report on alternative technology strategies for the isolation of nuclear wastes*, TIS-28818 (draft). Washington, DC: Department of Energy.

Johnstone, J.K., R.R. Peters, and P.F. Gnirk. 1984. *Unit evaluation of Yucca Mountain, Nevada test site: Summary report and recommendation*, SAND-83-0372. Albuquerque, N.M.: Sandia National Laboratories.

Landau, M. 1969. Redundancy, rationality, and the problem of duplication and overlap. *Public Administration Review* 29:350–357.

Montazer P., E.P. Weeks, F. Thamir, S.N. Yard, and P.B. Hofrichter. 1985. Monitoring the vadose zone in fractured tuff, Yucca Mountain, Nevada, *Proceedings of the NWWA Conference on Characterization and Monitoring of the Vadose (Unsaturated) Zone*, November 19–21, 1985, Denver, CO, National Water Well Association, pp. 439–469.

Montazer, P., and W. Wilson. 1984. *Conceptual hydrogeologic model of flow in the unsaturated zone, Yucca Mountain, Nevada*, USGS-WRI-84-4345. Lakewood, Colo: U.S. Geological Survey.

NAS (National Academy of Sciences). 1957. *Disposal of radioactive waste on land*. Washington, DC.

NAS (National Academy of Sciences). 1995. *The technical basis for the Yucca Mountain standard*. Washington, DC: National Academy Press.

Neff, J. 1984. Memorandum to B. Hewitt, April 16, 1984.

NRC (Nuclear Regulatory Commission). 1986. *NRC staff comments on the DOE final environmental assessments*. Division of Waste Management, December 22.

NWTRB (Nuclear Waste Technical Review Board). 1991. *Fourth report to the U.S. Congress and the U.S. Secretary of Energy*. Arlington, VA.

NWTRB (Nuclear Waste Technical Review Board). 1996. *Report to the U.S. Congress and the U.S. Secretary of Energy*. Arlington, VA.

NWTRB (Nuclear Waste Technical Review Board). 1997. *Report to the U.S. Congress and the U.S. Secretary of Energy*. Arlington, VA.

NWTRB (Nuclear Waste Technical Review Board). 1998. *1997 findings and recommendations*. Arlington, VA.

OTA (Office of Technology Assessment). 1985. *Managing the nation's commercial high-level radioactive waste*. Washington, DC.

PNL (Pacific Northwest Laboratory). 1992. *Example of post-closure risk assessment using the potential Yucca Mountain site*. Richland, WA.

Robertson, J.B., G.L. Dixon, and W.E. Wilson. 1982. Letter to M. Kunich, February 5.

Roseboom, E.H., Jr. 1983. *Disposal of high-level nuclear waste above the water table in arid regions*, U.S. Geological Survey Circular 903. Washington, DC.

Sandia National Laboratory. 1992. *An initial total-system performance assessment for Yucca Mountain*, SAND-91-2795. Albuquerque, N.M.: Sandia National Laboratories.

Sandia National Laboratory. 1994. *Total-system performance assessment for*

Yucca Mountain—SNL second iteration, SAND-93-2675. Albuquerque, N.M.: Sandia National Laboratories.

SEAB (Secretary of Energy Advisory Board). 1993. *Earning public trust and confidence: Requisites for managing radioactive wastes*. Washington, DC.

Sinnock, S., J.P. Brannen, and Y.T. Lin. 1984. *Preliminary bounds on the expected postclosure performance of the Yucca Mountain repository site, southern Nevada*, SAND-84-1492. Albuquerque, N.M.: Sandia National Laboratories.

Sinnock, S., Y.T. Lin, and M.S. Tierney. 1986. *Preliminary estimates of ground-water travel time and radionuclide transport at the Yucca Mountain repository site*, SAND-85-2701. Albuquerque, N.M.: Sandia National Laboratories.

Steinbrunner, J. 1974. *Cybernetic theory of decision-making*. Princeton, NJ: Princeton University Press.

Taubes, G. 1993. *Bad science: The short life and hard times of cold fusion*. New York: Random House.

Thompson, F.L., F.H. Dove, and K.M. Krupa. 1984. *Preliminary upper-bound consequence analysis for a waste repository at Yucca Mountain, Nevada*, SAND-83-7475. Albuquerque, N.M.: Sandia National Laboratories.

TRW. 1993. *Total systems performance assessment, 1993: An evaluation of the potential Yucca Mountain repository*. Cleveland: TRW, Inc.

TRW. 1995. *Total systems performance assessment, 1995: An evaluation of the potential Yucca Mountain repository*. Cleveland: TRW, Inc.

TRW. 1999. *License application design selection report*. Cleveland TRW, Inc.

USGS (U.S. Geological Survey). 1978. *Geologic disposal of high-level radioactive wastes—Earth sciences perspectives*, Geological Survey Circular 779.

U.S. House of Representatives. 1987. "Report: Establishing a Nuclear Waste Policy Review Commission, an office of the Nuclear Waste Negotiator, and for other purposes," 100th Cong., 1st sess., 100-425, part 1, Committee on Interior and Insular Affairs, November 5.

U.S. Senate. 1987. "Nuclear Waste Program," Hearing before the Committee on Energy and Natural Resources, 100th Cong., 1st sess., part 4, June 29.

U.S. Senate. 1992. "Department of Energy's Civilian Nuclear Waste Program," Hearing before the Committee on Energy and Natural Resources, 102nd Cong., 2nd sess., March 31.

UZFM (Unsaturated Zone Flow Model). 1997. UZFM Expert Elicitation Project, U.S. Department of Energy, Washington, DC.

Van Luik, A. 1994. Presentation to the Nuclear Waste Technical Review Board, Arlington, VA, January 12.

Vieth, D. 1984. Memorandum to DOE employees C. Hanlon and W. Hewitt, and contractor P. Gnirk, March 10.

Whitfield, S. (1984). Memorandum to E. S. Burton, March 14.

Williams, D., and B. Bodvarsson. 1996. "Conceptual model of flow in the unsaturated zone: New insights," presentation to the Nuclear Waste Technical Review Board, Arlington, VA, October 9.

Wilson, W. 1985. Letter to D. Vieth, director of the Nevada Waste Management Project Office. December 24.

Winograd, I., R.E. Anderson, T.C. Hanks, T.E. Reilly, and E.P. Weeks. 1998. "Viability assessment of a repository at Yucca Mountain: A report to the director, U.S. Geological Survey," November 25, Washington, DC.

PART FOUR

Cascades of Uncertainty: Prediction and Policy

How much longer will our oil reserves last? Well, that depends: on oil exploration and extraction technologies; on the future price of oil; on policy decisions about oil drilling in sensitive ecosystems; on geopolitics and the political stability of oil-rich nations and regions; on the rate of economic growth in the developing world; on the pace of innovation in alternative energy supply technologies.

Will acid rain continue to be a problem in the future? Well, that depends on, among other things, how fast we switch from coal-fired power generation to other sources of fuel, which in turn depends on . . . (see above paragraph).

How will the global climate change in the future? Well, that depends on many things, one of which is the amount of carbon dioxide emitted into the atmosphere by human activity, which in turn depends on . . . (see above paragraph).

An ecology of reality would reveal this: that every uncertainty is connected to every other uncertainty. This insight tells us not only that complex natural systems are difficult to predict, but also that the closer we get to what matters—that is, to the world as experienced by people—the more precipitous this cascade of predictive uncertainty becomes.

Oil and Gas Resource Appraisal: Diminishing Reserves, Increasing Supplies

Donald L. Gautier

Occasionally over the years knowledgeable people have announced that little oil remains to be found and that resource depletion is imminent. The reasons for such pessimism always seem logical and scientifically based. Who can argue with the geologists who explain the origin and distribution of petroleum? And who would argue that petroleum is not a finite, exhaustible resource that will one day be used up? The answer is: both historians and economists, who are quick to point out that the predictions have always been undone by new discoveries and new additions to reserves. In Ohio, in East Texas and Illinois, in California, Oklahoma, and West Texas, and more recently in Alaska, Russia, the North Sea, and the offshore Gulf of Mexico, big discoveries have toppled supply-driven prices. At the end of the twentieth century, oil is as cheap in real terms as it was prior to the 1973 OPEC embargo, and the commodities markets indicate stable energy prices into the foreseeable future. Today, most scientists, environmental activists, and politicians seem more concerned with the environmental consequences of fossil-fuel development, transportation, and consumption than with the threats to the availability of these fuels. What public concern remains regarding petroleum is focused on air pollution, global warming, or local employment rather than national petroleum dependency.

Resource availability has been seen as a global issue since at least the time of Thomas Malthus, but during the twentieth century, concern about oil shortage has periodically dominated popular thought and driven public policy. Early in the century, fears of shortage grew right along with the awareness that oil might be the engine of economic expansion. When petroleum became the preferred fuel for ships, cars, and stationary boilers, the possibility of energy exhaustion took on strategic significance. During the First World War, every seagoing nation planned ways to keep its ships sailing in the event of insufficient bunker fuel. By the early 1920s, virtually every promising geologic structure in the United States that could be identified from the surface had been drilled, and the future of crude oil supplies seemed gloomy. Demand for oil was increasing dramatically as Americans took to the highways, and the price of crude increased accordingly. The United States Congress set aside wide sections of the public domain in Alaska, Wyoming, and California as Naval Petroleum Reserves of last resort in the event of wartime shortfall.

The shortfall never came. Technological advances kept supplies ahead of demand, as the application of geophysical techniques such as gravity, magnetic, and seismic surveys allowed explorationists to identify subsurface structures that were potential reservoirs. Discoveries in California, Oklahoma, and Texas, not to mention Mexico and Venezuela, drove prices down, and with the discovery of the East Texas oil field, oil prices collapsed to just a few cents a barrel. Only two decades after the fears of shortage, the grim prospect of the Second World War found the United States unprepared for war, but with plenty of oil. Oil production was supervised by government agencies, and prices stabilized at around one dollar per barrel. Oil stayed cheap all through the fifties and sixties, even as demand rose exponentially.

By the early 1970s, demand was catching up with supply, and U.S. production had begun to fall. President Nixon felt compelled to declare Project Independence, to reverse the trend of increasing reliance on oil imports. The OPEC oil embargo of 1973, with its lines of cars and sharply rising prices, dashed the hopes of energy independence and ultimately led to the globalization of energy markets.

For better or worse, petroleum has proven to be the basis for economic development, and its consumption remains at the heart of the world economic order. Fears of shortage have given way to complacent reliance on market forces to resolve issues of resource availability. Now, at the opening of the twenty-first century, one hundred years after concerns about availability were voiced, the problem of oil supply may seem to be one of excess, not shortage. Even with four of the leading petroleum countries (Iraq, Iran, Russia, and Saudi Arabia) producing significantly below their capacity, huge multinational companies are stressed by the ferocious competition of abundant supply.

However, the complacency could be misplaced; discoveries of big new fields are now few and far between, while global demand is increasing. Almost every global economic scenario includes future increases in energy use, particularly in the developing countries, where energy consumption is a widely used, if indirect, measure of the standard of living, and most energy comes from fossil fuels. The difference between reserve additions from new discoveries and increasing global oil consumption is being accommodated by the application of technological innovations such as 3-D seismic imaging, computer-controlled directional drilling, deep-water technology, and efficient development practices. Voices of doom are again being heard, warning of the need to take dramatic action before radical price increases irreparably harm global economies (Campbell 1997; Ivanhoe 1996).

Few subjects better illustrate the intersection of science, public opinion, and political response than the prediction of oil and gas resource availability. The debate concerning the future of oil and gas resources and the appropriate role of government in establishing policy is emblematic of the larger debate concerning natural resources of all kinds. Like Thomas Malthus in the eighteenth century, neo-Malthusians point with alarm to the exponential growth of the human population and the accompanying demand for resources. The earth, they argue, is at or above its carrying capacity, and the government must act immediately to limit consumption, to conserve resources, and to bring population growth under control. If we do not take such steps, they assure us, ghastly consequences are just around the corner.

Meanwhile, the cornucopians point out that all such previously predicted shortages either never came to pass or were simply nullified by the actions of the market system. Resource substitution, they argue, will save the day, just as it did in the nineteenth century when whales were becoming scarce and whale oil was in short supply. The price of whale oil rose, and alternative fuels were found. Yes, it may be true that the big oil fields of the United States have been found, but changes in consumer habits, and inexpensive overseas production work together to keep fuel available and cheap. Government action, say the optimists, will only distort the market and lead to inflation, higher prices, and shortages. The debate continues, with the current U.S. government and most of the world firmly, if unconsciously, in the camp of the cornucopians. But this has not always been so.

With this context and background, we turn now to a review of the assessment work of the United States Geological Survey (USGS, the Survey) and its political motivations and consequences. The USGS's oil and gas assessment work serves to illustrate the role of resource prediction in development of government policy and the changing expectations of the role of a government-funded institution charged with making predictions.

Reserves, Resources, and Uncertainty

In the media and general literature the concepts of *reserves* and *resources* are rarely distinguished, but to those involved in resource appraisal work, the distinction is critical. To the analyst, oil and gas resources include virtually all naturally occurring concentrations of liquid or gaseous hydrocarbons in the earth's crust, some fraction of which is currently or potentially economically extractable. Reserves or proved reserves are that part of the resource that has been demonstrated, using engineering data, to be recoverable under existing economic and operating conditions. Thus reserves are that part of the resource that is expected to be produced at current economic conditions. Production can come only from reserves.

Oil and gas resource data (such as calculations of total oil or gas in place) are generally unavailable. Rather, analysts have access to drilling and production data and geological information. Estimates of resources commonly refer to undiscovered, conventional, economically recoverable resources—that is to say, that part of the resource base that has not been discovered, that which is accessible to currently available technology, and that which is recoverable at a profit.

To be useful, resource assessments need to be very specific in their scope and content. For example, for any particular area, an estimate of those resources that occur in large anticlinal structures—upwarped sedimentary rocks—visible at the surface of the earth could be completed today by most geologically trained resource analysts with very consistent and reproducible results. At the turn of the century, when the anticlinal theory of oil accumulation was new, such structures were the very frontier of exploration. A century later, oil in such structures either has already been produced or is counted in proved reserves. Today a resource assessment might have to evaluate oil accumulations in subtle stratigraphic traps, caused by transitions between different sedimentary rock types, hundreds of meters below the surface under deep water, in areas not even considered remotely prospective by the old-time oil finders.

The uncertainty surrounding a 1920s estimate of resources in surface anticlines probably reflects what was then known about such structures. Such uncertainty would have little to do with oil in the deep offshore structures being explored in the 1990s. Exploration proceeds in pursuit of particular concepts called exploration plays. Each play consists of a set of oil or gas accumulations that share similar geologic properties. Examples of plays might include the anticlines along the mountain front, stratigraphic traps in particular formations in the subsurface, or oil accumulations associated with salt domes in a certain area. Through time the largest, most obvious accumulations of each play are usually found first, and smaller more subtle accumulations are found later. Thus, if geologically and geographically related accumulations are considered within a single play, the history of drilling and the sizes and distribution of discovered fields can be used to make sensible predictions of the future of that play. Such an analysis, if done numerically, is a discovery process model. Similarly, unexplored plays that are geologically similar to well-known productive areas can be evaluated by means of analogy.

In the United States, oil plays have been pursued vigorously for more than one hundred years. Onshore, most sections of sedimentary rock with even remote prospects of oil production have been explored at some level. Although in recent years literally tens of thousands of new fields have been and continue to be discovered, in recent decades no really large fields have been found onshore in the forty-eight contiguous states, and the volume of oil in large fields greatly exceeds that in small fields. Consequently, estimates of undiscovered oil in the United States onshore have become more certain and volumetrically rather small. Much greater uncertainty surrounds the undiscovered resources in Alaska or in deep water offshore.

Moreover, the uncertainty surrounding natural gas resources is much greater than that surrounding oil. Until relatively recently, natural gas was found incidentally along with oil and few wells were drilled with the intention of producing natural gas. Natural gas exists not only in discrete, conventional accumulations with or like oil, but also in coalbeds, in gas hydrates offshore, in deep brines under enormous pressure, and in huge basin-center deposits. For example, in the Green River Basin of Wyoming alone, geologists estimate in-place gas resources in unconventional basin-center deposits to be present in amounts in excess of 5,000 trillion cubic feet (TCF), of which virtually none is currently being produced profitably. Contrast this with United States gas production, which hovers around 20 TCF per year.

A national oil and gas assessment must consider resources at a national level. This means the evaluation of resources of various kinds and at a variety of levels of uncertainty. Clearly, the terms of the assessment must be stated carefully. The areas, depths, and types of resources being evaluated have to be explicitly explained. If not, as we shall see, serious misunderstandings can arise.

Resource Assessment at the USGS

Prior to the OPEC embargo in 1973, while most of the public was unconcerned about oil availability, policy makers were privately worried about future supplies. Major oil companies were shifting their exploration budgets to overseas projects. Since the early 1950s the Malthusian view had been espoused by the clear and persuasive voice of M. King Hubbert, a research geologist with Shell Oil, who had been predicting the decline of United States oil production. Hubbert, by analyzing various oil and gas exploration, development, and production data, had correctly foretold the approximate year of maximum U.S. production (1971). He explained to everyone who would listen that the passing of the maximum signaled the beginning of a relentless decline from which United States production would never recover. By the early 1970s Hubbert had come to the USGS to conduct further research based on his studies at Shell and released the results in a series of influential publications (Hubbert 1972, 1974, 1979). Not only did he correctly forecast the year of maximum U.S. production, he confidently reported that no more than 55 billion barrels of oil (BBO) remained to be found in the United States (current U.S. consumption is on the order of 5 billion barrels of oil per year). Throughout the 1970s federal policies regarding oil and gas were consistent with the Hubbert view of the future. Even the major oil companies fully expected to see prices reach unprecedented levels and made investments predicated on $100/barrel oil.

Only a few dissenters remained. Many wildcat oil finders had always insisted that big discoveries were just around the corner, but within the United States government there was one highly visible and outspoken exception—the director of the USGS, Vincent McKelvey, who held a view radically different from that of Hubbert (McKelvey 1968, 1972, 1984). McKelvey believed that oil and gas resources are widely distributed and that the largely undrilled volumes of the earth's crust (then, as now, a significant fraction) could be expected to yield sizes and numbers of oil fields similar to those already discovered. To him, the most important resource was human ingenuity. McKelvey went on record saying that at least 200 and possibly more than 450 BBO remained to be found and developed within the confines of the forty-eight contiguous United States. The mutually exclusive views of Hubbert and McKelvey put the USGS in the awkward position of having recently endorsed both the most optimistic view and one of the most pessimistic views of the oil and gas future of the United States. The awkwardness sometimes turned bitter. McKelvey is said to have applied considerable pressure on Hubbert not to publish through the USGS, and indeed most of the Hubbert works of this period were released through Congress or through the National Academy of Science, of which Hubbert was a member. Hubbert, for his part, is said to have been influential in preventing McKelvey's election to the Academy. Ultimately,, the politics of resource assessment prevented King Hubbert from fully participating in Survey work and probably cost McKelvey his job. As Vince McKelvey continued to present his cheerful views, his optimism simply didn't fit with a national policy of strict controls on natural gas pricing and production and hefty federal subsidies for development of alternative fuels and nonhydrocarbon energy resources.

The National Assessments, Circulars 725 and 860

The USGS and its immediate predecessor agencies had been established during the frontier days of westward expansion for the purpose of evaluating the mineral and water resources of public lands in the United States and its territories. By the early twentieth century this mandate had come to include the evaluation of water and coal resources and estimates of the remaining oil reserves. From the early 1900s, the Survey had occasionally published resource estimates, which, like those of Hubbert and McKelvey, were largely the work of individual scientists. (The history of these early estimates has been summarized by McCulloh [1973], Miller et al. [1975], and Sheldon [1976].) However, the situation in the middle 1970s was new in terms of the economic importance of petroleum, the maturity and sophistication of the industry, and the global distribution of production and control. The urgency of the declining oil production, the critical dependency on imported petroleum, and the contradictory views being expressed by scientists of such high repute as Hubbert and McKelvey presented a clear problem. In the 1970s, developing new government programs was considered an appropriate response to problems, and money for research was relatively abundant. With such an obvious gap in knowledge, it seemed only natural that the USGS should be provided with the funds to determine the "truth" regarding oil and gas resources. Accordingly, in 1973, the Oil and Gas Resource Investigations Program was founded at USGS, utilizing a few traditional USGS scientists from the old Fuels Branch and new people straight out of industry research laboratories, oil exploration companies, and academia. The new program was to conduct geological, geophysical, and geochemical research into the origin and distribution of petroleum. A formal Resource Appraisal Group (RAG) was formed as well. Its purpose was to develop methodologies that could be employed systematically to predict future discoveries of oil and gas in the United States.

At about the same time, the USGS was approached by the Federal Energy Agency (FEA), which requested aid in fulfilling its urgent legal responsibility to generate an independent appraisal of the oil and gas resources of the United States. In that initial assessment for the FEA, the Resource Appraisal Group subdivided the United States into fifteen regions, further subdivided into 102 provinces. The provinces roughly approximated the petroleum basins of the United States and corresponded closely to the province boundaries established by the American Association of Petroleum Geologists (Cram 1971). A regional expert, typically a research geologist at the USGS, was called upon to gather the geological and engineering information necessary to evaluate the petroleum potential of each province. Assessments were conducted by means of a Delphi methodology (Miller et al. 1975), in which a group of resource assessment specialists reviewed the results of the province analysis, interviewed the regional experts, and made individual estimates of low (95 percent probability that actual resources exceed this number), most likely, and high (5 percent) volumes for each province. From these estimates, mean values were calculated and individual estimates were posted for review. If there were major differences among the estimates, the reasons for the differences were discussed and resolved, and a consensus was reached. The subjective probability estimates resulting from the Delphi assessments of each province were cast into log normal probability distributions and then the province-level distributions were aggregated to regional and national totals using standard Monte Carlo statistical techniques.

That initial assessment of the RAG was published as USGS Circular 725, prepared specifically for the FEA. In Circular 725, undiscovered oil resources were estimated to range from a low of 50 BBO at a 95 percent probability, to a high of 127 BBO at a 5 percent chance. The mean estimate of 82 BBO was somewhat higher than Hubbert's estimate of 67 BBO but dramatically lower than the USGS-McKelvey estimate of 400 BBO (U.S. Geological Survey 1974), not to mention the highest estimates (450 BBO) promoted by McKelvey and published by Theobald, Schweinfurth, and Duncan (1972). Corresponding RAG figures for gas were a statistical mean of 484 trillion cubic feet, with a range of 322 to 655 TCF. The report suggested that roughly one half of the undiscovered resources and one quarter of the gas resources occurred in offshore regions of the United States and in onshore frontier provinces of Alaska.

These estimates were done under contract and time constraints. The customers were understood to be U.S. government officials and only secondarily to be interested parties in industry, academia, and the public. There was some grousing from industry, but at the time such complaints were not considered to require government response. The assessment effort included analyses of undiscovered gas, but everyone knew that oil was the politically sensitive issue and the real reason for the study. When the results were presented to Director McKelvey, he knew immediately that his own organization had failed to support his view. Nevertheless, to his credit, he is said to have insisted on only minor editing changes in the text and to have immediately approved Circular 725 for release to the Federal Energy Agency and the public.

From the time of the 1973 oil embargo until 1981, crude oil prices rose from less than $3/barrel to more than $35/barrel. The domestic industry responded with intense exploratory and development drilling in the late 1970s, which yielded new geologic information, new technology, and a new economic climate. By 1981, in the midst of the drilling boom, the Survey was ready to release its second national assessment of oil and gas resources (Dolton et al. 1981). The study was published as USGS Circular 860, entitled "Estimates of Undiscovered Recoverable Conventional Resources of Oil and Gas in the United States." Circular 860 explicitly excluded whole categories of "unconventional" resources, partly because of their economic uncertainty, but also because there was no obvious way to evaluate them. The unassessed categories included natural gas in low-permeability ("tight") sandstone reservoirs, heavy (high gravity and viscosity) oil accumulations, tar deposits, oil shale, coal bed gas, gas in geopressured reservoirs and brines, and natural gas hydrates. The area under assessment had been expanded to include the continental slopes offshore, but otherwise the regions and provinces were similar to those of the 1975 study.

The 1981 study estimated undiscovered recoverable conventional oil resources of the United States to range from 64.3 to 105.1 BBO, with a mean estimate of 82.6 BBO of "conventional" oil. The oil estimate was thus very similar to that reported in the 1975 study. The authors emphasized that the pre-1975 disparity in estimates had been resolved and that a consensus had developed, with most new studies placing the range in estimated oil resources between 55 and 113 BBO. The Survey scientists felt they were now safely and unanimously in the middle of the consensus values concerning oil. There were some touchy subjects, but those were in the details. The authors explained that their assessment had been influenced by disappointing drilling results in the Gulf of Alaska, the Atlantic shelf, and the Southern California continental borderland.

Undiscovered natural gas, however, had changed significantly. Since 1975, the perceived role of gas in the energy mix had changed, and it was beginning to get real attention. Undiscovered conventional gas resources were estimated to be between 474.6 and 739.3 TCF, with a mean of 593.8 TCF. Thus, the 95 percent probability value of the new assessment was nearly identical to the mean of 1975.

In attempting to explain the differences between Circular 860 and previous work, the authors stressed the difficulty in making direct comparisons between estimates. They cited exclusion of natural gas from parts of the previous assessment, exclusion of the growth of reserves in known fields ("inferred reserves"), estimates of quantities other than recoverable resources, and differences in statistical reporting. Drilling in the Overthrust Belt of the Rocky Mountains and in the deep Anadarko basin of Oklahoma seemed to indicate previously unrecognized potential for gas. As a consequence, estimates of natural gas resources, especially in the Wyoming Overthrust, Oklahoma, and the Gulf of Mexico seemed optimistic indeed relative to previous work. These discrepancies didn't seem to bother the authors, who clearly regarded resource appraisals as works in progress. To some outside the inner circles, however, it was disturbing to see such large changes in the numbers.

Resource Assessment in an Era of Conservative Administration

The political environment changed drastically in 1981, as the election of Ronald Reagan brought a fiscal conservatism that squeezed the funding of many programs. In contrast to the previous decade, when federal budgets grew with regularity and government agencies were the instrument of choice for policy implementation, the new administration saw government regulations as problems in themselves, and regulatory agencies were singled out for special scrutiny. At USGS, it was the Conservation Division that had responsibility for leasing and royalty management on federal land. In 1982, Reagan's first secretary of the interior, James Watt, decreed that the Conservation Division of the USGS was to become the core of a separate agency, the Minerals Management Service (MMS), presumably subject to more careful scrutiny and tighter management than in the past. When the MMS was created, responsibility for estimating resources in the federal offshore areas became part of its leasing responsibility. That meant the end of national assessments conducted solely by USGS.

Meanwhile, the price of oil dropped precipitously. The domestic oil and gas industry feared for the future as it tried to compete with the cheap imported oil from Venezuela, Mexico, and the Persian Gulf. The new managers in the Department of the Interior sought to relax environmental policies that they said had locked up precious resources on federal lands. The message to oil and gas researchers at USGS was clear. The issue was no longer the imminent shortfall of domestic production relative to imports. Rather, they should determine the amount, quality, and distribution of untapped resources on federal lands, with an eye to making them available for development. Responding to the wishes of the new administration, a new assessment effort was launched in the mid-1980s called the Federal Lands Assessment Project (FLAP), designed to determine the quantity, quality, and distribution of oil and gas resources on federal lands. Additional studies were requested to determine the resource potential of certain roadless tracts being considered for wilderness designation. A special investigation was also funded to evaluate the oil and gas resources under the coastal plain of the former William O. Douglas National Wildlife Range, now renamed the Arctic National Wildlife Refuge (ANWR). The USGS conducted the geologic studies in support of the resource evaluation of ANWR (Bird and Magoon 1987).

Assessment methodology had significantly improved since Circular 860. The new play analysis approach in use at Exxon and at the Geological Survey of Canada provided a level of detail and analytical flexibility far superior to the old Delphi method of the USGS. In particular, the play analysis approach resulted in estimates of sizes and numbers of undiscovered accumulations. The sizes and numbers, in turn, accommodated economic analysis in a way not previously possible. Evaluation of the oil and gas potential of federal lands would require the type of detail attainable only through the new play analysis methodology. Still the resources on federal lands could not be analyzed outside of the broader context of national resource distribution. It would therefore be necessary to conduct an entirely new national assessment using the play analysis approach.

The new study was conducted in collaboration with the Minerals Management Service. The MMS would evaluate the federal offshore areas, and the USGS would confine itself to onshore and state waters. The work was labor intensive, fully utilizing for the first time computerized databases to evaluate resource potential on a play-by-play basis, developing detailed exploration maps, and statistically and spatially evaluating the trends of discoveries in each play as a function of time and drilling. The data available for the study were much greater than in the past. More than three million wells had been drilled in the United States by 1987, but oil prices had been falling steadily. Between the late seventies drilling boom and the new careful approach required by the competitive pressures of falling oil prices, virtually all available land with resource potential had been leased and drilled. In many cases drilling was done on even marginally prospective tracts. As a result, many ideas and areas had been tested for oil and gas potential.

The drilling frenzy had resulted in the discovery of many fields, but they were small and contributed little to the volume of discovered oil and gas in the United States. The areas that had been most promising in

1980 had been tested with disappointing results. The size of fields discovered had continued to drop. For example, prior to 1968, natural gas fields averaged about 600 billion cubic feet (BCF) per discovery; by 1987 this dropped to less than 50 BCF per discovery, and it continued to fall. The new methodologies, the computers, and the abundant statistical and geological information all pointed to the unavoidable conclusion that the estimated undiscovered conventional resources of the United States were actually somewhat smaller than had been reported in 1981. Undiscovered conventional oil resources were estimated to range from 33 to 70 BBO, with a mean estimate of 49 BBO, considerably less than the 82.6 BBO reported in 1981. And, rather than the fairly optimistic view of gas in 1981, the undiscovered conventional natural gas resource was now estimated to range from 307 to 507 TCF, with a mean of 399 TCF, a value less than the 95 percent probability estimate of 1981 (474.6 TCF).

As in past studies, the USGS and MMS had conducted their most recent assessment in a manner that explicitly and intentionally restricted itself to undiscovered *conventional* oil and gas resources. These conventional resources were the traditional fare of the oil industry—the discrete fields that had been explored for and discovered since the beginning of the petroleum age. During the same period, more than 22 BBO and 120 TCF of gas had been produced, reserves of oil had increased slightly, and gas reserves had held approximately steady. This seemingly paradoxical event, in which oil and gas were produced, but not discovered, while reserves held steady, was the result of growth of reserves in existing fields (the inferred reserves). This reserve growth reflected processes such as application of new technology in existing fields, discovery of new pools within already productive fields, and decreasing well spacing by in-fill drilling. In the comfortable light of retrospection, such processes, which were well known to the assessors, pointed to a serious flaw in the assessment from a public policy perspective. Even though growth of reserves in existing fields had proven to be the principal means of reserve addition, the resource assessment had given it short shrift, choosing instead to focus on potential discoveries of unknown fields. What was arguably the most important part of the assessment had thus been treated in only the most superficial and simplistic manner.

It was well known at USGS that, in addition to the conventional resources and inferred reserves, huge but economically problematic amounts of natural gas existed in broadly distributed continuous-type unconventional accumulations in coal beds and in low-permeability reservoirs. This was especially true in the Rocky Mountains, where the

USGS itself had been one of the leading research organizations releasing studies that showed natural gas to exist in basin-center deposits such as those of the Piceance and Green River basins (e.g., Law and Spencer 1989; Spencer and Mast 1986). But because the reserve growth phenomenon was largely the province of economists and petroleum engineers and because there was no reliable, reproducible way of evaluating the unconventional resources, the assessors simply elected not to attempt to include a full assessment of those categories. That decision turned out to be a terrible mistake.

Political Furor

At any time, such a downward revision in the best estimate of the nation's undiscovered resource would be the subject of considerable attention and criticism, but this time it caused a political explosion. Federally specified controls on the production and price of natural gas, which had been in place since the Carter presidency, were gradually being lifted by the Republican administration. While the Resource Appraisal Group at USGS was naively nearing completion of its assessment of conventional resources (Mast et al. 1989), the American Gas Association (AGA) had launched a major political campaign to expand the use of natural gas in the industrial sector. AGA was pressing Congress for deregulation of gas for electrical power generation. All the arguments for deregulation and expansion were based on the presence of large quantities of as yet untapped natural gas. Most analysts were using the old 1981 USGS assessment numbers as well as USGS studies of tight gas and coal bed gas resources, which indicated the presence of *thousands* of TCF of natural gas, mostly in the Rocky Mountain West. Unfortunately, this gas was part of the "unconventional" category that was specifically excluded from the new USGS-MMS assessment.

The directors of the USGS and MMS had for months been trying to schedule a briefing with Don Hodel, who succeeded James Watt as secretary of the interior, but they had not been able to get on his calendar. By the time a briefing was finally scheduled, on February 22, 1988, the text of the new joint USGS-MMS national assessment had been leaked to the DOE and to the AGA. The secretary evidently learned of the new assessment numbers from the American Gas Association, which had called demanding that he take appropriate action regarding the loose cannons at USGS. At the briefing, Hodel verbally skewered the Survey for changing methodology and questioned the credibility of the USGS scientists even to do such work (chief, USGS Office of Energy and Marine Geology, Charles Masters, personal communication, June, 1996). The secretary demanded a thorough review of the study and told those present that he was sure the numbers would be changed before they were released. It was clear to the director of the USGS, Dallas Peck, that his job was at stake (Dallas Peck, personal communication).

On March 10, 1988, UPI carried the following item:

> A study by the U.S. Geological Survey has reduced estimates of undiscovered U.S. oil and gas by about 40 percent, but the government has withheld the report for fear it could affect the volatile industry, the Washington Post reported yesterday. . . . Gas industry officials are particularly critical of the report. . . . "This is something our customers just don't need to see or hear at this time," Terry Uhl, a spokesman for the American Gas Association, told the Post.

The American Gas Association press release of March 11 unequivocally presented the AGA view:

> The U.S. Geological Survey's erroneous and unduly pessimistic preliminary estimates of the amount of natural gas that remains to be discovered in the United States . . . is highly inaccurate and clearly incomplete. The negative assessment by the USGS is the direct result of questionable methodology and assumptions, and is counter to projections by other respected organizations.

On March 18, 1988, the *Oil Daily* reported:

> "Preliminary U.S. crude oil and natural gas reserve (resource) estimates that are dramatically lower than their predecessors almost certainly will be changed," Interior Secretary Donald Hodel said Thursday.

The political heat was intense, and publication was briefly delayed. Importantly for the scientific credibility of the USGS, the numbers were not changed, but in a carefully phrased foreword, the directors of USGS and MMS emphasized those categories of resources that were not included in the assessment and carefully explained the possible shortcomings of the work. Reviews were requested and carried out by the American Petroleum Institute (1989), the American Association of State Geologists, the Energy Information Administration at DOE (1989), and finally in a study by the National Research Council (1988).

Most of the scientific reviews were supportive of the Survey work. Articles and editorials were published in *Science* magazine. The reviews by EIA and API generally endorsed the USGS methodology and findings. Several multinational oil companies discreetly informed Secretary Hodel that their own studies agreed with those of the USGS. The NAS committee made a number of relatively minor recommendations concerning tails of probability density functions, and combinations or subdivisions of plays, none of which resulted in significant changes to the numbers. Upon leaving his position as secretary of the interior, it is said that Don Hodel apologized to USGS Director Dallas Peck for all the fuss, admitting that the scientific work had been done properly.

But if the scientific work had been done properly, why the avalanche of stinging criticism? In this regard, the National Academy Committee was politically astute. Its report advised the USGS to include other categories of resources in addition to the conventional undiscovered and to actively seek scientific input and involvement from industry, state surveys, and others.

The USGS had consciously and carefully restricted its study to the discrete, conventional accumulations that had been the subject of previous assessments and that had been the objective of oil industry exploration efforts for most of the century. Little or no attention had been paid to the political context into which the work was being thrust. The USGS had focused on undiscovered oil onshore, even though virtually everyone knew that in recent years most additions to reserves had come from growth in previously discovered fields. The Survey scientists knew of the huge amounts of uneconomic natural gas present in unconventional reservoirs, but because there was no good methodology for assessing those resources, the USGS made the scientifically defendable but politically disastrous decision to exclude them from consideration, without giving much consideration to possible consequences.

The study had been done as if it were an end in itself, based on the long tradition of assessment work, but without really specifying its purpose. The potential users of the work, who probably neither knew nor cared about subtle distinctions in resource categories, had been blindsided. The brouhaha changed the outlook at USGS, which finally became cognizant of its audience and of the political context into which the numbers were being introduced.

The most recent national assessment, also conducted jointly with the MMS, was begun in January of 1991 and released in February of 1995. Industry experts were involved from the outset in identifying and documenting basic geologic concepts of exploration that could be used in developing play analyses. Methodology seminars and workshops were held involving prominent academic and industry scientists. The lesson had been learned about resource categories, and entirely new methodologies were developed to quantify the resources of oil and natural gas in the unconventional categories. More emphasis was placed on the growth of reserves in existing fields, which are the source of most additions to reserves in recent years. The new assessment was able to tap a much wider range of information and expertise than had been utilized before, hopefully without compromising the basic scientific independence necessary to produce results free from political interference.

The 1995 National Assessment reported recoverable resources of approximately 110 billion barrels of oil and 1,075 TCF of natural gas. In addition to 20 billion barrels of proved oil reserves and 135 TCF of proved conventional gas reserves, this large resource consisted of three distinct categories: (1) conventional resources, (2) reserve growth in existing fields, and (3) continuous-type (unconventional) accumulations. Undiscovered conventional oil resources, the principal commodity category of previous assessments, amounted to about 30 BBO. Anticipated reserve growth from existing oil fields accounted for 60 BBO, twice the conventional resource estimate. Continuous (unconventional) oil accumulations accounted for another 2 BBO. With respect to gas, the resource numbers were up significantly. Conventional gas was estimated at about 259 TCF. Growth of reserves in existing fields was estimated at 322 TCF, and continuous-type (unconventional) gas accumulations were estimated at about 358 TCF. Thus, although the estimates of conventional resources were similar to previous appraisals, the overall volume of domestic resources was reported to be much larger than in previous assessments. This increased volume was mainly the result of the more inclusive treatment of diverse resource categories.

Results were documented in enormous detail and organized in a completely digital format based on geographic information systems and published on CD-ROM (Gautier et al. 1995). The CD format took advantage of the digital age, containing the equivalent of 10,000 pages of paper documentation. More than twenty-five hundred copies were distributed, and the basic information documenting the assessment became as important as the results themselves. Instead of the previous narrow audience of resource analysts and economists, the 1995 National Assessment was intended for use by explorationists, teachers, land management agencies such as BLM and the Forest Service, and Wall Street arbitrageurs. This time academicians, industry groups, and various government agencies praised the work. Funding for the Energy Surveys Program at USGS seemed secure and protected even as the budget cutters were eyeing every aspect of federal spending. For the moment disaster had been averted. The new assessment may have been no truer than the previous one, but the audience and the political context were recognized and communication was improved.

Validity and Evolution of Oil and Gas Assessment

Just as the history of energy resource development has been one of geology and engineering, it has also been one of economics and policy. Careful analyses of explicitly defined resources have correctly predicted local depletion, but broad forecasts of future shortages based on extrapolation have been unreliable. Most actual shortages and associated price escalation have been linked to political events such as war or oil embargo, and to policies such as price regulation or production quotas. Instead of catastrophic price shifts, market forces have led to resource substitution. The long-term trend has been toward development of increasing quantities of energy at lower real cost. No doubt the earth is finite with regard to the distribution of fossil fuels, but the end of fossil fuel use is not in sight.

How then does one measure the success or failure of oil and gas resource assessment? Probably not in terms of accuracy, even though that is the desired goal. The prediction of the quantity, quality, and distribution of essentially unknown entities such as oil and gas accumulations is, by definition, an attempt to narrow uncertainty surrounding decisions. The value of an assessment must be in its usefulness. King Hubbert's logistic equations were useful to the managers at Shell Oil with whom he collaborated in planning his original work. If the Hubbert curves have not been especially helpful in applications far beyond their original purpose, they have certainly served the unintended purpose of framing a national scientific dialogue on resource availability.

We probably should not attempt to separate the resource predictions from their political context, and we probably should not attempt predictions without having clearly defined policy objectives. The broadly stated policy goals satisfied by the earlier assessments no longer suffice. General but naive questions concerning strategic shortages have given way to more specific issues. As the detail of available information has increased and the predictive reliability has improved, questions have evolved to become much more specific. Instead of policy questions concerning the desirability of imposing natural gas price controls, land managers want to know the environmental consequences of opening a roadless area or wildlife refuge to exploratory drilling, or the expected income from leasing a particular tract of federal land. Resource predictions are useful in developing economic models of oil and gas markets. The ability to predict whether or not a particular new pipeline will be economically viable depends on the reliability of our understanding of available but untapped resources. Prediction of future greenhouse emissions depends on which fossil fuels will be available for consumption. We may never know if any such prediction is correct. Only if decision makers understand the assumptions, methods, and uncertainties associated with a prediction will they be able to use it effectively in the development of national energy policies.

References

American Petroleum Institute. 1989. *Ad hoc review of national assessment of undiscovered conventional oil and gas resource.* Unpublished report. Washington, DC: American Petroleum Institute.

Bird, K.J., and L.B. Magoon, eds. 1987. *Petroleum geology of the Arctic National Wildlife Range.* U.S. Geological Survey Bulletin 1778. Washington, DC.

Campbell, C. J., and J. H. Laherrère 1998. The end of cheap oil. *Scientific American* 278 (March): 78–83.

Cram, I.H., ed. 1971. Future petroleum provinces of the United States—Their geology and potential. *American Association of Petroleum Geologists Memoir* 15(1): 1–803.

Dolton, G.L., K.H. Carlson, R.R. Charpentier, A.B. Coury, R.A. Crovelli, S.E. Frezon, A.S. Khan, J.H. Lister, R.H. McMullin, R.S. Pike, R.B. Powers, E.W. Scott, and K.L. Varnes. 1981. *Estimates of undiscovered recoverable conventional resources of oil and gas in the United States,* U.S. Geological Survey Circular 860. Washington, DC.

Energy Information Administration. 1989. *An examination of domestic natural gas estimates.* EIA Special Report 89-01. Washington, DC: U.S. Department of Energy.

Gautier, D.L., G.L. Dolton, K.I. Takahashi, and K.L. Varnes, eds. 1995. *1995 national assessment of United States oil and gas resources—Results, methodology, and supporting data,* U.S. Geological Survey Digital Data Series 30 (CD-ROM). Washington, DC.

Hubbert, M.K. 1972. Estimation of oil and gas resources. In *Proceedings of Workshop on Techniques of Mineral Resource Appraisal,* pp. 16–50. Denver, Colorado: U.S. Geological Survey.

Hubbert, M.K. 1974. U.S. energy resources, a review as of 1972. In *U.S. Senate Committee on Interior and Insular Affairs, U.S. Energy Resources: A Review as of 1972, a Background Paper,* part I, 1–201. 93rd Cong., 2nd sess., Committee Print, serial no. 3-40 (92-74).

Hubbert, M.K. 1979. Hubbert estimates from 1956 to 1974 of U.S. oil and gas. In *Methods and models for assessing energy resources, proceedings of* the First IIASA Conference on Energy Resources, Luxembourg, Austria, M. Grenon, ed., pp. 370–383. Oxford, UK: Pergamon Press.

Ivanhoe, L.F. 1995. Future world oil supplies: There is a finite limit. World Oil, October, pp. 77–88.

Law, B.E., and C.W. Spencer, eds. 1989. Geology of the tight gas reservoirs in the Pinedale anticline area, Wyoming, and at the multiwell experiment site, Colorado. U.S. Geological Survey Bulletin 1886, chapters A–O, unpaginated. Washington, DC.

Mast, R.F., G.L. Dolton, R.A. Crovelli, D.H. Root, and E.D. Attanasi (U.S. Geological Survey); P.E. Martin, L.W. Cooke, G.B. Carpenter, W.C. Pecora, and M.B. Rose (Minerals Management Service). 1989. Estimates of undiscovered conventional oil and gas resources in the United States—A part of the nation's energy endowment. Washington, DC: U.S. Geological Survey and Minerals Management Service.

McCulloh, T.H. 1973. Oil and Gas. In United States mineral resources, D.A. Brobst and W.T. Pratt, eds. U.S. Geological Survey Professional Paper 820, pp. 477–496. Washington, DC.

McKelvey, V.E. 1968. Contradictions in energy resource estimates. In Proceedings of the Seventh Energy Gas Dynamics Symposium, L.B. Holmes, ed., pp. 18–26. Evanston, IL: Northwestern University Press.

McKelvey, V.E. 1972. Mineral resource estimates and public policy. American Scientist 60(11): 32–40.

McKelvey, V.E. 1984. Undiscovered oil and gas resources: Procedures and problems of estimation. In Proceedings of the 27th International Geological Congress 13: 333–352.

Miller, B.M., H.L. Thomsen, G.L. Dolton, A.B. Coury, T.A. Hendricks, F.E. Lennartz, R.B. Powers, E.G. Sable, and K.L. Varnes. 1975. Geological estimates of undiscovered recoverable oil and gas resources in the United States, U.S. Geological Survey Circular 725. Washington, DC.

National Research Council, 1988. Energy Related Research in the USGS. Washington, DC: National Academy Press.

Sheldon, R.P. 1976. Estimates of undiscovered petroleum resources—A perspective. U.S. Geological Survey Annual Report Fiscal Year 1975, pp. 11–22. Washington, DC.

Spencer, C.W., and R.F. Mast, eds. 1986. Geology of tight gas reservoirs. American Association of Petroleum Geologists Studies in Geology #24. Tulsa, Okla.

Theobald, P.K., S.P. Schweinfurth, and D.C. Duncan. 1972. Energy resources of the United States, U.S. Geological Survey Circular 650. Washington, DC.

Predictive Modeling of Acid Rain: Obstacles to Generating Useful Information

Charles Herrick

In the early 1970s, a number of U.S. scientists gained national attention by asserting that lakes in the Adirondack Park and other pristine areas of upper New York, Vermont, and New Hampshire were dying from "acid rain." Environmental groups joined the debate, claiming that acid rain was damaging forests and lakes across the northeastern United States and Canada.[1] The possibility that acid rain was responsible for observed damages to agricultural crops was also a matter of concern. It was argued that acid rain was caused by sulfur dioxide emissions from large, coal-burning power plants, mostly located in the Midwest and along the Ohio River Valley. Situated to take advantage of the region's high-sulfur coal reserves, these thermoelectric facilities provided much of the power for the U.S. industrial heartland. Emissions from these power plants were vented through extremely tall smokestacks to help diffuse combustion by-products, thereby reducing local pollution. The use of tall stacks caused sulfur dioxide to be lofted into the atmosphere, where it could be transported by prevailing winds, mixed with precipitation and other "acid precursor" pollutants, and deposited in the Northeast in the form of acid rain.

Congressional leaders reacted to the public outcry and demanded large reductions in emissions of sulfur dioxide from midwestern power plants. However, the cost of reducing emissions was predicted to be very high, which raised the specter of increased electricity costs and severe unemployment in the high-sulfur coal industry. The acid rain debate was therefore regionally divisive, pitting states in the Northeast against those of the Midwest and the Ohio River Valley. The debate was punctuated by scientific uncertainties concerning the scope, nature, and magnitude of the acid rain phenomenon.

Efforts to assess the causes and effects of acid rain were initiated late in the Carter administration; research was continued and expanded under the Reagan administration, which endorsed the Acid Precipitation Act of 1980 (Title VII of the Energy Security Act of 1980, P.L. 96-294), creating a ten-year research program, the National Acid Precipitation Assessment Program (NAPAP). The NAPAP research program was directed by an Interagency Task Force composed of senior officials from the Departments of Agriculture, Energy, and the Interior, the National Oceanic and Atmospheric Administration, the Environmental Protection Agency, and the White House Council on Environmental Quality. NAPAP coordinated its research, monitoring, and assessment activities with states, academia, key international acid deposition research programs, and other nonfederal research organizations.

The scientific justification for NAPAP's research program was based on numerous uncertainties associated with acid deposition causes and effects. The policy justification for NAPAP was based on the need to better characterize the acid rain problem before committing the nation to a potentially costly and regionally divisive control and mitigation regime. This improved understanding of acid precipitation and its effects was to be utilized in a predictive context, specifically to derive and evaluate alternative policy scenarios. In other words, NAPAP was a clear example of predictive science being marshalled to establish a foundation for environmental policy formulation.

Evolution of NAPAP

When considering the applicability of predictive science in environmental policy, the case of acid rain must be understood in terms of two especially salient factors: (1) NAPAP research and modeling activities were undertaken to evaluate alternative policy prescriptions for acid rain; and (2) acid rain is a multidisciplinary construct, which is to say, acid rain does not exist as an empirical phenomenon within the bounds of a single discipline. Each of these factors is discussed below.

NAPAP's Policy Context

Congress intended NAPAP to conduct research to guide policy action. The stated purposes of Title VII of the Acid Precipitation Act of 1980 were:

1. to identify the causes and sources of acid precipitation;

2. to evaluate the environmental, social and economic effects of acid precipitation; and

3. based on the results of the research program . . . [to] take action to the extent necessary and practicable (A) to limit or eliminate the identified sources of acid precipitation, and (B) to remedy or otherwise ameliorate the harmful effects which may result from acid precipitation.

Title VII thus mandated a wide variety of scientific and analytical activity, ranging from basic causal research (objective 1), to socioeconomic impact assessment (objective 2), and including applied policy analysis and a mandate for prescriptive policy development (objective 3). The possibility that such objectives might be difficult to integrate or even mutually inconsistent was not addressed. Moreover, the entire program was placed within an unforgiving ten-year time frame and placed under the joint management of five federal agencies and the White House Council on Environmental Quality. In other words, at its inception NAPAP was challenged with a compressed schedule, a ponderous management structure, and a potentially volatile blend of objectives.

Difficulties aside, the NAPAP interagency management team went through several exercises intended to articulate linkages between policy needs and the program's research objectives. The policy focus embodied in P.L. 96-294 is reflected in NAPAP's early organizational structure, which included eight scientific and technological research categories and a ninth category called Assessment and Policy Analysis. Articulated in NAPAP's initial assessment plan, the goal of the assessment and policy analysis function was to "provide an objective basis for establishing sound energy production, resource management and environmental protection policies" (NAPAP 1981). The plan stated that such information is "urgently required because: (1) there is a growing national and international concern about acid precipitation and its effects; (2) projected changes in energy policies . . . could result in the emission of more precursors of acid precipitation; (3) present information on acid precipitation is insufficient to support the development of models capable of predicting its occurrence and assessing its consequence; and (4) a number of potentially irreversible effects of acid deposition have been postulated, but if, when, and where these may occur cannot now be predicted with confidence."

NAPAP's early policy focus is also reflected in the program's first annual report, published in January 1982, which summarized NAPAP's Research Policy Focus in terms of three questions about acid precipitation: "(1) What is it? (2) What are its impacts? [and] (3) How can it be effectively managed?" (NAPAP 1982a). During the program's first annual review meeting, agency scientists and outside observers (from nongovernmental organizations, and industry) developed a list of "Task Force Outputs Which Answer Policy Questions." In this exercise, specific policy questions were formulated and specific (expected) research outputs identified that would help to answer each question. The meeting was structured around four topics, one of which was entitled "Research and Assessments to Address Policy Concerns." It is clear from the meeting report that NAPAP scientists were aware of the need to address specific policy issues and to do so quickly (NAPAP 1982b).

But NAPAP's initial policy focus was soon to dissipate. As mentioned above, NAPAP's initial organization structure included an Assessment and Policy Analysis category, funded at $1.36 million (8 percent of the NAPAP budget). By the time the 1983 Annual Report was published, the "Policy Analysis" research category had been eliminated, and replaced simply by "Assessments" (NAPAP 1983). NAPAP's 1984 Annual Report indicates a pervasive turn away from policy objectives as a source of program direction. For instance, a four-page section of the report entitled "Highlights of Program Accomplishments" does not address or even allude to the policy relevance of NAPAP's research and monitoring accomplishments; indeed, it does not even use the word "policy." Moreover, a section entitled "The Role of Science" circumscribes NAPAP's research domain and admonishes policy officials to make the "subjective judgments" necessary to determine an appropriate response to the acid precipitation issue:

> Decision makers, not researchers, must decide the level of scientific information necessary for decision making. Scientists attempt to define at any point in time what is known, with what level of certainty, and what is not yet known. . . . Some of the issues which decision makers address require making a host of subjective judgements for which scientists have no special expertise. For example, scientists have the task of relating the response of ecosystems to the amount of acid deposition they receive; but it is the role of the policy maker to determine the acceptable level of response—whether emissions should be limited further, and by how much, considering the social costs and benefits in addition to other factors. (NAPAP 1984)

Clearly inconsistent with two of the three original program directives articulated in Title VII, this statement also removes economics and policy analysis from the "appropriate" sphere of scientific activity and seeks to draw a well-defined boundary between the research and policy processes. After 1984, economics and integrated assessment virtually disappeared as program priorities, replaced by an emphasis on the primacy of basic, "curiosity-driven" science (ORB 1991). After the appointment of a new NAPAP director in 1988, the program again turned toward the development of integrated, policy-focused assessments. The plan for NAPAP's final assessment described twenty-five diverse policy and control scenarios, most of which were never formally modeled. There are two primary reasons for this: (1) NAPAP's near militant focus on disciplinary science between 1983 and 1987 made after-the-fact integration technically and practically difficult; and (2) by 1989, the Bush administration had assumed leadership for reauthorization of the Clean Air Act and discouraged policy-focused studies.

The Multidisciplinary Character of Acid Deposition

Studies conducted under NAPAP covered a broad range of disciplinary terrain, focusing on the following areas: emissions of acid precursor pollutants; atmospheric transport and chemical transformation dynamics; acid deposition regimes and air quality monitoring; effects on surface waters, aquatic life, forests, agricultural crops, building materials, and cultural artifacts; human health and visibility degradation; and economic analysis of alternative control and mitigation strategies. NAPAP also reviewed the status of acid deposition control technology development programs, such as those conducted by the Electric Power Research Institute and the Department of Energy–sponsored Clean Coal Technology Program.

Much of NAPAP's effort focused on the development of predictive models, perhaps as much as one-third to one-half of the program's total budget. NAPAP either developed or adopted thirty-nine numerical models to enhance understanding of the acidification process and its effects (NAPAP 1989b). More specifically, models were developed in almost all of NAPAP's disciplinary categories to assist in (1) the characterization of future emissions based on projected energy and economic activity, (2) the simulation of atmospheric processes leading to deposition of acidic substances, (3) the characterization of the impacts of changes in deposition or air concentrations on aquatic and terrestrial systems, (4) the characterization of biological effects on aquatic systems due to changes in water chemistry, (5) the assessment of effects on materials and visibility due to changes in deposition and air concentrations, and (6) the evaluation of the costs of control strategies and the economic benefits of effects mitigation. The predictive nature of this process was complicated by the broad time frame of acidification effects. For example, episodic surface water acidification can occur over a matter of days following a snowmelt event or heavy rain, whereas the acidification of some forest soils can take as long as a century, due to buffering by vegetation, soil, and bedrock.

As illustrated in figure 12.1, NAPAP's final assessment was conceived as an exercise in integrated modeling. Disciplinary models were to be linked for application in a stepwise process. First, output information from economic drivers was to be incorporated into an emissions model set to produce projections of alternative emissions scenarios. The emissions projections were then introduced into atmospheric models, producing deposition estimates through a system of transport, chemical transformation, and removal mechanisms. This information was to be input into various effects models to predict changes in aquatic and terrestrial systems, agricultural crops, health, materials, and visibility. Finally, output from these models was to be used to evaluate the economic benefits associated with alternative policy scenarios.

The Results of NAPAP

Most participants and observers agree that NAPAP was successful in meeting its scientific goals. This characterization is supported by at least five criteria: (1) the program funded and produced hundreds of successfully peer-reviewed articles and studies; (2) its major research components conducted regular programmatic peer review sessions, with a very high degree of positive evaluation; (3) evaluations performed by independent review bodies consistently cited high-quality science; (4) there was a near complete lack of charges of scientific fraud; and (5) program critics point to factors such as political manipulation of program reports but never claim scientific incompetence. Specific areas of scientific accomplishment are shown in table 12.1.

Despite the wealth of scientific understanding created by NAPAP, almost all retrospective evaluations of the program are critical of its lack

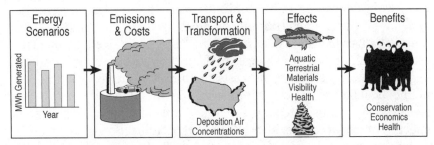

Fig. 12.1   Schematic of NAPAP model areas and linkages. (MWh = megawatt hours. Source: NAPAP 1989b)

TABLE 12.1

Scientific accomplishments of NAPAP (Cowling 1992).

Inventories	Development of the first continental-scale inventories of sulfur dioxide (SO_2), nitrogen oxides (NO_x), and volatile organic compounds (VOCs).
Trends	Development of spatial and temporal trends in the chemistry of precipitation across the United States from 1980 through 1990.
Transformations	Improved understanding of the chemical transformations of airborne SO_2, NO_x, VOCs, and their reaction products during transport from their point of origin to their ultimate deposition on soils, vegetation, surface waters, and materials.
Modeling	Development and partial validation of a prognostic, nonlinear model—the Regional Atmospheric Deposition Model (RADM).
Gradients	Description and analysis of temporal, spatial, and elevational gradients in the chemical exposure of forests in a region extending from southeastern Ontario to western North Carolina.
Status surveys	Development of methods to conduct regional surveys of the acid-base status of surface waters. These methods were applied to lakes and streams in several regions of the nation.
Linkages	Increased understanding of linkages among both current and cumulative deposition of airborne sulfur and nitrogen compounds and the interactions of those substances with: (a) the acid-neutralizing capacity and other chemical properties of bedrock and soils; (b) the acidity and especially the alkalinity of surface waters; and (c) the nature, health, and productivity of aquatic ecosystems.
Impacts	Improved understanding of the effects of sulfur and nitrogen compounds on many natural and economic resources including the following: (a) agricultural crops, (b) natural and commercial forests, (c) agricultural and forest soils, (d) aquatic ecosystems, (e) human health, (f) engineering and construction materials, (g) cultural artifacts, and (h) the aesthetic quality of scenic vistas in various regions of the country.

of direct input into the policy process (Roberts 1991). At base, much of this criticism springs from the program's failure to articulate an unambiguous policy prescription ahead of the political schedule to reauthorize the Clean Air Act Amendments. Consider the following chronology: Vice President George Bush announced his support for acid rain reduction during the presidential campaign in 1988 and proposed legislation soon after his inauguration in 1989. However, there was no formal, publicly available statement of Principal Findings from NAPAP until June 1990, when the *1989 NAPAP Annual Report* was issued (NAPAP 1989a). The Senate and House of Representatives passed their respective versions of the 1990 Clean Air Act Amendments in April 1990, and final passage of the Clean Air Act Amendments of 1990 occurred on October 22, 1990. President Bush signed the Amendments into law on November 15, 1990 (P.L. 101–549). By contrast, the initial draft of the *NAPAP Integrated Assessment* was completed in September 1990 and the document published in November 1991.

NAPAP was clearly behind schedule, at least in terms of the political acid rain debate. But there are two other ways in which it failed to meet its policy goals. In the first place, little of NAPAP's plan to conduct an integrated, predictive modeling exercise was actually realized. In most cases, NAPAP's predictive models were exercised in stand-alone disciplinary studies, integrated only in the sense that they employed deposition or air concentration estimates derived from the transport and transformation model as a basis for effects calculations. In most cases, NAPAP's models were not designed with integration in mind—they utilized discipline-specific constructs, assumptions, and variables. In many cases, models in the same discipline area (e.g., forestry effects) were not constructed to "nest" together, leading to a situation in which researchers could speak confidently about acidic deposition impacts on individual tree branches but not about impacts to whole trees, tree stands, or forests. NAPAP's early focus on disciplinary science left the program ill prepared to conduct its final, integrated assessment. In the end, NAPAP staff combined model-based information with monitoring data, findings from mechanistic and controlled exposure studies, survey information, and other research results to derive a composite picture of acid deposition causes and effects. NAPAP program managers referred to this as a "weight of evidence" approach. However, it proved exceedingly difficult to generate consensus regarding what may or may not have counted as evidence.

In the second place, NAPAP's research program did not address the policy questions that actually animated the acid rain debate of the late 1980s. For instance, NAPAP was virtually silent on issues such as the impacts of acid rain controls on regional coal mining employment (Rubin, Lave, and Morgan 1992); and it said little about the cost-benefit differences among the contending proposals for eight, ten, and twelve million tons of sulfur dioxide reduction. Likewise, NAPAP's aquatic and terrestrial effects studies provided little illumination on issues surrounding the timing, geographic dispersion, and sectoral focus of proposed control regimes. NAPAP research also failed to address issues pertinent to emissions trading, such as the problem of regional hot spots and the potential for economic savings.[2] Instead, NAPAP's *1990 Integrated Assessment* report answered questions such as: "How does acidic deposition affect red spruce at high elevation?" "How do simulated changes in ozone affect crop yields?" and "How much does light scattering by particles contribute to visibility reduction?" (NAPAP 1991). In other words, NAPAP's findings were not transparently pertinent to the policy issues being debated prior to the formulation and passage of Title IV of the Clean Air Act Amendments of 1990.

The Problem of Integration

Congress's stated objective in establishing NAPAP was to determine whether acid precipitation was a problem and, if so, what could be done about it. The problem with such a charge is that acid rain can be characterized from numerous valid perspectives: chemical, meteorological, biological, ecological, social, and others. Within each of these perspectives, further specialization is not only possible but, from a research point of view, absolutely necessary. However, technical characterizations developed under one disciplinary regime may be quite irrelevant or even substantively inconsistent with approaches utilized under another discipline. They are likely to involve different spatial and temporal scales, different chemical species, different degrees of analytical detail due to different levels of expertise or funding, different assumptions concerning nonmeasured parameters, broadly divergent (and sometimes inconsistent) modeling approaches and statistical testing procedures, and so on. Indeed, constructs and methodologies can vary substantially within disciplines. The following examples illustrate the sort of intradisciplinary pluralism that came to typify NAPAP research (Herrick and Jamieson 1995).

- Aquatic damage from acid deposition can be characterized in several ways. If damages are stated in terms of the *number of lakes affected,* then projections of decreased deposition appear to provide a substantial decrease in damages. If the same projection is expressed in terms of *percentage of affected lakes*, then the decrease in damage appears less significant. If acidity is characterized in terms of pH (the standard chemical measure of acidity) rather than acid neutralizing capacity (ANC, a different measure), then future gains would be smaller still. Moreover, the choice of a reference pH value can radically alter the number of "acidic" surface waters. Numerous decisions must be made: A national scale assessment discounts regional variation;[3] a focus on chronic acidity produces a different perspective than one including short-term episodes; analyses dealing with the current situation may inadvertently miss longer-term processes threatening future degradation; and monitoring for direct effects may not reveal possible indirect or synoptic effects. Still another consideration is whether chemical acidification has actually harmed aquatic life. All of these measures are valid, but no one (or combination) of them is intrinsically more correct than the others.

- A major impetus for acid rain control was the fear of damage to aquatic resources. NAPAP's aquatic effects research team determined that 4 percent of U.S. lakes and 8 percent of streams were acidic, with values that varied from region to region. Florida had the highest percentage of acidic surface waters (33 percent of lakes and 39 percent of streams). In the mid-Atlantic Highlands, mid-Atlantic Coastal Plain, and the Adirondacks, 6–14 percent of lakes and streams were chronically acidic; about three times that number were subject to periods of temporary acidification (days to weeks) due to storms and snowmelt conditions. Yet as soon as these facts were reported, other scientists argued that many lakes were naturally acidic because forest soils and organic decomposition produced a "normal" acidic condition, especially for smaller lakes. In other words, lakes and streams may indeed have been acidic, but not necessarily because of air pollution. Indeed, some scientists argued that the dominant theory of surface water acidification was wrong, that it failed to address natural factors controlling surface water chemistry.

If edification is one's objective, then it does not matter whether surface water acidification is characterized in terms of pH or ANC. Once the research question has been appropriately bounded and documented, the matter is settled. However, the policy context does not accommodate the "luxury" of a bounded, or disciplinary, perspective. This dynamic is especially pertinent to science-policy-assessment initiatives predicated on the development and exercise of an "authoritative" modeling capability. If the science characterizations and mechanisms underlying the models are disciplinary and, hence, bounded in scope, arguable, or otherwise subject to interpretation, then the models will inherit this basic contestability. Thus, while disciplinary boundaries are helpful in "normal" science, they present a difficulty in science-based analysis of policy issues, especially when predictive modeling is involved. Disputes tend to arise over the boundaries between disciplines, resulting in a lack of consensus concerning which rules, constructs, and methods are most relevant or "weighty" in a given situation. This, in turn, leads to disputes concerning the strength and validity of scientific evidence. Disputes such as these can bedevil the process of integrated research; in particular, such disputes can prolong the assessment process and create opportunities for policy entrepreneurs to "game" the research process, leading to the lack of closure exhibited under the NAPAP process.

Alternative Approaches

NAPAP's policy context, when combined with the multidisciplinary nature of acid rain, necessitated significant alteration of the traditional approach to academic research—an approach characterized by increasing specialization coupled with a fundamental understanding that research is serendipitous, evolutionary, and not subject to the imposition of bureaucratic management regimes. Instead, NAPAP research was intended to address policy issues and demanded an integrated perspective rather than the specialization associated with academic and basic research (although, during the period from 1983 to 1987, the program director placed a low priority on integrated assessment, economic research, and policy analysis). What is more, NAPAP called on researchers to converge upon a state-of-the-science consensus within a stipulated time frame, a factor that made it necessary to augment findings and data with informed judgment. Such a situation is by no means exceptional in the context of policy-driven research and modeling; indeed, this can and should be viewed as the normal—and generally unavoidable—state of affairs. The result is a value-laden dynamic in the core of a process presumed to be objective and value free. In other words, just as research and monitoring results become infused with informed judgment, so too will the outputs of predictive modeling exercises.

As argued above, the policy process can rarely accommodate the luxury of a disciplinary perspective on environmental issues, making it difficult to exercise an authoritative modeling capability. NAPAP was unable to address acid rain as a coherent policy problem. Instead, it remained many separate issues, each bounded by disciplinary and methodological conventions. Defining its activity in terms of disciplinary science, NAPAP lacked the nondisciplinary perspective that would have allowed it to characterize acid rain in terms of a set of policy options (Herrick and Jamieson 1995; Rubin et al. 1992). In other words, NAPAP was able to produce a "banquet" of data and findings but unable to help nonspecialist decision makers determine which "dishes" should weigh most heavily or, indeed, even be considered in the policy choice (Herrick and Jamieson 1995).

As suggested before, predictive modeling is tied to its science base. And if the bounds of the science base are arguable, this fundamental contestability will almost certainly "infect" subsequent modeling activities. Three things can help to mitigate the contestability and increase the applicability of predictive modeling in a policy context: (1) when possible, attempt to combine modeling with an adaptive management regime; (2) recognize and emphasize the heuristic (as opposed to prescriptive) value of models; and (3) make sure that basic research, modeling, and policy analysis are designed and conducted in parallel, not sequentially. These three recommendations are discussed below.

Seek Adaptive Applications for Model Outputs

One alternative would involve a combination of adaptive management, *ex post* program evaluation, and environmental monitoring. Such an approach would draw on both the theory of policy sciences (Lasswell 1971; Brunner and Ascher 1992) and the growing use of adaptive approaches to ecosystem management. Rather than predicating action on the reduction of uncertainty, the adaptive approach to environmental policy acknowledges that both our ability to craft public policy and our scientific understanding of nature are imperfect. As explained by Kai Lee (1993), perhaps the most articulate proponent of the adaptive management of ecosystems:

> [An] adaptive policy is one that is designed from the outset to test clearly formulated hypotheses about the behavior of an ecosystem being changed by human [use or intervention]. . . . If the policy succeeds, the hypothesis is affirmed. But if the policy fails, an adaptive design permits learning, so that future decisions can proceed from a better base of understanding. . . . Adaptive management is highly advantageous when policy makers face uncertainty, as they almost always do in the environmental arena.

This fact may help to illuminate an effective role for science in policy making—not to attempt to predict and dictate policy choices but to monitor the effectiveness of those choices after they have been made. If models are designed, calibrated, and applied to characterize changes associated with the imposition of specific policy regimes, and if policy regimes are viewed as temporary and experimental rather than final and authoritative, then the issue of disciplinary boundaries loses salience.

As an alternative to the large-scale, long-term research and modeling program played out under Title VII and NAPAP, Congress and the administration could have acknowledged the uncertainties associated with the acid rain debate and initiated an "experimental" series of spatially and temporally restricted control regimes. Indeed, by the mid-1980s, several states and Canadian provinces had already initiated acid rain control programs.[4] Such regimes could have been federally subsidized and monitored for comparative outcome. Regimes that appeared effective could have been replicated and ineffective approaches modified or terminated. Modeling could be very useful in planning, evaluating, and designing alterations for such an approach.

Ironically, Title IX of the Clean Air Act Amendments of 1990 did include provisions for policy evaluation, ongoing monitoring, and progress assessment. Title IX reauthorized NAPAP as a mechanism to evaluate the effectiveness of the acid rain provisions of the Clean Air Act, including the effectiveness of the emissions trading program mandated under Title IV. To their credit, congressional authors of the 1990 amendments recognized an opportunity to establish an adaptive policy regime. However, most observers agree that implementation of Title IX has been weak and sporadic. Acidic deposition monitoring networks have shrunk due to funding cutbacks, model validation activities have been postponed, and research on mechanisms has been curtailed. In short, it appears that an opportunity to use predictive models in an adaptive context is being lost.

Avoid Conflation of Heuristic and Prescriptive Prognostic Applications

Democratic government is based on the presumption that an informed citizenry is able to make sound decisions. The Acid Precipitation Act of 1980 could have mandated a program of research, education, and communication. In such a case, policy advocates would be free to draw upon scientific findings and model-based projections to develop whatever policy scenarios they saw fit. Can scientific information and predictive modeling activities add value in a policy context even if they do not address specific policy prescriptions? Most of NAPAP's critics have focused on the program's high-visibility reports and studies (Roberts 1991; Rubin et al. 1992). Few acknowledge or seriously consider the fact that NAPAP staff responded to numerous congressional queries and requests for information. Many inquiries were also received from governors and other state and local officials, NGOs, and interest organizations. In addition, NAPAP and task force staff fielded hundreds, perhaps thousands, of telephone queries and requests for information. Many of those phone conversations were with key congressional staffers. NAPAP staff were also generous in their dealings with the media, answering hundreds of questions in real time. Much to the chagrin of the task force, some of the program's data and findings actually debuted in the popular media, with a series of articles by the syndicated columnist Warren Brookes gaining particular notoriety. Far from being absent from the Clean Air Act debate, a case might be made that NAPAP findings were actually ammunition for frontline troops in the Clean Air Act debates. While NAPAP's assessments did not settle the acid rain policy debate, at least they assured that the public debate was fueled by solid information, not "junk science."

Some observers take this perspective a step further, arguing that NAPAP data and findings set the stage for the acid rain policy decisions embodied in the Clean Air Act Amendments of 1990 (Russell 1992). In the late 1970s and early 1980s some scientists and much of the public believed that acid rain posed a grave, immediate, and irreversible threat to the health and productivity of aquatic and terrestrial ecosystems. By contrast, other scientists and some of the public believed that air pollution was predominantly an urban problem, that the acids present in precipitation were much too dilute to have significant effects on any but the most sensitive lakes and streams, and that acid rain was unlikely to affect crops or forests. Given this polarization and lack of a defining majority on either side, political action was probably impossible. Even if an acid rain control policy could have been forced through Congress, the scars would have made compliance difficult and tainted the atmosphere for cooperation on other environmental issues.

By the time NAPAP concluded its decade of research, these extreme views had converged into a more moderate middle ground. The flow of information from NAPAP may thus have educated the public and decision makers throughout the program's life. By the time NAPAP's final assessment reports were released, a large proportion of the U.S. population was aware of the issue, the poles of the debate had collapsed, and parties on all sides recognized that there was enough merit on the other sides that a compromise solution was both possible and legitimate. Under that perspective, predictive modeling may have provided value not so much through the accurate portrayal of future events and conditions, but as a heuristic device to help highlight issues and areas of scientific congruence, disagreement, and uncertainty.

In this sense, it could be argued that NAPAP was quite successful, setting the stage for a political solution to the acid rain debate. However, we should not let the fact that NAPAP produced and disseminated a great deal of valuable information distract us from the program's failure to provide and evaluate solutions to the acid rain policy debate. Congress did not ask NAPAP merely to set the stage, it asked NAPAP to help cast, direct, and produce the play. Through NAPAP, Congress invited the science community into the policy-making process; research was funded not simply to broaden and deepen the acid rain knowledge base but to "provide . . . a basis for policy" (NAPAP 1981). Predictive modeling may be used in a prescriptive policy context, and it may also be used as a heuristic device to facilitate education and communication. Models can be used in either context, but care should be taken to assure that results and applications are not conflated.

Create Appropriate Organizational Structure and Managerial Approach

As mentioned earlier, NAPAP was managed through a cumbersome interagency mechanism consisting of six "leading agencies," each with a different mission, focus, and basis of research expertise. This organizational structure had the constructive effect of coordinating the research activities of the different federal agencies. The program's research funding, workshops, and joint publications helped create an "epistemic community" surrounding the acid rain issue, and this in turn helped to nourish the dialogue that led eventually to a policy solution. However, NAPAP had no true "lead agency" with both budgetary and administrative authority over program activities. Instead, the program evolved a consensus management approach that required all six lead agencies to approve NAPAP's reports, testimony, research plans, and other official documentation prior to their release to Congress and the public.[5] It is argued that this led to compromise solutions in which research and assessment results were frequently watered down to suit the differing positions and missions of the lead agencies and their stakeholders (Cowling 1992).

While NAPAP may indeed have published compromise solutions, I believe it is a mistake to pin this on the dilution of research results. Such an argument presumes that NAPAP's findings were politically salient, but to be blunt, that was not the case. As I have already argued, NAPAP did not address many of the policy questions that animated debate over reauthorization of the Clean Air Act. Instead, NAPAP remained hunkered in mainstream, disciplinary science, answering questions that were restricted in scope (e.g., "What is the effect of acidic deposition on

high-elevation red spruce?"), confined to specific disciplinary para-
digms, and not readily amenable to application in a broader policy
environment.

With the exceptions of the White House Council on Evironmen-
tal Quality (CEQ) and EPA, NAPAP was run by old-line science
agencies. Those agencies never demanded that NAPAP break out of
its disciplinary research program and pursue integrated assessments
of alternative policy scenarios. This suggests the need for an alternative organizational structure or management approach. Specifically, a new institutional arrangement may be necessary to help
create and sustain a match between social issues, the research
agenda, and the methodological needs of policy analysis. Apart
from the (now defunct) Office of Technology Assessment (OTA), the
federal environmental regime includes no institution for the ongoing
integration of "what policy analysts want to accomplish and what
researchers can reasonably achieve" (Sarewitz 1996). I am suggesting that the policy application of predictive models might best be
accomplished through an organization separate from those that
conduct the scientific research on a particular issue. Such an
arrangement would create an organization with a clear responsibility for policy analysis, an organization with no wiggle room to claim
partial success for broadening the scientific knowledge base with its
results, even if they weren't policy relevant. Perhaps such an organization could "commission" research agencies to develop state-of-
the-science reports, use delivery orders to specify the models to be
built and their capabilities, and maintain channels of constructive
negotiation with the science community over the timing and plausibility of specific tasks. Under such an arrangement, scientific
research, model development, and policy analysis would be conducted in parallel, not sequentially. As demonstrated by NAPAP, the
typical sequential application—basic research, followed by model
development and integration, followed by policy analysis—allows
too few opportunities for policy analysts to steer and influence
research and prediction activities.

Notes

1. Acid rain was discovered originally in England in the latter decades of the
   nineteenth century. However, it did not become a policy issue until 1967,
   when the Swedish scientist Svante Oden and a scientific committee

appointed by the Parliament of Sweden succeeded in calling the world's attention to the phenomenon at the United Nation's Conference on the Human Environment. Norway was the first European country to initiate a national program of research. Much of what was learned about the phenomenon and effects of acidification was discovered through European, Scandinavian, and Canadian national research programs.

2. The Clean Air Act Amendments of 1990 instituted an innovative "cap and trade" system under which total SO_2 emissions were limited, but individual sources were offered the option of either controlling emissions or purchasing excess emissions allocations under the trading program. During congressional debates over the acid rain provisions of the act, it was feared that strong emissions gradients, or hot spots, would result if sources in close geographic proximity to one another decided to purchase excess allotments rather than opting for SO_2 controls. It was argued that such a scenario was plausible even if the overall cap was not exceeded.

3. Surface water acidity depends on many factors, including the pH of precipitation, the chemical composition of watershed soils and bedrock, the flushing rate of a water body, the presence of nondeposition sources of acidity (e.g., organic acidity or mine drainage), and a variety of other watershed-specific factors.

4. These states include Wisconsin, Minnesota, New York, and Massachusetts.

5. In the case of congressional testimony or response to congressional queries, Office of Management and Budget approval was also required.

References

Brunner, R., and W. Ascher. 1992. Science and social responsibility. *Policy Sciences* 25: 295–331.

Cowling, E. 1992. The performance and legacy of NAPAP. *Ecological Applications* 2(2):111–116.

Herrick, C., and D. Jamieson. 1995. The social construction of acid rain: Some implications for science/policy assessment. *Global Environmental Change* 5(2): 105–112.

Lasswell, H.D. 1971. *A pre-view of policy sciences*. New York: American Elsevier.

Lee, K.N. 1993. *Compass and gyroscope: Integrating science and politics for the environment*. Washington, DC: Island Press.

NAPAP. 1981. *National Acid Precipitation Assessment Plan*. Washington, DC: Interagency Task Force on Acid Precipitation.

NAPAP. 1982a. *Annual report: National Acid Precipitation Assessment Program*. Washington, DC: National Acid Precipitation Assessment Program.

NAPAP. 1982b. *Report of the first annual meeting of NAPAP*. Washington, DC: National Acid Precipitation Assessment Program.

NAPAP. 1983. *Annual report: National Acid Precipitation Assessment Program*. Washington, DC: National Acid Precipitation Assessment Program.

NAPAP. 1984. *Annual report: Interagency Task Force on Acid Precipitation*. Washington, DC: Executive Office of the President Publications.

NAPAP. 1989a. *Annual report: National Acid Precipitation Assessment Program*. Washington, DC: National Acid Precipitation Assessment Program.

NAPAP. 1989b. *NAPAP Assessment Plan update*. Washington, DC: National Acid Precipitation Assessment Program.

NAPAP. 1991. *1990 Integrated Assessment*. Washington, DC: National Acid Precipitation Assessment Program.

ORB. 1991. *The experience and legacy of NAPAP*. Washington, DC: Oversight Review Board for the National Acid Precipitation Assessment Program.

Roberts, L. 1991. Learning from an acid rain program. *Science* 251: 1302.

Rubin, E., L. Lave, and M. Morgan. 1992. Keeping climate research relevant. *Issues in Science and Technology* 13(2): 47–55.

Russell, M. 1992. Lessons from NAPAP. *Ecological Applications* 2(2): 107–110.

Sarewitz, D. 1996. *Frontiers of illusion: Science, technology, and the politics of progress*. Philadelphia: Temple University Press.

Prediction and Other Approaches to Climate Change Policy

Steve Rayner

Climate Change, Policy, and Uncertainty

Science provides abundant reasons policy makers and the public *ought* to be concerned about climate change (e.g., Watson, Zinyowera, and Moss 1996; OTA 1993; NAS 1992). Although some outcomes of climate change are thought to be potentially positive, such as carbon fertilization of plants or ice-free Baltic ports, public and scientific attention are mostly focused on the potential negative impacts of temperature and precipitation changes on human populations and the ecological systems on which humans depend. Examples of the former include direct impacts such as more floods and increased energy demand for heating and cooling, as well as indirect impacts such as changes in nutrition due to agricultural shortages and new patterns of vector-borne diseases. Ecological impacts could result from changes in the duration and timing of growing seasons, the availability of fresh water supplies, and sealevel rise.[1]

The extensive scientific literature is supplemented by an equally voluminous corpus of economic analysis suggesting that policies designed to mitigate the onset of climate change would either enhance or inhibit global productivity (e.g., Repetto and Austin 1997; Pearce et al. 1996; Fankhauser 1995). Whatever the benefits and costs of climate change, they are likely to be unevenly distributed geographically and socioeconomically, with the greatest burden falling on poor people living in vulnerable regions. Hence, differential impacts on development and issues of international and intergenerational equity have recently come to the fore (e.g., Toth 1999). In both the scientific and economic literatures, uncertainty about the future is a pervasive issue.

Uncertainty is compounded by the active role played by so-called skeptics. For example, some governments, such as that of Saudi Arabia, fearing the impact of reductions in fossil fuel use on their economies, have expressed doubt about the scientific basis for climate change concerns. Such doubts are supported by certain industrial lobbies, including those of U.S. coal companies. Scientists who support such skepticism (for example, Seitz, Jastrow, and Nierenberg 1989; Balling 1992; Michaels 1992) are often reviled by climate modelers.[2]

In the midst of uncertainty about the consequences of climate change, policy makers have sought to act. The 1992 United Nations Framework Convention on Climate Change (FCCC) articulates as its goal—its "ultimate objective"—"to achieve, in accordance with the relevant provisions of the Convention, stabilization of greenhouse gas concentrations in the atmosphere that would prevent dangerous anthropogenic interference with the climate system." While the language of the convention indicates widespread acceptance of the idea that human tinkering with the climate is undesirable, it provides little insight into what concentrations should be considered "dangerous" or to whom or what danger is presented. Popular culture (e.g., the movie *Waterworld*) and even the "serious press" have presented catastrophic climate change images that suggest that human life as we know it would be changed beyond recognition.[3] But most scientists and policy makers consider these doomsday scenarios unlikely. Scenarios for the twenty-first century, as projected by the Intergovernmental Panel on Climate Change, suggest that climate change is most likely to exacerbate the challenge facing human and natural populations that are already existing in marginally sustainable conditions. However, climate is far from the determining factor in the fate of those populations. Poverty, urbanization, unsustainable resource management strategies, and so forth are at least as much to blame for the dangers that already confront the poor.

Danger is not solely a function of the state of the climate, but of the ability of a population and environment to respond to that state. Hence, the causes and distribution of climate change are not the only sources of policy uncertainty inherent in the goals of the convention. Yet the FCCC, and Conferences of Parties to the FCCC that have followed, have focused policy efforts exclusively on emissions reduction policies and very little (if at all) on improving the resilience of populations through measures designed to help human and natural systems adapt to climate change.[4] Uncertainty thus permeates the climate change issue at every level, from atmospheric dynamics to societal impacts to human response. The conventional scientific response to uncertainty is to try to reduce it through more research. Among policy makers and corporate decision makers (particularly in the United States), this response has been encouraged by a strong demand for "accurate" climate predictions. At the same time, uncertainty about the consequences of action is usually invoked in the political sphere to support the status quo. Thus, the role of prediction and its attendant uncertainties has been central to the relationship between science and policy on the climate change issue. This chapter argues that predictive modeling, which is often portrayed as the necessary foundation for action on climate change, in fact provides an insufficient basis for sustainable policies related to species loss, habitat degradation, declines in human health, and loss of human lives. This suggests the need for policy makers (and scientists) to focus more attention on research and policies that do not depend so heavily on prediction for their success.

The Role of Prediction in Climate Change Science and Policy

Scientific interest in the role of the atmosphere in maintaining the surface temperature of the earth can be traced back to the work of Fourier in the first half of the nineteenth century. Arrhenius (1896) explicitly linked the combustion of fossil fuels to atmospheric CO_2 and calculated an expected value for the mean annual temperature rise at various latitudes that would result from an expected doubling of atmospheric CO_2—a scenario that is still used as a benchmark by contemporary climate modelers. Without even a pocket calculator, Arrhenius arrived at figures only a couple of degrees higher than what we have been able to achieve with the latest Cray supercomputers. Arrhenius's apparent success on the back of an envelope does not begin to reflect the enormous increase in scientific knowledge about climate over the past century, because his result was partly the outcome of compensating errors in his calculation. However, his work is politically interesting, in that it suggests a significant increase in societal sensitivity to reasonably consistent scientific predictions of climate change over the last hundred years.

In the middle of the twentieth century, Callendar (1940) revisited the issue of global air temperature increases resulting from industrial emissions of carbon dioxide. However, the scarcity of reliable measurements of atmospheric concentrations and uncertainty about the absorptive capacity of the oceans made it difficult for scientists to determine whether there were long-term trends in atmospheric CO_2 concentrations. The first reliable time-series measurements of atmospheric CO_2 were begun as a result of the efforts of Roger Revelle, who had demonstrated that ocean uptake of CO_2 was a very slow process and that CO_2

would accumulate in the atmosphere (Revelle and Suess 1957). Revelle persuaded the United States to begin continuous monitoring of CO_2 concentrations at Mauna Loa (Hawaii) as part of its contributions to the International Geophysical Year of 1957–58 (Keeling 1960). Thus, in recent decades, scientists have obtained direct measurements of rapidly rising CO_2 concentrations in the atmosphere, as well as ice-core records indicating that those increases are indeed part of a trend dating from the Industrial Revolution.

At the end of the twentieth century, prediction in climate change science and policy is inextricably tied to computer model simulations. General circulation models (GCMs) that atmospheric scientists use to explore the global climate lie at the heart of climate prediction. Manabe and Wetherald (1967) produced the first modern analysis of the heat balance of the atmosphere and its responses to anthropogenic changes in CO_2, water vapor, and other factors affecting temperature. Meanwhile, early models of the ocean carbon cycle were refined and, beginning in the late 1970s, were coupled to models of the atmosphere and the climate. These models have been developed to explore additional climate factors, such as soil moisture, and to incorporate feedback mechanisms that can alter the sensitivity of the climate to changes in greenhouse gas concentrations.

The number of climate models grew rapidly in the late 1980s, from about a half dozen built and operated in the United States and Britain. The Atmospheric Model Intercomparison Project, published in 1992, listed twenty-seven atmospheric GCMs originating in the United States, Europe, Asia, and Australia (Gates 1992).

GCMs have been the focus of most of the past decade's political controversy about the extent, and even the reality, of climate change. The underlying philosophy of national and international research and assessment programs such as the $2-billion per-year United States Global Change Research Program (USGCRP) and the Intergovernmental Panel on Climate Change (IPCC) has been "getting the science right," on the assumption (explicitly stated by President Bush in 1990, and annually thereafter in the USGCRP budget reports) that we cannot develop sound policy without substantially reducing scientific uncertainty about the future climate via an enhanced understanding of basic earth system science processes.

This idea has been elaborated as the "cascade of uncertainty," the notion that uncertainties inherent in our understanding of basic earth systems are exacerbated by uncertainties over emissions (see figure 13.1). In turn, this situation makes anticipation of impacts even more difficult to determine, especially in the context of global socioeconomic

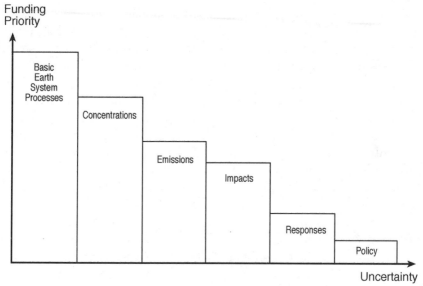

Fig. 13.1   "The cascade of uncertainty."

uncertainty and uncertainty about how people, communities, nations, and markets will respond to such impacts. The cascade or accumulation of uncertainty has been invoked to justify a wait-and-see approach to climate policy, as well as to promote the earth sciences research agenda. Some have argued that cascading uncertainties are an impediment to sound policy making, and others have argued that they justify stringent precautionary measures. These opposing viewpoints each serve to elevate the role of GCMs in climate policy debates.

GCMs: The Quest for Prediction

Since 1990, the field of coupled climate modeling, i.e., linking three-dimensional representations of the atmosphere, ocean, ice caps, and land surface, has developed rapidly. Such models are evaluated both by assessing how well they "predict" past climates and by running them alongside each other to identify convergences, divergences, and overall coherence. While such evaluations lead modelers to profess confidence in their simulations, they readily acknowledge that the ability of current atmospheric models to simulate the observed climate varies with scale and variable, e.g., temperature, precipitation, or cloud cover (e.g., Gates et al. 1996).

Modelers generally assume that erroneous results can be corrected by comparing specific model predictions about climate with the actual climate. A common validation technique is to run climate models "backwards" to see if they can reproduce past climate conditions. But there are some practical obstacles to this method of validation. The observational data to which the model is compared are themselves informed by theoretical commitments and assumptions that may not be transparent to the modeler or the user of model-based information. As Jasanoff and Wynne (1998) note, the reconstruction of past climates is an act of "heterogeneous archaeology" that combines analysis of physical data (e.g., of polar ice cores, fossilized pollen, tree rings, and the like) with at least superficially incommensurable social artifacts, such as parish records. Translating such data and artifacts into measures of past climate, in turn, depends heavily upon theory and inference.

At least some of the data used to validate climate models are themselves model outputs. GCMs are one of the important resources used to construct time-series data sets of past climates accessible only through the kind of tentative, incomplete, and scattered proxies described above. Moreover, the past climate, which models are expected to reproduce, is defined by data such as sea-surface temperatures, pressures, and precipitation. But such data themselves reflect implicit choices of what is important to measure, how and where data are collected, what standardization methods are used, and so forth. Thus, the data sets used as the standard against which a model's performance is evaluated may not be entirely independent of the model or, at least, of the assumptions shaping it.

Model validation thus contains an inevitable element of circularity. Moreover, an important difference between climate predictions and, say, weather predictions is that climate predictions of the distant future cannot be evaluated in terms of what transpires. Thus, policy makers and other users of climate models must necessarily rely on unverifiable evaluation techniques.

The stability of some of the model-based projections may be due as much to institutional factors influencing the modelers as to properties of the models themselves (for examples of this phenomenon, see chapters 8 and 10). It has been observed of models generally that they are creative endeavors in which some properties ascribed to objects will be genuine properties of the objects modeled, but others will be merely properties of convenience (Cartwright 1983) or of necessity (Jasanoff and Wynne 1998). One example, which has been explicitly recognized by the IPCC is the practice of "flux adjustment." The coupling of ocean and atmosphere models highlighted the situation in which very small discrepancies in the surface fluxes (ocean-atmosphere heat exchanges) caused models to drift away from the observed climate.[5] Some researchers have intervened in the models by adding a correction or adjustment to modify fluxes before they are imposed on the ocean. "The term 'flux adjustment' is not meant to imply a knowledge of the 'right' answer for the fluxes, since they are only imprecisely known" (Gates et al. 1996, p. 237). In other words, they are "guesstimates" derived using the craft skills of the modeler. There is nothing inherently wrong with this practice, and the IPCC found that "there is no evidence . . . that the use of flux adjustments *per se* is substantially distorting the response to increases in greenhouse gases" (Kattenberg et al. 1996, p. 311). However, their presence is not necessarily transparent to users of modeling information outside of the modeling community itself. Other modelers have chosen not to make flux adjustments and thus to accept the resultant drift. Recently, several models that run without flux adjustment have yielded good portrayals of current climate without flux imbalances. The debate over whether or not to adjust may thus fade into the history of science. However, the flux adjustment decision remains a clear example of how expert judgment may play an important role in the modeling process that is not easily visible to policy-oriented users of modeling information.

A second example of possible institutional factors underlying the stability of climate science concerns the range of global average temperature changes for a doubling of atmospheric carbon dioxide concentrations from preindustrial levels. The range usually given is 1.5–4.5°C. These figures are remarkably close (within 2°C) of Arrhenius's original back-of-the-envelope calculations over a century ago. They have remained noticeably stable throughout the past decade, which has seen some fundamental changes in modeling approaches and scientific understanding. If this estimate were a product of the GCMs, it would be reasonable to expect a quantified probability distribution across the range, with a most likely value (say, 2.5°C) somewhere in the middle, tailing off toward low values at the extremes. In practice, however, the range is not derived deterministically from the formal models but is the result of diffuse expert judgment and negotiation among climate modelers (van der Sluijs 1997; Shackley et al. 1998).

One GCM expert involved in the IPCC process observed (Jasanoff and Wynne 1998, p. 70):

> What they were very keen for us to do at IPCC, and modelers refused and we didn't do it, was to say we've got this range 1.5–4.5 degrees, what are the probability limits of that? You can't do it. It's not the same as an experimental error. The range is nothing to do with probability—it is not a normal distribution or a skewed distribution. Who knows what it is?

Commentators disagree about the extent to which the negotiated stability of the projected temperature range represents a consensus about the scientific credibility of the values (S. Schneider, Stanford University, personal communication, January 1998) or a more hybrid consensus that takes account of what policy makers would find credible (Jasanoff and Wynne 1998). In either case, the emergent stability has helped to domesticate climate change as a seemingly manageable problem for both science and policy.

The chair of the scientific working group of IPCC, Britain's Sir John Houghton, reflected on the need for pragmatic limits on the framing of scientific forecasting when he observed:

> There are those who home [in] on surprises as their main argument for action. I think that this is a weak case. No politician can be expected to take on board the unlikely though possible event of disintegration of the West Antarctic ice sheet. What the IPCC scientists have been doing is providing a best estimate of future climate under increased greenhouse gases—rather like a weather forecast is a best estimate. Within the range of possibility no change of climate is very unlikely. Sensible planning I would argue needs to be based on the best estimate, not on fear of global catastrophe or collapse. (Jasanoff and Wynne 1998, p. 71)

Thus, what may appear to be the natural approach to producing climate knowledge is a complex exercise in which scientific judgment interacts with policy makers' needs for sensible and usable planning instruments.

Clearly, the goals of climate modelers and the expectations of policy makers converge. Both seek more accurate prediction on a finer scale. But while the scientific community is rigorous in its attempts to deal with explicit scientific uncertainties and to communicate them to policy makers, policy makers continue to operate with an unrealistic expectation of scientific capabilities. Furthermore, the nature of what we refer to as "uncertainty" often lies outside of the GCM modeling framework—for example, inherently unpredictable thresholds for rapid climate destabilization, as well as unpredictable extrinsic effects, such as volcanoes—and remains diverse and problematic. This is true not only of GCMs, but of the whole suite of earth systems and ecological models that are used for predicting hydrological, ecological, agricultural, and other impacts of climate change on natural systems that support human life on earth (see Oreskes, Schrader-Frechette, and Belitz 1994).

Climate Change Damage Cost Estimates

If GCMs are conventionally considered the "front end" of climate prediction, then the "back end" might be the prediction, using economic models, of costs and benefits from climate change and its mitigation.

Even the best predictive models of climate and other earth systems processes do not provide policy makers with information that they can readily use. The outputs of such models are expressed in biophysical units such as tons of forest biomass or wheat per hectare. Economic analyses attempt to translate diverse biophysical impacts of climate change into monetary terms.

IPCC economists recognize that "[the] level of sophistication in socioeconomic assessments of climate change assessments is still rather modest. Damage estimates are tentative and based on a number of simplifying assumptions" (Pearce et al. 1996, p. 183).

In most impact studies, the main variable that drives the impact functions is the globally averaged change in annual surface temperature for an equilibrium doubling of the preindustrial CO_2 concentration equivalent of greenhouse gases. Nonmarket damages are estimated using willingness-to-pay measures that proved particularly controversial when applied to the costs of human life (Pearce et al. 1996). According to the IPCC, best-guess estimates of annual worldwide damage costs on this basis range from 1.5 to 2 percent of world GNP. However, Fankhauser (1995, p. 54) writes that predictions of economic impact:

> are of course far from exact and one should allow for a range of error of probably at least +/- 50 percent. We should also remember that several greenhouse impacts have not been quantified. These are probably predominantly harmful, with the possible exception of climate amenity. Overall, the results are thus clearly in the upper quadrant of the . . . range of 0.25 percent to 2 percent of GNP. *A more reasonable range is probably 1 percent to 2 percent of world GNP,* at least for developed countries and the world as a whole.

The original emphasis in this passage tells us a lot about the state of the art in economic damage assessment. Whatever the model results, the central estimates seem to be stabilized by the expert judgment of the community, in a fashion not dissimilar to the central estimates of global temperature rise due to doubling of CO_2. That is to say, despite the technical apparatus of prediction, the central estimate is no more or less than a consensual, subjective judgment.

Furthermore, just as the global average equilibrium temperature rise for doubled CO_2 is an arbitrary artifact from a scientific standpoint—and a largely meaningless one in terms of the real world, in that neither people nor ecosystems actually experience global average temperature—its aggregate impact on world GNP is equally "demonstrably unimportant in leading to actual impacts" (Rotmans and Dowlatabadi

1998, p. 311). The regional variation in predicted damage is substantial. While the worldwide central estimates of 1–2 percent are believed to be typical for developed countries, for less industrialized nations the range expands from a minimum of 2 percent to a maximum of 9 percent—even more if alternatives to willingness to pay are used to estimate nonmarket impacts, particularly the value of a statistical life (that is, the potential costs that can arise from protecting against an increasing risk of mortality; Pearce et al. 1996).

More detailed treatment of market impacts requires disaggregation by sector. Usually, climate-dependent sectors such as agriculture, coastal defense, forestry, and water resources, as well as energy and transport are chosen as the focus of study. However, researchers' knowledge about even the *existing* climate impacts on these sectors is severely deficient. The uncertainties are enormous, and we know only that climate impacts will be part of other social, economic, and environmental changes that may influence society. For example, uncertainties in population projections have at least the same influence on estimates of world food supply as climate change uncertainties (Toth 1994).

Prediction of the costs of *preventing* climate change through greenhouse gas emissions reductions is based on two kinds of economic models. One kind, referred to as top-down, includes aggregate models of the entire economy. Top-down models are macroeconomic models based on statistical observations of past behavior. Bottom-up models examine the technological options for energy savings and fuel switching that are available in various sectors of the economy. In contrast to top-down models, in which the scope for substituting technologies is based on past experience, bottom-up models estimate substitution potential on the basis of the actual technologies that individuals and firms could profitably adopt at various price levels.

The predictions of various models for the costs of climate change mitigation policies diverge rather spectacularly. Some suggest that stabilizing CO_2 emissions at 1990 levels could require a tax of up to $430 per ton of carbon by 2030 and impose total costs of up to 2.5 percent of annual GDP (Charles River Associates 1997), while others predict similar goals could be reached with much smaller taxes and negligible, even beneficial, impacts on the overall economy (Gaskins and Weyant 1993).

A comparison of 162 runs of sixteen leading mitigation cost models revealed that they are universally sensitive to a handful of key structural features and assumptions (Repetto and Austin 1997). These are:

- whether the model assumes that the economy adjusts efficiently in the long run or suffers persistent transitional inefficiencies;

- the assumed scope for interfuel and product substitution;

- whether the model assumes a backstop nonfossil energy source is available at some constant cost;

- the time frame in which the reductions are to be achieved;

- whether the model assumes that costs from climate change impacts will be avoided;

- whether reductions in fossil fuel combustion reduce other damages from air pollution;

- whether tax revenues are returned to the economy through reduction of a distorting tax rate or lump-sum rebates; and

- whether the model assumes that international burden-sharing options are available.

Repetto and Austin (1997, pp. 13–14) conclude:

> Surprisingly, these eight assumptions (along with the size of the CO_2 emissions reduction) account for fully 80% of the variation in the predicted economic impacts. This is remarkable because it assumes that all other modeling assumptions—hundreds of assumed parameter values and relationships—are comparatively unimportant.

That is the good news. The bad news is that knowing the sensitivity of the model does not reduce uncertainty and that potential variability for the eight critical assumptions remains very high indeed. Any apparent consensus in the models "does *not* imply that their predictions are accurate but only that most modeling exercises have employed similar assumptions" (Repetto and Austin 1997, p. 11, original emphasis). If the search for accurate predictive capability in global climate modeling represents a major challenge to human scientific ingenuity, it is nevertheless one that many scientists deem to be feasible. By comparison, any desire for accuracy in long-term (100-year) economic prediction seems cruelly misplaced. And, in fact, sophisticated practitioners actually disown the goal of accurate prediction in favor of using models as heuristic tools for organizing inquiry, identifying interdependencies, and developing a better overall understanding of complex issues (Rotmans and Dowlatabadi 1998). This is especially the case in the field of integrated assessment.

From Prognosis to Diagnosis: Approaches to Integrated Assessment
In recent years, scientists have sought to link computer models of a wide range of relevant phenomena into so-called integrated assessment (IA) models (Rotmans and Dowlatabadi 1998). Integrated assessment is defined as "an interdisciplinary process of combining, interpreting, and communicating knowledge from diverse scientific disciplines in such a way that the whole set of cause-effect interactions of a problem can be viewed from a synoptic perspective" (Rotmans and Dowlatadbadi 1998, p. 292). According to the IPCC (Watson et al. 1996), the component models within IA may (but do not invariably) include:

- general circulation models (GCMs), which are designed to simulate climate dynamics and to predict patterns of change in temperature and precipitation over long time periods;

- greenhouse gas emissions models, which predict global emissions scenarios resulting from assumptions about economic activity and technology, including economic impacts arising from climate change and from policies designed to reduce emissions;

- carbon cycle models, which simulate the fate of emissions in the atmosphere, oceans, and terrestrial biosphere; and

- ecological models that simulate the responses of managed and unmanaged ecosystems, agriculture, etc., to predicted changes in temperature, precipitation, soil moisture, atmospheric carbon concentrations, and so on (see figure 13.2).

IA modelers are often at pains to point out that their goal is diagnostic rather than prognostic. That is to say, IA models should be regarded as a means for ordering information and guiding inquiry rather than as predictive "truth machines." However, it is fair to note that such careful caveats about the scope and purpose of IA models tend to melt into the background when both IA practitioners and users confront the apparent but misplaced concreteness of tables and graphs representing various model runs.

Furthermore, not all of the model makers whose work may be incorporated in IA frameworks share such modest aspirations. Although, currently, predictability decreases as finer scales are modeled, climate

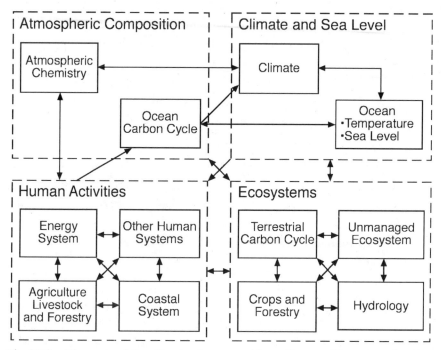

Fig. 13.2   Elements of the climate change problem. (Source: Watson et al. 1996)

modelers, for the most part, do present their ultimate goal as developing a suite of simulation models that will predict such factors as temperature and precipitation at increasingly finer scale, over longer periods, and with greater precision than has been possible to date.

> Major improvements in climate model simulation and projection come from increasing spatial resolution, that is representing more details in space, and from improving process parameterizations that describe small scale dynamic and thermodynamic processes. Better projections of climate will result from ensembles composed of multiple simulation runs. (ACPI 1998, p. 3)

Indeed, at its best, integrated assessment is an iterative process in which scientific insights are communicated to decision makers while decision makers' experiences and needs simultaneously inform scientific assessment. "This complex, intuitive and value-laden process operates at a variety of levels and scales, so researchers cannot address the process by only one, unique approach" (Rotmans and Dowlatabadi

1998 p. 294). Diverse approaches are required, ranging from formal and experimental methods such as models, to heuristic and exploratory methods such as expert judgment or policy exercises.

Current integrated assessment techniques include, in rough order of increasing formality:

- qualitative assessments based on limited, heterogeneous data sets without formal models;

- scenario development for exploring a variety of possible images of the future;

- simulation gaming, which represents a complex system by a simpler one with relevant behavioral similarity, and represents complex decision management or policy situations through role-playing by human participants; and

- computer-aided integrated assessment models that analyze the behavior of complex systems, reveal interrelationships and feedbacks, make uncertainties (and their accumulation) explicit, and compare the implications of alternative policy strategies.

Of these four methods, modeling again receives the most attention from both scientists and policy makers. Integrated assessment models necessarily consist of simpler versions (metamodels, also called reduced-form models) of the more complex or expert models that describe each domain of the interlinked climate-ecology-society system and that have previously been tested, calibrated, validated, and documented in the literature. In general, the interpretative and instructive value of an integrated assessment model is far more important than its very limited predictive capability.

Rotmans and Dowlatabadi (1998) provide an extensive catalogue of the sources of predictive limitations in integrated assessment. I will focus on three of these: limited validation, the absence of stochastic behavior, and the treatment of uncertainty in the models.

Validation of integrated assessment models of the global system is severely limited because available data are hardly sufficient to adequately characterize the processes being modeled, and because once these data have been used to calibrate models, there is insufficient additional information to conduct meaningful validation. At present, the best practitioners have used their models in backcasting exercises to parameterize and calibrate key processes. But there can be no guarantee that historically validated models will continue to apply in the future (Oreskes et al. 1994). Furthermore, such models offer no guidance for dealing with phenomena outside the bounds of historical experience.

Stochastic behavior is not dealt with in the models. Climate itself is a stochastic process, with significant impacts arising when extreme (i.e., rare) weather events are experienced. In addition, technological innovations and social movements are strongly stochastic. But most integrated assessment models are developed using a continuous formulation of the underlying processes. This is rarely the case in the real world.

Several attempts have been made to classify the different types and sources of uncertainties in models (cf. chapter 3). For example, Morgan and Henrion (1990) distinguish three kinds of uncertainty: uncertainty about empirical quantities, uncertainty about the functional form of models, and disagreements among experts. An alternative classification is the distinction of Funtowicz and Ravetz (1985) between technical uncertainties (e.g., as found in different measurement techniques, such as surface versus satellite temperature records), methodological uncertainties (the right choice of analytical tools), and epistemological uncertainties (the subjective conception of a phenomenon). But despite the important implications of these distinctions for decision making, they have not been incorporated into integrated assessment models.

In other words, disagreement among experts arises not simply from different technical interpretations of the same available scientific evidence, but also from different values and perspectives brought to the problem by decision makers portrayed in the models—as well as the values and perspectives of the modelers themselves. With the possible exception of targets (van Asselt et al. 1995, Janssen 1996), Currently available methods are unable to make uncertainties associated with disagreement and subjectivity explicit within models. Yet assumptions about values are especially important for integrated assessment. For example, different attitudes about the future may lead to the selection of high, low, or even negative discounting for long-term environmental damages (Rayner, Thompson, and Malone 1999).

Despite this diversity and the resultant uncertainty about the future that results from it, the conventional approach to creating future scenarios for integrated assessment modeling is to extrapolate from the present to posit a future that is more of the same. The future world of the IPCC First and Second Assessment Reports (IPCC 1990; Houghton et al. 1996) is essentially today's world but more so: more people, more economic growth, and more technology (although largely of the same sort). History suggests that such assumptions are unrealistic. An analyst or decision maker at the turn of the last century would have been hard pressed to envisage even the broad outlines of the changes in technological capacity and its distribution over the succeeding one hundred years. Indeed, a late nineteenth-century British parliamentarian expressed his concern that at the prevailing rate of growth of emissions, London would be buried several feet deep in horse manure by the 1950s (Rayner and Malone 1998).

The rapid rate of socioeconomic and technical change relative to climate change contrasts with the slower background rate of change of the natural world. Ecologists frequently warn us that it is not so much the amount of climate change that is dangerous but the possibility that it will occur faster than the rate at which ecosystems can adapt. On the other hand, society is changing at an accelerating, if uneven, rate. The implications of the rate of climate change for society may therefore be quite different from its implications for unmanaged ecosystems. Not only may societies adapt to climate impacts, but technological change may lead to a more rapid displacement of fossil fuels than is conceivable today (Rayner and Malone 1998). Although there is no simple technological fix, there can be no fix without technological change. The problem is that *there is no way of telling today* whether rapid social and technological change will prove to be a saving grace or yet another factor compounding the challenge of global environmental governance.

Although scientifically important efforts are currently underway to effect qualitative improvements in climate models, there is no practical likelihood that models of the remaining strongly stochastic elements of the climate-ecology-society system will ever produce reliable long-term predictions. Whatever improvements in understanding we can derive from integrated assessment models—and they may well be considerable—their *predictive* performance, like the sound reproduction of stereo systems, will never exceed that of their weakest components. In any case, for long-term predictions, it is impossible to reliably assess their accuracy, because the predicted events lie far in the future.

Alternatives and Complements to Prediction as the Basis for Climate Policy

Reduction of uncertainty about the future is a prerequisite for achieving both the political momentum and the technical capacity necessary to implement the current policy framework on global climate change. Yet prediction in climate policy is characterized by fundamental uncertainties that are subject to various interpretations based on the technique and the sociopolitical milieu. Under such conditions, it would seem that predictions alone are an inadequate basis for policy. Sometimes climate modelers argue that imperfect predictions are better than nothing as a basis for policy, but "nothing" is not the only alternative to prediction.

If prediction across the whole field of climate change issues is not a sufficient basis for sound policy, the question remains: What types of policies can respond to the threat of global climate change without having to depend on predictions? If decision makers cannot predict the unpredictable and policy makers cannot know the unknowable, how can society face the prospect of profound change occurring at an accelerating pace? The answer may be to focus on building responsive institutional arrangements that monitor change and maximize the flexibility of human populations to respond creatively and constructively to it (also see chapter 14).

Approaching the climate issue from the starting point of assessing human vulnerability and societal adaptation may seem to be far less amenable to concerted rational action by national governments than the current focus on implementing emissions reduction targets. But it also opens the space for discussion of the adaptive strategies that inevitably will be required, even to tackle the likelihood of climate change resulting from past and present emissions. Adaptation also may be more directly relevant to stakeholders, as it allows for a variegated response to local conditions. For instance, an adaptation measure designed to protect a coastal community from sea-level rise may have no feature or characteristic in common with measures designed to stem desertification. That is to say, adaptation is a bottom-up strategy that starts with changes and pressures experienced in people's daily lives. This characteristic contrasts with the top-down approach of national targets for emissions reductions. The connections between emissions targets and people's everyday behavior and responsibilities seem less direct, even abstract. Designing adaptation strategies may be more sensitive to the real tradeoffs made by real people in a way that top-down emissions reduction strategies such as carbon taxes may not be.

Viewing climate change as an issue of societal choices opens the range of possible actions consistent with the commitment of nations to the United Nations Framework Convention on Climate Change. Instead of tracing a narrow line of causality from emissions to climate to impacts, we can explore how emissions-producing activities are embedded in social institutions and ways of life—and, then, what alternatives are possible within those institutional arrangements. Instead of trying to force institutional change through taxing or regulating outputs, we can plot multiple pathways to satisfying human needs for the goods and services provided by emissions-producing activities. Policy makers can take sound steps that do not depend on increasingly accurate predictions of climatic or social change. In a commentary on the Kyoto process published in *Nature*, Rayner and Malone (1997) made five such recommendations reiterated here:

1. *Design policy instruments for real-world conditions rather than try to make the world conform to a particular policy model.* Much of the policy proposed to deal with climate change is underlain by the assumption that there are well-behaved markets with a large number of traders who have perfect information; deviations from the ideal are viewed as barriers that can be removed by inserting the right information (e.g., the right price signal). Unfortunately, even highly industrialized countries exhibit significant variation from this ideal model of information flows and barriers. In less industrialized countries, these variations are often so large as to render the model useless. Many less industrialized countries are unable to carry out even the most elementary functions of government, let alone implement climate change policies such as those addressing land and water use. In these conditions, the issue of optimizing across regulations, taxes, permits, education, and demonstration projects becomes academic. And when conventional development approaches treat these fundamental structural differences as mere barriers to implementation, policy instruments cannot fail to fail.

   The solution is to design information to fit the everyday perspectives of diverse actors and to design policy instruments to suit specific conditions. For instance, Nepal has recently returned significant tracts of forest, nationalized in the 1950s, to the control of local communities that previously exercised guardianship of the forests through indigenous institutions. This move to neotraditional resource management reflects the central government's lack of monitoring and enforcement capacity to regulate the forests. It also recognizes that inequalities of market power and the weakness of markets for land and forest products in Nepal militate against purely private forest ownership as a sustainable strategy.

2. *Incorporate climate change concerns into other, more immediate issues such as employment, defense, economic development, and public health.* Without a major policy stimulus (such as a carbon tax, which seems to be a dead political option, at least in the United States) or an unmistakable signal that climate change is imminently threatening, any country is likely to delay the kinds of behavioral changes that would be necessary to mitigate or adapt to climate change. Climate change will stay at the policy periphery, while attention will stay focused on policy core issues like national economic policy or corporate manufacturing strategy.

   Under such conditions, an appropriate strategy is to build climate concerns into the everyday concerns of people at the local level and the larger-scale concerns of policy makers at the national level. For example, domestic end-use energy efficiency seems to be more appealing when presented to consumers as a money-saving measure or to increase home comfort than when it is touted for its environmental benefits. At the national level, the success of Germany in meeting its climate policy goals was entirely due to widespread acceptance of the need for industrial restructuring to secure economic development. Joining climate change to issues of societal resilience opens the agenda to a broad range of focus areas, including economic development, institutional restructuring, fostering civil society, and strengthening indigenous arrangements (e.g., land tenure) that are already working. Resilience encompasses not just preservation from harm (where that is possible) but also strengthening or establishing alternative economic activities (both market and nonmarket) and social structures, as illustrated by the economically independent and environmentally sensitive energy and industrial programs of the Indian nongovernmental organization Development Alternatives.

3. *Take a regional and local approach to climate policy making and implementation.* In the day-to-day lives of most people in the world, local government is a more salient political actor than the central government. It delivers or withholds essential services; it mediates between the citizen and the nation–state through local officials, such as police officers who may have to monitor vehicle emissions, or building inspectors who are responsible for seeing that new construction meets energy-efficiency standards. Furthermore, over 50 percent of the world's population now lives in urban areas, where the density, mixture, and physical layout of residential and commercial neighborhoods all influence the energy intensity of the community. Many of these factors are directly under the control of community governments.

However, almost all of the climate change policy research and analysis is aimed at high-level policy makers. Funding agencies tend to be those of national governments or of interest groups and organizations seeking to influence national government policy or international negotiations. While this research is important, it is not very helpful to a city manager, the general manager of an aluminum smelter, the operator of a regional reservoir system, or a householder seeking guidance on how to do the right thing for the climate at the same time as doing the best for his or her citizens, stockholders and employees, consumers, or family members.

Policy makers can seek out and encourage local-level activities in many ways. For example, the Municipal Leaders Summit for Climate Change in New York in 1993 established the Cities for Climate Protection program. This program was an extension of an earlier initiative linking fourteen cities in the United States, Canada, Europe, and Turkey, designed to strengthen local commitment to reduce urban greenhouse gas emissions, to research and develop best practices in pilot communities, to share planning tools and experiences, and to enhance ties among municipalities across national boundaries, especially among those in industrialized and less industrialized countries (cf. Brunner 1996).

4. *Direct resources to identifying vulnerability and promoting resilience, especially where the impacts will be largest.* Whatever the level at which decisions are made, sustainability is about being nimble, not about being right. Policy makers should balance their current emphasis on linear goal setting and implementation by paying more attention to promoting societal resilience through enhancing the capability to switch strategies as conditions change. This is particularly urgent where populations are vulnerable to the early impacts of climate change.

Vulnerability includes risks to people, land, and infrastructure— but just as important are political and economic systems and other institutional arrangements (as well as the environment). Changes in regional patterns of habitability would exacerbate existing problems for poor populations living in environmentally vulnerable areas, such as low-lying tropical regions.

No standard framework exists for identifying the many complex sources of vulnerability. Poverty is generally recognized as one of the most important correlates of vulnerability to hazard, but it is neither necessary nor sufficient for it. The very young and the old are often identified as especially vulnerable. Other variables widely invoked are differences in health, gender, ethnicity, education, and experience with the hazard in question. Empirical, local-level studies reveal such complex mosaics of vulnerability as to cast doubt on attempts to describe patterns and estimate trends at the global or even the regional scale.

The IPCC Second Assessment Report (1995) has made a preliminary identification of regions and societies where climate change impacts are likely to be most severe, for example, coastal zones and areas that are already warm and dry (Watson et al. 1996). However, social science warns us of broad pronouncements about relative vulnerability. Some researchers argue that the industrialized world is more vulnerable because of increasing interdependencies and rigidities in the industrial system and its supporting infrastructures. Other researchers have argued that the vulnerability of the less industrialized world is greater because of its immediate dependence on agriculture. When all is said and done, building both the social and the financial capital of the poor may be their best defense.

5. *Use a pluralistic approach to decision making.* The Framework Convention provides an important symbolic edifice expressive of worldwide concern about climate and about the persistent issues of global development that are inextricably bound up with it. However, the real business of responding to climate concerns may well be through smaller, often less formal, agreements among states; states and firms; and firms, nongovernmental organizations (NGOs), and communities. This process is likely to be messy and contested, but the potential exists to make the most of diversity and the variety of decision strategies available to decision makers.

What will connect the diverse elements of a plural policy approach? The goal of creating resilience provides the theme, with resilience defined by a society's capacity to draw upon multiple ways of using resources and distributing goods and to switch from strategies that are not working to ones that will work, at least until they are replaced by still better strategies. Each society needs to have within it multiple working methods of resource management, as illustrated in the *longe durée* by Putnam's (1993) contrast between the stagnation and decline of southern Italy from the twelfth century onward and the rise of the more institutionally diverse *communes* of the north. A society that uses one strategy only (say, only authoritarian or hierarchical management) will be extremely vulnerable to disruption. Complex, overlapping, plural, interdependent civic institutions embodying diverse combinations of several basic strategies extend a society's capabilities to develop in a sustainable fashion, even—or, rather, especially—when confronted with surprise. Hence, international development efforts are increasingly focused on nurturing institutions of civil society for resource protection, such as the water users associations promoted by the World Bank in Pakistan.

Policy making that links the local and global levels requires extension of civic life, both as civic science (linking scientific and technical knowledge with local knowledge and craft skills) and civic society (associational links outside of governments and markets), at all levels to complement the market and government. Similarly extending integrated assessment analysis and inquiry will enable scientific efforts to provide information useful to decision makers at all levels—not only global and national, but also at the levels of firms, NGOs, and households.

The focus on targets and timetables for emissions reduction simplifies and bounds climate change as a distinct problem. In so doing, it domesticates a large, complex, and unruly set of life's circumstances as being capable of solution through the application of scientific prediction, rational analysis, goal setting, and policy implementation by technocratic elites. But targets and timetables essentially represent a top-down, pollutant-by-pollutant, media-specific approach to environmental management on a global scale. This approach is increasingly recognized as obsolete at the local and national levels; there is even less reason to believe it should dominate at the global scale.

The first essential for policy in a complex world is to resist the urge to declare one viewpoint true and to reject others. For example, in welcoming delegates from seventeen governments to the April 1990 White House Conference on Science and Economics Research Related to Global Change, President Bush claimed that political decision makers were being asked to choose between "two diametrically opposed schools of scientific thought" on the reality, severity, likelihood, and timing of global environmental change. *Since that time, the role of predictive models in global change policy has all too often been distilled to the search for a particular answer: Is temperature going up? By how much? What will it cost to mitigate this change? What will it cost if we don't?* Abandoning such an approach in the face of enormous natural and societal complexity is neither mindless relativism that says one idea is just as good as any other, nor a recipe for passivity and the abdication of choice. Where people argue about the way the world (natural and social) actually works or the way the world ought to work, we are likely to find ourselves facing competing partial truths. To commit oneself, family, firm, community, or nation to just one of these viewpoints about how the world works is to gamble that it will turn out to be right and the others wrong. It is far more likely that all will be partly right and all will be partly wrong. Recognizing this and stewarding the kind of institutional pluralism necessary to maintain multiple viewpoints and a rich repertoire of policy strategies from which to choose is what promoting societal resilience, sustainable development, and climate change governance is all about.

Conclusion

The story told by integrated assessment modeling is that from a policy-making standpoint, the imprecision and uncertainties inherent in the current generation of GCMs represent only a small fraction of the uncertainty associated with climate change, its impacts, and the consequences of alternative policies (see figure 13.1). Although climate change threatens to worsen the lot of poor people in vulnerable geographic locations, it is unlikely to be the deciding factor in whether human society as a whole prospers or suffers. In fact, for much of the world's population, technological, economic, social, and political change is likely to occur at such a rate that changes in the global climate regime of the order anticipated by the IPCC may be barely noticeable.

At the level of national politics, scientific prediction of climate change provides opportunities to advance diverse environmental, technological, economic, and political agendas. For example, in Britain, the Conservative government was able to marshal concerns about climate change to justify its policy to break the political power of the National Union of Mineworkers by switching from coal to nuclear electricity and the so-called dash for gas (Everest 1988). The German government adopted an activist stance on climate change in large part because it reinforced economically motivated policies designed to bring about restructuring of German industry. It is not coincidental that these are the only economically healthy countries to have achieved voluntary reductions of greenhouse gas emissions to 1990 levels. Elsewhere, concern about climate change has provided a new audience for advocates of renewable energy technologies who had been out of favor since the oil price shocks of the 1970s receded.

At the level of international relations, I have elsewhere suggested that if the threat of climate change did not exist, we would have had to invent it, or something very much like it, to respond to the challenges of global governance at the end of the twentieth century, when widespread recognition of global economic and ecological interdependence is accompanied by powerful drives to establish strong independent ethnic, local, and regional identities (Rayner 1994). The existence of a potentially catastrophic consequence of business as usual is a powerful incentive for change; among other things it is an opportunity to revive the issues of equity in international development that dropped off the international agenda with the demise of the New Economic Order of the 1970s and 1980s. Climate change is particularly effective in this respect because the threat appears to be an automatic consequence of human action—a deterministic response of nature—not a legal or coercive sanction that can be imposed or stayed by the discretion of any party. This places climate in a longstanding worldwide tradition of "natural" sanctions on human behavior (Douglas and Wildavsky 1982).

Scientific prediction, as distinct from prophecy, is meant to provide a value-free application of inductive reasoning to the material world that is distinguishable in its essence from the morally charged revelations of oracles, prophets, and politicians. Prophesy, at least in the Judeo-Christian tradition, was for the most part explicitly aimed at the goal of changing social behavior. Individuals and civilizations came to a grisly end by fire, flood, or the sword because they would not heed the warnings of prophets that could have turned away God's wrath. In contrast with the divine origin and the moral conditionality of prophesy, scientific prediction is supposed to be based on rational observation of deterministic systems.

As the world enters the twenty-first century, the distinction between prediction and prophesy seems to be increasingly blurry, another manifestation of the unsustainable firewall separating facts and values in science and social policy. Scientists and policy makers are deeply engaged in a hybrid science and policy discourse where predictions become warnings of dire consequences if lifestyles characterized by profligacy in both consumption and consummation are not changed.

But predictions have not and cannot transform the technical and socioeconomic complexities of climate change into a series of politically manageable choices. While scientists may reasonably strive for a breakthrough improvement in the quality of their long-term modeling of the climate system, the broader field of societal risk management or practical politics suggests that policy will likely be driven by negotiation among competing values rather than probabilistic forecasting. This suggests that policy makers need to extend their search for guidance beyond even integrated assessment modeling to the development of a participatory "vernacular" (O'Riordan and Rayner 1991) or "civic science" (Lee 1993; Rayner and Malone 1998) that taps into the everyday concerns and wisdom of citizens as well as the expertise of climatologists.

Notes

1. The anticipated rate of change is often stated to be likely to exceed the capacity of ecosystems to migrate or adapt. The same has been said of the adaptive ability of socioeconomic systems. However, a recent survey of the relevant social sciences (Rayner and Malone 1998) suggests that the background rate of socioeconomic and technological change will continue to outpace climate changes.

2. For instance, a climate modeler anonymously reviewing this paper insisted that Balling and Michaels "certainly do not qualify as physical climate scientists." In fact, both are trained and hold university appointments in the field.

3. For example, the famous cover of *Der Spiegel* showing water lapping around the top of the twin spires of Cologne Cathedral.

4. Despite this omission from the international climate policy agenda, the goals of adaptation and building resilience to climate impacts are receiving increasing attention within the research communities associated with the science and human dimensions of climate change. This line of research suggests a broader policy goal than emissions reductions: that of securing human welfare under the prospect of environmental change.

5. This was shown by Gleckler et al. (1994), who demonstrated that the surface fluxes in many uncoupled atmospheric models would imply a northward oceanic heat transport in the Southern Hemisphere, inconsistent with empirical observations. This resulted from errors in the cloud radiative forcing in the models, which allowed excessive solar radiation to reach the surface.

References

ACPI (Accelerated Climate Prediction Initiative). 1998. *The accelerated climate prediction initiative: Bringing the promise of simulation to the challenge of climate change*. Pacific Northwest National Laboratory, PNNL-11893. Richland, WA.

Arrhenius, S. 1896. On the influence of carbonic acid in the air upon the temperature of the ground. *Philosophical Magazine* S5, 41(251):237–277.

Balling, R. 1992. *The heated debate: Greenhouse predictions versus climate reality*. San Francisco: Pacific Research Institute for Public Policy.

Brunner, R.D. 1996. Policy and global change research: A modest proposal. *Climatic Change* 32: 121–147.

Callendar, G.S. 1940. Variations in the amount of carbon dioxide in different air currents. *Quarterly Journal of the Royal Meteorological Society* 66:395–400.

Cartwright, N. 1983. *How the laws of physics lie*. Oxford, UK: Oxford University Press.

Charles River Associates. 1997. *World economic impacts of U.S. commitments to medium term carbon emissions limits*. Prepared for the American Petroleum Institute, CRA no. 837-06. Washington, DC.

Douglas, M., and A. Wildavsky. 1982. *Risk and culture*. Berkeley: University of California Press.

Everest, D. 1988. *The greenhouse effect*. London: Royal Institute of International Affairs and Policy Studies Institute.

Fankhauser, S. 1995. *Valuing climate change: The economics of the greenhouse*. London: Earthscan.

Funtowicz, S.O., and J.R. Ravetz. 1985. Three types of risk assessment: A methodological analysis. In *Risk analysis in the private sector*, C. Whipple and V. Covello, eds. New York: Plenum.

Gaskins, D.W., and J.P. Weyant. 1993. Model comparisons of the costs of reducing CO_2 emissions. *American Economic Review* 83(2):318–323.

Gates, W.L. 1992. AMIP: The atmospheric model intercomparison project. *Bulletin of the American Meteorological Society* 73:1962–1970.

Gates, W.L., A. Henderson-Sellers, G.J. Boer, C.K. Folland, A. Kitoh, B.J. McAvaney, F. Semazzi, N. Smith, A.J. Weaver, and Q-C. Zeng. 1996. Climate models—Evaluation. In *Climate change 1995: The science of climate change*, J.T. Houghton et al. eds. Cambridge, UK: Cambridge University Press.

Gleckler, P.J., D.A. Randall, G. Boer, R. Colman, M. Dix, V. Galin, M. Helfand, J. Kiehl, A. Kitoh, W. Lau, X-Z. Liang, V. Lykossof, B. McAvaney, K Miyakoda, and S. Planton. 1994. *Cloud-radiative effects on implied ocean energy transports as simulated by atmospheric general circulation models*. Report 15, PCMDI, Lawrence Livermore National Laboratory, Livermore, Calif.

Houghton, J.T., L.G. Meira Filho, B.A. Callender, N. Harris, A. Kattenberg, and K. Maskell, eds. 1996. *Climate change 1995: The science of climate change*. Cambridge, UK: Cambridge University Press.

IPCC (Intergovernmental Panel on Climate Change). 1990. *IPCC first assessment report*. Cambridge, UK: Cambridge University Press.

Janssen, M. 1996. *Meeting targets: Tools to support integrated assessment modeling of global change*. Doctoral dissertation, University of Maastricht.

Jasanoff, S., and B. Wynne. 1998. Science and decisionmaking. In *Human choice and climate change. Volume 1: The societal framework*, S. Rayner and E.L. Malone, eds. Columbus, OH: Battelle Press.

Kattenberg, A., F. Giorgi, H. Grassl, G.A. Meehl, J.F.B. Mitchell, R.J. Stouffer, T. Tokioka, A.J. Weaver, T.M.L. Wigley, and contributors. 1996. Climate models—Projections of future climate. In *Climate change 1995: The science of climate change*, J.T. Houghton et al. eds. Cambridge, UK: Cambridge University Press.

Keeling, C.D. 1960. The concentration and isotopic abundances of carbon dioxide in the atmosphere. *Tellus* 12:200–203.

Lee, K.N. 1993. *Compass and gyroscope: Integrating science and politics for the environment*. Washington, DC: Island Press.

Manabe, S., and R.T. Wetherald. 1967. Thermal equilibrium of the atmosphere with a given distribution of relative humidity. *Journal of Atmospheric Sciences* 24:241–259.

Michaels, P. 1992. *Sound and fury*. Washington, DC: Cato Institute.

Morgan, G.M., and M. Henrion. 1990. *Uncertainty: A guide to dealing with uncertainty in quantitative risk and policy analysis*. New York: Cambridge University Press.

NAS (National Academy of Sciences). 1992. *Policy implications of green-house warming: Mitigation, adaptation, and the science base.* Washington, DC: NAS.

Oreskes, N., K. Schrader-Frechette, and K. Belitz. 1994. Verification, validation and confirmation of numerical models in the earth sciences. *Science* 263:641–646.

O'Riordan, T., and S. Rayner. 1991. Risk management for global environmental change. *Global Environmental Change* 1(2):91–108.

OTA (Office of Technology Assessment). 1993. *Preparing for an uncertain climate.* Washington, DC, Government Printing Office.

Pearce, D.W., W.R. Cline, A.N. Achanta, S. Fankhauser, R.K. Pachauri, R.S.J. Tol, and P. Vellinga. 1996. The social costs of climate change: Greenhouse damage and the benefits of control. In *Climate change 1995: Economic and social dimensions of climate change,* J.P. Bruce, H. Lee, and E. Haites, eds. Cambridge, UK: Cambridge University Press.

Putnam, R. 1993. *Making democracy work: Civic traditions in modern Italy.* Princeton, NJ: Princeton University Press.

Rayner, S. 1994. Governance and the global commons. Discussion paper 8, Centre for the Study of Global Governance, London School of Economics. Reprinted in *Global governance: Ethics and economics of the world order,* M. Desai and P. Redfern, eds. London: Pinter.

Rayner, S., and E. Malone. 1997. Zen and the art of climate maintenance. *Nature* 390:332–334.

Rayner, S., and E.L. Malone. 1998. *Human choice and climate change. Volume 4: What we have learned.* Columbus, OH: Battelle Press.

Rayner, S., M. Thompson, and E. Malone. 1999. Equity issues and integrated assessment. In *Fair weather: Equity concerns in climate change.* F.L. Toth, ed. London: Earthscan.

Repetto, R., and D. Austin. 1997. *The costs of climate protection: A guide for the perplexed.* Washington, DC: World Resources Institute.

Revelle, R., and H.E. Suess. 1957. Carbon dioxide exchange between the atmosphere and ocean and the question of an increase in atmospheric CO_2 during the past decades. *Tellus* 9:18–27.

Rotmans, J., and H. Dowlatabadi. 1998. Integrated assessment modeling. In *Human choice and climate change. Volume 3: Tools for policy analysis,* S. Rayner and E.L. Malone, eds. Columbus, OH: Battelle Press.

Seitz, F., R. Jastrow, and W. Nierenberg. 1989. *Scientific perspectives on the greenhouse problem.* Washington, DC: George C. Marshall Institute.

Shackley, S., P. Young, S. Parkinson, and B. Wynne. 1998. Uncertainty, complexity and concepts of good science in climate change modelling: Are GCMs the best tools? *Climatic Change* 8:159–205.

Thompson, M., R. Ellis, and A. Wildavsky. 1990. *Cultural theory.* Boulder, CO: Westview Press.

Toth, F.L. 1994. Practice and progress in integrated assessments of climate change. In *Integrative assessment of mitigation, impacts, and adaptation to climate change*, N. Nakicenovic, W.D. Nordhaus, R. Richels, and F.L. Toth, eds. CP-94-9 Vienna, Austria: International Institute for Applied Systems Analysis.

Toth, F. 1999. *Fairness and climate change*. London: Earthscan.

van Asselt, M.B.A., J. Rotmans, M. den Elzen, and H. Hilderink. 1995. *Uncertanity in integrated assessment modeling: A cultural perspective based approach*. GLOBO Report Series 9. Bilthover: RIVM.

van der Sluijs, J. 1997. *Anchoring amid uncertainty: On the management of uncertainties in risk assessment of anthropogenic climate change*. Leiden: Ludy Feyn.

Watson, R.T., M.C. Zinyowera, and R.H. Moss, eds. 1996. *Climate change 1995: Impacts, adaptations and mitigation of climate change: Scientific-technical analysis*. Cambridge, UK: Cambridge University Press.

PART FIVE

Prediction in Perspective

For any of the complex environmental phenomena discussed in the preceding case studies, a perfect predictive model would include every possible variable associated with the phenomena, and every possible relationship among those variables. Such a model would capture the exact current state of each variable and thus accurately predict its future state. Such a model would equal—would be indistinguishable from—reality. It would have to contain as many molecules as the reality that it was representing, and the events that it was predicting would unfold at the same rate as reality. This model—the perfect prediction machine—is both impossible and useless.

The minute we step back from the perfect, and perfectly useless, model, we introduce approximations, assumptions, uncertainty, and error into the prediction process. Here is a crucial question: to what extent are approximations, assumptions, uncertainty, and error challenges to be overcome by science, and to what extent are they realities to be accommodated by society? Expressed differently, perhaps the apparent limitations of scientific prediction are the very essence of human judgment and freedom.

Alternatives to Prediction

Ronald D. Brunner

Perhaps it is time for scientists and policy makers to reconsider the emphasis on predictions in policy decisions.[1] Over past decades, scientists and to a lesser extent policy makers have often acted as if predictions based on scientific laws were prerequisites for major policy decisions. But there is growing evidence that in the future we may rely more on flexible policies that are based on experience at local or regional scales. Such "distributed" policies are particularly promising for the cases in this volume, each of which discusses the interaction of a complex natural event and a complex policy process. As has been shown repeatedly, decision making based on predictions is problematic in such situations. This chapter reviews some evidence on a possible transition from an emphasis on science-based predictions to an emphasis on action based on distributed policies. It also considers the conditions under which such a transition might occur, and the implications of the transition for science policy and leadership.

Converging Trends

Confidence that the natural or social sciences can predict with precision, scope, and accuracy through systems of generalizations is based on the reductionist hypothesis. According to Philip Anderson (1972), a Nobel prize–winning solid-state physicist, this is the hypothesis that "[the] workings of our minds and bodies, and of all the animate or inanimate matter of which we have any detailed knowledge, are . . . controlled by the same set of fundamental laws. . . ." Anderson accepted the reductionist hypothesis but rejected the corresponding constructionist hypothesis as a fallacy: "The ability to reduce everything to simple fundamental laws does *not* imply the ability to start from those laws and reconstruct the universe" (Anderson 1972; emphasis added). Instead, at each level of complexity in the aggregation of elementary particles, scientists find new fundamental properties. Each level requires fundamental research of its own. Thus more complexity entails qualitatively different behavior—a conclusion dramatized in the often-repeated statement, "Psychology is not applied biology, nor is biology applied chemistry" (Anderson 1972). "More is different" thus became a rallying cry for research on "complex adaptive systems" (Horgan 1996), which are what the ten cases in this volume surely represent.

Research on such systems calls into doubt the possibility of finding systems of generalizations that predict with precision, scope, and accuracy—especially insofar as humans and other living forms are concerned. The behavior of a complex adaptive system emerges from interactions among large numbers of component parts, some of which may be people who have their own, subjective perspectives on themselves and the rest of the system. Holland (1992) observes that people use "internal models" to *anticipate* the future, basing current actions on expected outcomes" (emphasis in the original). Because humans are not omniscient, the internal models are invariably incorrect in many ways, but *they can be improved through hindsight* (Holland 1992). The internal models are also the distinguishing characteristic of complex adaptive systems, explaining why their behavior is so intricate and difficult to understand.

> Because the individual parts of a complex adaptive system are continually revising their ("conditioned") rules for interaction, each part is embedded in perpetually novel surrounds (the changing behavior of other parts). As a result, the aggregate behavior of the system is usually far from optimal, if indeed optimality can even be defined for the system as a whole. For this reason, standard theories in physics, economics, and elsewhere, are of little help because they concentrate on optimal end-points, whereas complex adaptive systems "never get there." They continue to evolve, and they steadily exhibit new forms of emergent behavior. History and context play a critical role.... (Holland 1992)

Computer models can deduce behavior that is comparable to that of any complex adaptive system composed of many adaptive agents. But computer models cannot reliably predict the behavior of any such system if the context and rules of interaction of the agents in the system are always open to new experience and insight and the system never reaches equilibrium (Horgan 1995). Anderson, at least, accepts the perspective on natural history that "life is shaped less by deterministic laws than by contingent, unpredictable circumstances" (Horgan 1996). For instance, because of the limited predictability of open systems, computer models cannot accurately integrate the impacts of hurricanes into predictions about beach behavior (see chapter 8), the effect of faults and fractures on fluid flow in rocks (chapter 10), or the influence of changing patterns of energy consumption on hydrocarbon consumption (chapter 11) or carbon emissions (chapter 13).

These insights from complex adaptive systems were anticipated long ago by social scientists who *observed* how humans behaved in order to understand how to improve the rationality of their decisions. They typically concluded that observed acts are best understood on the basis of simplified "internal models" that are not universal but contingent on local context. For example, Harold Lasswell observed that "living forms are predisposed to complete acts in ways that are perceived to leave the actor better off than if he had completed them differently. This observation draws attention to the actor's own perception of the alternative act completions open to him in a given situation" (Lasswell 1971). But no one is omniscient: The actor's own perception is somewhat mistaken, somewhat different from the perceptions of others, and subject to change through new experience and insight. Similarly emphasizing human cognitive constraints relative to the complexity of real-world problems, Herbert Simon recognized that "the intended rationality of an actor requires him to construct a simplified model of the real situation in order to deal with it. He behaves rationally with respect to this model, and such behavior is not even approximately optimal with respect to the real world" (Simon 1957). Both Lasswell and Simon emphasized that "internal models" are not fixed or given, nor can they be assumed.

Observations interpreted in light of insights such as these, rather than the reductionist hypothesis, challenge the conventional emphasis on science-based predictions in policy decisions. Consider the following alternatives to the conventional emphasis on predictive science in policy making:

1. The potential to predict human behavior with precision, scope, and accuracy is *limited* because human behavior is not determined by fundamental laws. Instead, human behavior stems most directly from choices and decisions that are contingent upon individual, unavoidably subjective perspectives, or "internal models." The internal models are open to new human experience and insight, and systems significantly affected by human behavior are therefore open systems as well. For example, the total earth system at the focus of research under the U.S. Global Change Research Program (USGCRP) is an open system, as human activities powered by Industrial Age technologies caused the buildup of carbon dioxide in

the atmosphere and thereby contributed to global climate change (see chapter 13). But validation and verification sufficient to qualify generalizations as scientific laws depend on replications in closed systems, such as those constructed in the laboratories of the natural sciences. Of course, few natural systems are closed: floods are influenced by changing land-use patterns, radioactive waste by energy production technology and political trends. None of the cases considered in this book treat perfectly closed systems, although asteroids might be considered approximately closed in the short term.

2. Prediction with precision, scope, and accuracy is *insufficient* for rational policy decision. Prospective or retrospective assessment of the consequences of a policy alternative also depends on policy goals, which cannot be taken as fixed or given in an evolving situation. Policy goals need to be adjusted in light of new experience and insight to add degrees of freedom in the evolutionary process. For example, assessments of the scientific literature indicate that some climate change induced by human activities is unavoidable. As this expectation has intensified among scientists and spread to policy makers, adaptation to climate change has become more important as a policy goal relative to mitigation (Pielke 1998). It is hardly rational to ignore the vulnerabilities of particular communities to unavoidable climate changes, given our proven ability to adapt to climate impacts like floods on a local level (chapter 5).

3. Prediction with precision, scope, and accuracy is *unnecessary* because rational policy decision is inevitably a process of trial and error to some degree. It is procedurally rational to act on the most promising alternative in a situation, despite projected uncertainties in costs and benefits—provided the alternative is modest enough to assess in an appropriate period of time, to fail gracefully if it does fail, and to learn from the experience (Brunner 1996). Making progress on complex problems depends on factoring the overall problem into its geographic or functional parts, working on them in parallel and in series in the appropriate situations, and integrating any partial solutions back into the evolving whole (Simon 1969). For example, manufacturers such as Johnson & Johnson have reduced their carbon dioxide emissions *and* their energy costs, and cities such as Fargo, North Dakota, have reduced their vulnerabilities to floods (Brunner and Klein 1999; Pielke 1999). Such distributed policies have helped decision makers to mitigate and adapt to climate change on small scales. This approach can be generalized.

Moran (see chapter 9) suggests that insurance and bonding can address the environmental risks of hard-rock mining, without resort to flawed hydrogeochemical models. Herrick (chapter 12) showed how tradeable emissions permits helped address the acid rain problem, while predictive models had little direct impact on policy.

4. Science-based predictions are *not politically independent*, even when the research is carried out as objectively as possible. It is important to understand that multiple models, each simplified in different ways, exist for any decision situation. The selection of one simplified model over all the others that are equally valid or invalid entails a subjective purpose that goes beyond science. Hence, opposition to policies based on the model selected may (and often does) reflect normative as well as technical considerations. For example, Gautier (chapter 11) showed how a prediction of future oil and gas reserves by the U.S. Geological Survey was greeted with harsh criticism because of the way the scientists who carried out the prediction chose to define the problem and its parameters—not because of any inherent flaw in the science itself.

Scientific prediction, in other words, is inherently problematic as a basis for decision making in the face of complexity. As this difficulty becomes more widely appreciated, it stimulates a search for new approaches—a search that appears to be converging on a more or less common outlook on decision making. Consider several examples that emphasize responsiveness to experience in concrete situations in environmental policy. Paul Stern has envisioned a "second environmental science," in which "[the] most important point is probably that human beings are continually responsive to interventions . . . so that it will never be possible to write a cookbook for behavior change. It is absolutely essential to treat interventions as dynamic and to monitor and revise them continually" (Stern 1993). In the adaptive management of ecosystems, C. S. Holling observes that, "the essential point is that evolving systems require policies and actions that not only satisfy social objectives but also achieve continually modified understanding of the evolving conditions and provide flexibility for adapting to surprises" (Holling 1995).

In the case of global climate, Rayner and Malone (1997) suggested that, rather than try to predict the unpredictable, policy makers should "focus on building responsive institutional arrangements that monitor change and maximize the flexibility of populations to respond creatively and constructively to it." The design of responsive institutional arrangements might begin with small-scale policies that have already succeeded in mitigating or adapting to climate change in particular situations, and with the harvesting of experience from those successes for adaptation in other situations through policy research (Brunner and Klein 1999; see also Dyson 1988).

Conditions and Projections

What conditions account for the recent rise of alternatives to the conventional emphasis on science-based predictions in major policy decisions? The search for answers to this question begins with problems inherent in models of open systems. The problems of omission and misspecification, discussed above, are compounded by problems of application for policy purposes. A mathematical or numerical model in itself can be abstract and complete, however much it must omit from the open system it represents. But the rules for applying the model cannot possibly account for all real-world contingencies and are therefore incomplete. The rules define the class of phenomena for which the model can be taken as an adequate representation—for example, the behavior of sand on a beach. The rules also specify the correspondence between variables in the model and observable phenomena—for example, between the parameter in the model representing wave height and actual waves on the beach (chapter 8). Consequently, according to Friedman (1953), "In seeking to make science as 'objective' as possible, our aim should be to formulate rules explicitly in so far as possible and continually to widen the range of phenomena for which it is possible to do so. But, no matter how successful we may be in this attempt, there inevitably will remain room for judgment in applying the rules." How much data is "enough"? What level of uncertainty is "satisfactory"? What is the "appropriate" time frame for measurement? There would be no need for peer review if there were no room for judgment in applying scientific theory or method, just as there would be no need for courts of law if there were no room for judgment in applying substantive and procedural rules of law. In the real world, "room for judgment" in practice entails limits to the objectivity that is supposed to exist in principle.

Because of such problems, predictive science seldom meets the expectations created by its proponents. It is expected to promote consensus on policy through objective forecasts, but different groups of scientists produce quite different forecasts for the same situation. For example, modeling groups working under the auspices of the National Research Council's Committee on Nuclear and Alternative Energy Systems in 1978 produced quite different energy forecasts for 1990—despite their use of the same initial conditions, assumptions about economic growth, and policy scenarios. Shortly after these forecasts were produced, William Ascher (1981) observed: "Some of the models must be wrong in quite significant ways on policy-relevant issues. It is also discouraging that even agreement across models need not be an indication of validity; they could all be wrong." There was a precedent for the possibility that they could all be wrong: All multiyear predictions of U.S. energy consumption for 1975 had turned out to be higher than actual consumption because none of the forecasting models incorporated the possibility of an oil embargo and much higher energy prices in 1973–74. The oil embargo is a vivid illustration of the potential significance of factors omitted from a predictive model of an open system. Moreover, as Gautier (chapter 11) illustrates, assessments of oil and gas *supply* are also highly sensitive to factors that are extrinsic to the "closed" geological systems that contain the oil—factors such as advances in hydrocarbon extraction technologies and the politics of oil exploration on public lands.

Predictive science is expected to improve the accuracy of policy-relevant forecasts. But empirical limits on forecast accuracy have been documented in systematic appraisals across many policy areas and over time, and plausible explanations have been offered for why improvements in forecast accuracy are unlikely (Ascher 1981). Consider the judgment of Alice Rivlin (1984) on economic forecasting, the area of forecasting in the social sciences for which we have the most experience:

> The poor showing of the forecasters is not due to any lack of effort or ingenuity. . . . The real problem is that the economic system is extremely complicated, that our own economy is battered by forces outside itself which are inherently unpredictable, such as the weather and foreign wars. I doubt we will ever improve the accuracy of forecasting very much, especially the forecasting of economic turning points. Instead, we will have to learn to live with the uncertainty.

If predictive science is to provide the foundation for rational policy, it is frustrating to discover that the National Acid Precipitation Assessment Program was driven by science, not policy, and turned out to be somewhat irrelevant to the Clean Air Act Amendments of 1990, which it was created to support (Roberts 1991; chapter 12, this volume); or that similar concerns have been raised about climate change research (Economist 1994; chapter 13, this volume). In the aggregate, the expected benefits from predictive science have been frustrated often enough to call into question fundamental assumptions about the relation between science and society (Pielke and Byerly 1998). Will the conventional emphasis on science-based predictions for policy purposes be rejected by decision makers in the future? Emerging experience suggests that alternatives to prediction will begin to play a greater role in policy for the following three reasons:

- First, the probability of rejection increases to the extent that such predictions become more problematic for scientists or policy makers. Only those who are frustrated or concerned are receptive to alternatives; those who are contented are prejudiced against alternatives to the status quo. Many of the cases in this book document the problematic nature of predictions as a policy tool.

- Second, the probability of rejection increases to the extent that promising alternatives come to the attention of those who are receptive. Their frustrations or concerns only motivate changes in behavior; a promising alternative is necessary to shape and direct those changes successfully. Such alternatives are available in the majority of the cases presented here. Monitored, retrievable storage of nuclear waste, for example, offers an alternative to geologic isolation that would obviate the need to predict the behavior of the waste package tens of thousands of years into the future.

- Third, the probability of rejection increases to the extent that action on a promising alternative by its early adherents does in fact succeed in alleviating their frustrations or concerns. Nothing succeeds like success in diffusing and adapting an alternative more broadly; no one knowingly copies a failed alternative. California appears to have increased its earthquake preparedness despite (or even because of) the failure of earthquake prediction.

Recall, as well, that Johnson & Johnson's mitigation policy succeeded in reducing carbon dioxide emissions and energy costs, and that Fargo, North Dakota's adaptation policy succeeded in reducing the city's vulnerability to floods. Such policies are likely to be diffused and adapted by other companies and communities but only if the policies come to others' attention, and only if the others are frustrated or concerned enough to be receptive. Where such conditions are met, the conventional emphasis on science-based predictions will tend to be replaced by an alternative emphasis on distributed policies based on experience in particular contexts.

The pace of change can be expected to accelerate in periods of crisis and slow down between crises. Congressional attention focuses on the earthquake or flood problem after an earthquake or flood disaster. In climate change policy, for example, the heat wave, drought, and other extreme climate events that afflicted North America in 1988 functioned as a mini-crisis. Prior to those extreme events, reports of global warming were buried in congressional testimony and the back pages of newspapers. Amid those extreme events, quite similar reports became major news stories, spreading public awareness of global warming and stimulating public concerns. According to a climate scientist, Stephen Schneider, "In 1988, nature did more for the notoriety of global warming in fifteen weeks than any of us [scientists] or sympathetic journalists and politicians were able to do in the previous fifteen years" (Ungar 1995). Decision makers alleviated public concerns through increased support for predictive research, including establishment of the USGCRP in 1989. But after more than a decade of support for predictive research, it might take more than the promise of more predictive research to alleviate public concerns. It might take action on distributed mitigation and adaptation policies to meet public demands to "Do something!" Or, as Byerly (1998) sardonically asks: "If sea level rises will the National Governors Association say 'you [scientists] continue developing your computer models and we will lobby for extra billions so you can tell us what to do about our flooding'?"

Science Policy

What are the policy implications of a recognition that scientific prediction is not performing as promised? The overall goal of science policy is unproblematic in itself: No one seriously disputes that benefit to the public good, in exchange for public support of science, is the preferred outcome. But the overall goal must be continually specified down to the level of individual policies, programs, and projects, and the action alternatives for realizing specific goals are problematic. Basic research aimed at improving the precision, scope, and accuracy of predictions appears to be insufficient to deliver the more specific and immediate benefits expected and demanded from science. Three changes in science policy would help ameliorate this problem and better adapt scientific research to the emergent needs of society.

- The first change is to distinguish between predictive research and policy research. Policy research applies available knowledge to improve the integration of ends and means, of goals and alternatives, as experience unfolds in particular contexts. The basic task is not prediction but increased freedom through insight. This means bringing to the attention of policy makers factors that may have been overlooked or misconstrued in existing "internal models" so that policy makers are free to take them into account in making choices and decisions. From a policy perspective, there is no point in merely refining predictions of climate change and impacts as if human choices and decisions were determined by scientific laws or were irrelevant. However, there is a point in invalidating those predictions through further insight and action on more enlightened policies that minimize loss of life, property, environment, and other values and exploit beneficial opportunities that may arise from climate change.

The specific values to be protected and advanced through policy research should not be hidden under a cloak of objectivity. Instead, they should be clarified openly by scientists in consultation and collaboration with others in particular decision contexts. It is a mistake to assume that ignoring values in the design of research is socially responsible, or that taking values into account necessarily compromises objectivity in the conduct of research. "The policy approach does not mean that the scientist abandons objectivity in gathering or interpreting data, or ceases to perfect his tools of inquiry. The policy emphasis calls for the choice of problems which will contribute to the goal values of the scientist, and the use of scrupulous objectivity and maximum technical ingenuity in executing the projects undertaken" (Lasswell 1951). The policy approach also calls for assessment of the social consequences of research in terms of goal values. The open clarification of goal values and social consequences tends to culminate in specifications of the public good or the common interest for particular situations. When the common interest is specified, it is difficult for special interests to make a persuasive case that they ought to be served at net cost to common interests, and even more difficult to make that case prevail in a open political arena in which many interests may participate.

• The second change in science policy is to expand opportunities to practice policy research on behalf of common interests, as needs or opportunities arise in specific areas of policy. For this to happen, research proposals and results must be evaluated broadly—not only by scientific, disciplinary criteria, but also by criteria relevant to the policy issue addressed. One way to implement this change is to include decision makers along with scientists and science-policy administrators in the process of evaluating research. The decision makers should be representative of the diverse social interests most directly at stake in the geographic or functional area of the research. One modest proposal along these lines (Brunner 1996) can be summarized in five steps:

1. Support a number of research teams to work concurrently for specified periods of time on improvements in the policy information readily usable by policy makers.

2. Give each team broad discretion to define its own approach to such improvements, in order to encourage diversity in specializations, methods, science and policy interfaces, and so forth.

3. Evaluate and compare the teams' improvements in policy information at an annual policy research conference composed of representative policy makers, scientists from the teams, and administrators.

4. Select for continued support over the next funding period the teams that have made satisfactory progress and terminate support for the teams that have made the least progress.

5. Repeat the cycle of support, adding new teams based on competitive proposals, so long as it is justified by improvements in information readily usable by policy makers.

It is not necessary to presume that anyone knows enough to prescribe standards of "information readily usable" on behalf of the "common interest" that are suitable for all situations. It is sufficient to maintain a process that encourages a diversity of plausible approaches toward these goals, selects those that have made the most progress in some situations, and builds cumulatively on those successes. This approach is fundamentally different from centralized, prediction-focused research programs described in earlier chapters, such as USGCRP and the National Acid Precipitation Assessment Program. It has been successfully implemented in various degrees in the cases of earthquake and weather prediction.

This five-part proposal is designed to encourage integration of the scientists' knowledge with the policy makers' expertise in particular geographic or functional areas. Area expertise is no less important than scientific knowledge for policy purposes because, for example, each river basin is a unique combination of the characteristics of many earth-process and human dimensions relevant to a range of environmental policy concerns. Similarly, each functional area differs from the others, so that insurance, public health, and manufacturing, for example, are not equivalent for policy purposes. Enlightened policy must be adapted to these differences, and to the policy processes already established in each area (Brunner 1999). In the case of climate change, for example, the president of a corporation and the mayor of a city must still worry primarily about the vitality of the enterprises for which they are accountable. Reducing greenhouse gas emissions or vulnerabilities to climate change are merely additional emerging interests that eventually might have to be integrated or balanced with the traditional interests served by the decision processes. Successful promotion of the new interests depends on an understanding of the traditional interests—marketing a successful product, protecting citizens from hazards—and of the decision processes for reconciling them.

As scientists gain experience in providing information readily usable by policy makers, they will tend to move beyond their initial "internal models" of themselves as providers of predictions in policy situations. Regardless of their disciplinary origins, scientists will tend to converge on a contextual, problem-oriented, and multimethod outlook as an adaptation to the requirements of practice (Lasswell 1971). The beginnings of such a convergence can be seen in such diverse fields as social psychology (Stern 1993), ecology (Holling 1995), hydrology (Baker 1998), and anthropology (Rayner and Malone 1997).

- The third change recommended for science policy is to support education (including continuing education) designed to implement a contextual, problem-oriented, and multimethod outlook. Well-designed "tools" distilling a vast amount of relevant experience are typically necessary to implement such an outlook systematically and successfully; personal experience alone is not normally sufficient. Indeed, the limitations of personal experience sometimes motivate progress toward a broader policy outlook. For example, Stern (1993) found that four variables typically overlooked by energy economists—promotion, simplification, reliability, and trust—explain differences of an order of magnitude or more in consumer responsiveness to energy conservation programs with identical financial incentives. Holling (1995) found that resource managers overlooked the gradual loss of resilience in their ecosystems, institutions, and communities by focusing too narrowly on controlling a target variable, such as timber production, through the use of modern technologies. Apparent "success" with respect to the target variable diverted attention from impending collapse until it was too late. Baker (1998) found that "increasingly sophisticated predictive modeling schemes . . . are now the principal operating tools both for applied management hydrology and for basic geophysical hydrology," even though the modeling schemes in themselves are not satisfactory for spurring action.

Conclusion

The preceding sections have sketched a possible transition from reliance primarily on science-based predictions in major policy decisions to reliance primarily on action on distributed policies based on experience in particular situations. If nothing else, this sketch may be used to stimulate and clarify other expectations about whether the future of science will differ significantly from its past for policy purposes, and if so, how? But suppose this sketch has enough plausibility to be taken seriously—what are the implications for leaders in the many specialized areas of science and policy? Leadership is often the critical factor in resisting change or in shaping change in preferred directions—even though it cannot be effectively included in predictive models.

Leaders might begin by acknowledging that public support for science ultimately depends on the public benefits science provides, as assessed by the public and its representatives. From this perspective, concerns that science has not fulfilled the public expectations it creates are problems and opportunities to be addressed through changes in the practice of science for policy purposes. Thus, leaders might distinguish predictive research from policy research; only the latter is adapted to applications of scientific knowledge and area experience in open systems. They might provide more opportunities to practice policy research on behalf of common interests in specific areas of policy, in order to provide information readily usable by policy makers. They might nurture a contextual, problem-oriented, and multimethod outlook through support of education in appropriate intellectual tools.[2] To the extent that such leaders succeed, they will help some scientists avoid the unwanted complications of policy research, help others develop their predispositions toward policy research, and help policy makers ameliorate policy problems in particular areas.

An open question is whether scientists and policy makers will be flexible enough to move beyond the boundaries of predictive science to embrace a more inclusive concept of "science" insofar as policy is concerned. Sarewitz and Pielke (chapter 1) suggest that prediction as applied to policy rests on a misapplication of "traditional" scientific prediction, which was aimed not at predicting the future, but at testing hypotheses. Oreskes (chapter 2) further reveals that prediction—of whatever sort—was not always a central rationale for science. We might agree with Kaplan (1963) that "policy must be scientific to be effective. . . . But to say scientific is not to speak [only] of the paraphernalia and techniques of the laboratory; it is to say realistic and rational—empirically grounded and self-corrective in application. Policy is scientific when it is formed by the free use of intelligence on the materials of experience." This evolution would reintegrate modern science with a much older science of the concrete based on experience and judgment in particular environments since the dawn of civilization. According to the anthropologist Claude Levi-Strauss (1966), "This science of the concrete . . . was no less scientific and its results no less genuine [than the predictive natural sciences]. They were achieved ten thousand years earlier and still remain at the basis of our own civilization."

Notes

1. This material is based on work supported by the National Science Foundation under Grant no. SBR-9512026. Any opinions, findings, and conclusions or recommendations are those of the author and do not necessarily reflect the views of the National Science Foundation. Daniel Sarewitz, Roger A. Pielke, Jr., and Radford Byerly, Jr., provided helpful comments. Parts of this chapter are reprinted in revised form from *Technological Forecasting and Social Change* 62, Ronald D. Brunner "Predictions and Policy Decisions," 73–78 Copyright 1999, with permission from Elsevier Science.
2. In this regard, the "policy sciences" offer a compelling set of appropriate intellectual tools (see especially Lasswell and Kaplan 1950; Lasswell 1971; Lasswell and McDougal 1992). The policy sciences also offer the benefit of being well developed and well used over more than half a century.

References

Anderson, P. W. 1972. More is different. *Science* 177:393–396.

Ascher, W. 1981. The forecasting potential of complex models. *Policy Sciences* 13:247–267.

Baker, V. R. 1998. Hydrological understanding and societal action. *Journal of the American Water Resources Association* 34:819–825.

Bolin, B. 1994. Next steps for climate change analysis. *Nature* 368:94.

Brown, K. S. 1999. Taking global warming to the people. *Science* 283:1440–1451.

Brunner, R. D. 1996. Policy and global change research. *Climatic Change* 32:121–147.

Brunner, R. D. 1999. A third way for the climate change regime. Revision of a paper presented at the 17th Policy Sciences Annual Institute, Yale Law School, New Haven, CT, October 2–4.

Brunner, R. D., and R. Klein. 1999. Harvesting experience: A reappraisal of the U.S. climate change action plan. *Policy Sciences* 32:133–161.

Byerly, Radford, Jr. 1998. A path beyond the ecology of science. In *AAAS science and technology policy yearbook 1998*, A. Teich, S. Nelson, and C. McEnaney, eds., p. 84. Washington, DC: American Association for the Advancement of Science.

Dyson, F. J. 1988. Quick is beautiful. In *Infinite in all directions*, pp. 135–157. New York: Harper & Row.

Economist. 1994. A problem as big as a planet. *Economist* (November 5): 83–85.

Friedman, M. 1953. The methodology of positive economics. In *Essays in positive economics*, pp. 3–43. Chicago: University of Chicago Press.

Holland, J. H. 1992. Complex adaptive systems. *Daedalus* 121:17–30.

Holling, C. S. 1995. What barriers? What bridges? In *Barriers and bridges to the renewal of ecosystems and institutions*, L. H. Gunderson, C. S. Holling, and S. S. Light, eds., pp. 3–34. New York: Columbia University Press.

Horgan, J. 1995. From complexity to perplexity. *Scientific American* 272 (June): 104–109.

Horgan, J. 1996. *The end of science: Facing the limits of knowledge in the twilight of the scientific age*. Reading, MA: Addison-Wesley.

Kaplan, A. 1963. *American ethics and public policy*. New York: Oxford University Press.

Kaplan, A. 1964. *The conduct of inquiry: Methodology for behavioral science*. San Francisco: Chandler Publishing.

Lasswell, H. D. 1951. The policy orientation. In *The policy sciences*, D. Lerner and H. D. Lasswell, eds., pp. 3–15. Stanford, CA: Stanford University Press.

Lasswell, H. D. 1971. *A pre-view of policy sciences*. New York: Elsevier.

Lasswell, H. D., and A. Kaplan. 1950. *Power and society: A framework for political inquiry*. New Haven, CT: Yale University Press.

Lasswell, H. D., and M. S. McDougal. 1992. *Jurisprudence for a free society: Studies in law, science, and policy*. New Haven, CT, and Dordrecht: New Haven Press and Martinus Nijhoff.

Levi-Strauss, C. 1966. *The savage mind*. Chicago: University of Chicago Press.

Pielke, R. A., Jr. 1998. Rethinking the role of adaptation in climate policy. *Global Environmental Change* 8:159–170.

Pielke, R.A., Jr. 1999. Who decides? Forecast and responsibilities in the 1997 Red River flood. *Applied Behavioral Science Review* 7:83–101.

Pielke, R.A., Jr., and R. Byerly, Jr. 1998. Beyond basic and applied. *Physics Today* (February):42–46.

Rayner, S., and Malone, E. L. 1997. Zen and the art of climate maintenance. *Nature* 390:332–334.

Rivlin, A. M. 1984. A policy paradox. *Journal of Policy Analysis and Management* 4:17–22.

Roberts, L. 1991. Learning from an acid rain program. *Science* 251:1302–1305.

Simon, H. A. 1957. *Models of man*. New York: John Wiley & Sons.

Simon, H. A. 1969. *The sciences of the artificial*. Cambridge, MA: M.I.T. Press.

Stern, P. C. 1993. A second environmental science: Human-environment interactions. *Science* 260:1897–1899.

Ungar, S. 1995. Social scares and global warming: Beyond the Rio Convention. *Society & Natural Resources* 8:443–456.

Prediction in Society

Dale Jamieson

The problem of prediction in society arises because the following propositions all appear to be true:

1. prediction is essential to science;

2. science is essential to good decision making in a broad range of cases (such as those discussed in this book); yet

3. prediction is not essential to good decision making in the same range of cases.

Although each of these premises seems independently plausible, it is hard to see how they could all be true, since the negation of 3 would seem to follow logically from 1 and 2. What has gone wrong? I argue that 1 and 2 are both false, so there is no problem with believing 3. However, it is important to understand why 1 and 2 are false and also why they are plausible.

Prediction and Science

The idea that prediction is essential to science is part of a picture of science and nature that has a long history, going back to Hobbes, Descartes, Newton, and the birth of the scientific revolution. The great successes of seventeenth-century science were in mechanics, both earthly and celestial, and at the heart of the scientific revolution was the metaphor of nature as a machine. This metaphor became a guiding idea in subsequent attempts to understand nature. It reached its apex in 1747 when the French philosopher and physiologist Julien La Mettrie published *Man a Machine*, a systematic attempt to extend the model of the machine to human beings. This mechanistic model of humans and other animals remained influential up through twentieth-century behaviorism, and there are still echoes of this model in contemporary cognitive science and molecular genetics.

The idea that nature is an orderly mechanism has been accompanied historically by a particular view about how to study nature. According to this view, nature can best be understood by decomposing it into its proper parts, describing them carefully, and identifying the causal laws that connect them. Once nature has been analyzed in this way, future states can be predicted and past and present states explained. A past or present state is explained when it is deduced from a conjunction of past states and laws of nature. A prediction is a deduction about future states on the basis of past or present states in conjunction with laws of nature. This view is made vivid by contemplating what has come to be known as Laplace's demon.

Lecturing in Paris in the 1790s, Laplace asked his audience to consider a being with the following characteristics:

> An intellect which at any given moment knew all the forces that animate nature and the mutual positions of the beings that comprise it, if this intellect were vast enough to submit its data to analysis, it could condense into a single formula the movement of the greatest bodies of the universe and that of the lightest atom: for such an intellect nothing could be uncertain, and the future just like the past would be present before its eyes. (Stewart 1993, pp. 25–26)

Laplace's demon knows the complete state of the world at any given time, and from this can infer any future state of the world.

In these skeptical times there may be a tendency to scoff at this grand vision. Indeed James Gleick (1988, p. 14) writes, "In these days of Einstein's relativity and Heisenberg's uncertainty, Laplace seems almost buffoon-like in his optimism." Yet modern science has traditionally aspired to the role of Laplace's demon, and it is important to appreciate why prediction has been regarded as central to science.

First, that a theory makes predictions is a mark of its being science. Imagine someone asserting that "God is good" but also asserting that this predicts nothing about any future state of the world (perhaps she also believes that God moves in mysterious ways). Even if we want to say that this belief is part of a theory, we certainly would not say that it is part of a scientific theory. For its truth or falsity makes no difference in the world. Second, making predictions that turn out to be true is part of what confirms a scientific theory. This is such a deep part of our common sense about scientific reasoning that it may be difficult for us even to notice that we sort theories in part on the basis of whether what they predict in fact occurs. Finally, science is directed toward discovering laws of nature rather than mere empirical generalizations, and it is essential to laws of nature that they support predictions. The power of physics, for example, is that it is supposed to tell us about the basic structure of the universe and not just how some corner of the universe appears to someone at a certain place and time. Nicholas Rescher (1997, p. 160) puts this point well:

> A law like "Copper conducts electricity" is not an experiential inventory ("Wherever we've looked into the matter, we have found that . . ."), nor is it an historical report ("Heretofore, copper has always . . ."). If the thesis at issue is indeed a *law*, then it must, as such, characterize how things stand always and everywhere—and thereby look to the future every bit as much as to the past-&-present. And this means that natural science, as an inherently law-oriented enterprise, is unavoidably involved in the business of prediction.

As Oreskes points out in her contribution to this volume (chapter 2), there have always been embarrassments to this view, geology and biology being prime examples. Traditionally, there have been three ways of dealing with these problem children.

The first is to say that they really do make predictions; it is just that the predictions they make are very general and not very specific. Again, Rescher writes (1997, p. 161): "While evolutionary theory does not predict specific outcomes by way of forecasting the modifications of particular species, it does, nevertheless, provide for predictive inferences at the general level of trends and statistical tendencies." While this may be true, Rescher's weak sort of predictability would not be enough for those who are in thrall of Laplace's demon, for it is the ability to predict specific outcomes that is supposed to be what scientific knowledge delivers, and that is not what is delivered by knowledge of geology and biology.

The second approach has been to distinguish systematic sciences like physics and chemistry from historical sciences like geology and biology. Whatever underwrites the historical sciences as sciences is different from what underwrites the systematic sciences as sciences. They may equally be sciences, but what makes them such is different.

Finally, there is the tough love approach to the problem children: "If you want to grow up and be sciences, then you'd better figure out how to support predictions." According to this view, biology and geology, at least insofar as they do not make predictions, are not sciences at all.

One hears quips that express this point of view: "There is science and then there is stamp collecting," "The plural of anecdote is not data," and so on.

In recent years the view that prediction is essential to science has been breaking down, in part because internal changes within science have increasingly made successful prediction appear to be less of a reasonable goal. Gleick has already reminded us of Einstein and Heisenberg. But we have also become aware of chaos, and various threshold and nonlinear phenomena. Some people even think that, because of various quantum phenomena, the universe may not be deterministic. Let me try to explain these views by going back to Laplace's demon.

The possibility that there could be a demon who knows the complete state of the world at any given time, and from this can infer any future state of the world, requires both a metaphysical and an epistemological assumption. The metaphysical assumption is the truth of determinism: that for all states of the world, earlier states are sufficient for later ones. The epistemological assumption has two parts: first, that the demon could have complete knowledge of the state of the world; and second, that the demon could in fact perform the inferences that link earlier states of the world to later ones.

Even if we put aside the metaphysical assumption and accept that determinism is true, it is amazing that anyone could believe the epistemological assumption. Ironically for those who champion predictability in earth system science, weather and climate are often cited as paradigm examples of chaotic systems that cannot be predicted. Chaotic systems are highly complex, irregular, and, although they are deterministic, often give the appearance of randomness. Extreme sensitivity to initial conditions is one characteristic of some chaotic systems. In a 1972 lecture to the American Association for the Advancement of Science, Edward Lorenz dubbed this sensitivity the Butterfly Effect. Imagine the world as it is at this instant, and then imagine a world exactly like this one but with an additional butterfly flapping its wings. Such small differences in initial conditions cascade so rapidly that the gross states of these worlds relevant to determining weather quickly diverge. The Butterfly Effect explains why even the best weather forecasts are speculative beyond two days and virtually worthless beyond a week. Predicting weather and climate is made even more difficult by the fact that the earth is not a closed system. Its states are influenced by many exogenous variables such as solar flux. Even if the internal structure of the earth's systems were well behaved and well understood, prediction still might not be possible due to those external influences.

In the light of these changes within science, there is a move toward understanding science not as the search for laws of nature, but as the attempt to construct adequate models of nature (e.g., Cartwright 1989; see also Casti 1997). The view that science is principally concerned with modeling nature makes the search for laws less central, thus moving science away from a single-minded concern with prediction. There are various important similarities and dissimilarities between these perspectives, but for our purposes what matters is that models are said to forecast or to produce scenarios rather than to predict, and a bad forecast or a crazy scenario is not regarded as falsifying a model.

Consider, for example, the case of economic models. Robert McNown (1986, p. 363) summarized the fate of macroeconomic forecasts in the first half of the 1980s in the following way:

> The final issue of *Business Week* provides some casual, but revealing, information on the predictive performance of the major modeling services over the previous year. . . . The December 28, 1981 issue offers a brief column entitled, "How the Forecasters Went Wrong in 1981," and this should have been taken as a disclaimer for the feature story, "Scanning a Brighter but Hazy Future." The "brighter future" for 1982 turned out to be the most severe recession of the post–World War period. . . . At the end of 1982 we had "Why the Forecasters Really Blew It in 1982" and the main fare offered "Slow Motion Recovery . . . " as a prediction for 1983. Growth in real GNP for 1983 turned out to be a robust 6.5% versus an average prediction from the models of only 3.7%.

In the light of such failed forecasts people do not fire the economists and throw away their models. Instead, they argue that we need more economists and better models.

This is also what happens with climate models. Most climate models are subject to what is called "climate drift": through time, they tend to move away from realistic states to states that are clearly unrealistic. There are various reasons for climate drift, but when atmospheric and oceanic models are coupled and incompatible heat and salt transfers are implied, climate drift is inevitable. One approach to the problem of climate drift is to ignore it: the "actual" climate response in a perturbation experiment is treated as the difference between a climate change simulation and a drifting control run that both proceed from the same initial state. But as an authoritative report points out, "this strategy may lead to erroneous results if the climate system is drifting into a state with basically different response characteristics" (Houghton et al. 1996, p. 417). The other approach is "flux adjustment." This involves changing the values of fluxes so as to avoid climate drift. There is generally little empirical basis for these adjustments, and in some cases they are plainly inconsistent with what is known. Yet climate models continue to be used, and virtually everyone advocates their improvement, not their rejection.

As the focus on modeling shows, prediction is not essential to science, though prediction and science are closely linked both historically and epistemologically. There are areas of science in which investigators are not primarily focused on producing true predictions, and the nature of reality and our cognitive limitations may be such that even the best description of the world would not produce such predictions. For these reasons proposition 1 is false.

Science and Decision Making

My second proposition—that science is essential to good decision making—gains its plausibility from a combination of confusion and bias. Language invites the confusion of supposing that a policy decision about science is (or should be) a scientific policy decision, and that what scientists do must be science. But a policy decision about science is no more a scientific policy decision than a policy decision about alcohol is an alcoholic policy decision. Indeed, much of what scientists do is not science at all. They make breakfast, walk the dog, teach students, fix lab equipment, write grant proposals, review manuscripts, gossip about their colleagues, and so on. These days scientists are called upon to study and render opinions about many problems, but those problems are not thereby transformed into scientific problems. Most of the problems discussed in this book have been studied by scientists and have important scientific dimensions, but they are not scientific problems, and there is little reason to suppose that science provides the right instruments for making sound decisions in these cases. Science bears on predicting earthquakes or asteroid impacts, but assessing those hazards wisely involves an enormous amount of knowledge about human beings and society that one does not gain from reading geology or astrophysics journals. What scientists know isn't always relevant to decision making, and what is relevant to decision making isn't always what scientists know. There are differences between information, data, and predictions, and it is far from clear what makes some information "scientific information" in the first place. Moreover, the need for good judgment cannot be eliminated from decision making. Good judgment in assessing such hazards includes knowledge and imagination about the possibilities for mitigation and prevention, some acquaintance with different conceptions of fairness in distributing benefits and burdens, respect for the importance of process, and keen sensitivity to public concern about both the hazard and various prevention and mitigation strategies.

In addition to these considerations, there is also a story to be told about the long-standing American obeisance to science and technology and the way that it underpins the widespread belief that science is essential to good decision making. Even our homegrown philosophical tradition, pragmatism, which saved us from many of the absurdities visited upon the continent of Europe by its indigenous philosophical traditions, has tended to elide science, progress, the American Dream, and the welfare of humanity as if these were all the same projects pursued under different descriptions. Pragmatists such as Dewey and Quine avoided the often obscure abstractions of Heidegger and Sartre and the authoritarian politics that they masked, but their enthusiasm for science and technology was much too uncritical. Even Rorty (see, e.g., 1998), who elevates edifying humanistic discourse to the same level as scientific communication, has trouble distinguishing America and "progress."

Part of the attraction of scientific decision making has also been the promise of value-free decisions about public policy, thus allowing us to bypass the messy business of dialogue and negotiation. As I have argued elsewhere, Americans tend to be allergic to the open discussion of fundamental moral and political differences (Jamieson 1992). Pretending that our differences can be washed away in the solvent of scientific decision making, we instead conduct these arguments in the language of decision theory, economics, or whatever technical discourse is currently in fashion. This conflation of scientific and policy discourse, foreshadowed in the enthusiasm for science characteristic of founding fathers such as Franklin and Jefferson, has been extremely powerful in American life, and this is surely part of why proposition 2 has seemed plausible to many people.

Of course, in many cases scientific information is useful for making good decisions, and this is part of why people embrace 1 and 2. If you are deciding whether to bomb someone (for example), it is important to know whether the bomb will really explode. But sometimes knowing what religion people practice is also useful in decision making. It does not follow from the usefulness of either scientific or religious information in some cases that either is essential to good decision making. The fact that science is often only marginally relevant to good decision making is not something to be mourned, nor should we blame scientists for failing to solve policy problems. The logic and discourse of science and policy are different; they are brought together at risk of injury to both. In the old days (perhaps the 1980s Reagan White House), policy decisions were informed by astrology; today they are informed by science. Science may be more useful for decision making than astrology, but one theoretical discourse is no more intrinsically suited to decision making than another. This is because science and policy have fundamentally different aims: Science aims at truth, while decision making aims at right action.

Interlude: Science, Society, and Self-Interest

If what I have said is correct, 3 is true while 1 and 2 are false. There are some further, perhaps more mundane, reasons for the widespread acceptance of 1 and 2. The rhetoric that supports 1 and 2 is both ubiquitous and strong in part because believing 1 and 2 serves the interests of a broad range of constituencies. Scientists, like everyone else in a highly economically rationalized society, are under pressure to earn their keep. Presenting themselves as real-world problem solvers is one way of trying to convince people that they deserve to be fed. The focus on prediction brings with it big science and big money. What most scientists seem really to want to do is (relatively) basic science: they want to discover the most fundamental particle, understand the human genome, the atmospheric system, the immune system, and so on. Of course, they may believe (like many other people) that this will have all sorts of good effects: that what's good for science (or business, or investors, or families, etc.) is good for America. But basically scientists do (relatively) basic science because that is what they have been trained to do, and most people (and other animals) like to express their training. Moreover, the activities that they have been trained to perform are (unsurprisingly) the activities that are most highly valued in their communities. So saying that you can't prevent genetic diseases, AIDS, or climate change unless you can make predictions from basic science is a way of maintaining business as usual in a shifting rhetorical environment. For their part decision makers and their constituencies are quite happy to approach various complex problems in this way since they want the technical fixes that predictive science promises. Everyone wants to reproduce, safe sex is a bore, and no one wants to stop driving. And oddly enough, while almost no one these days believes that you can solve the problem of poverty by giving poor people money (which on the face of it sounds like a pretty plausible solution), they do seem to believe that you can solve all sorts of complicated environmental problems by giving scientists money (which on the face of it sounds like a pretty implausible solution). This obviously tells us something about the broader cultural context in which science is done.

Concluding Remarks: Brave New (Policy) World

In conclusion, it should be said that the kinds of problems that we face are quite complex, and they do not respect traditional boundaries between various systems of knowledge. I can now watch the ozone hole develop in (more or less) real time on my PC. Twenty years ago I didn't know that there was an ozone hole. My father didn't know that there was such a thing as ozone. Our ability to monitor global systems is increasing at high velocity, but that doesn't translate into "solutions," or even into understanding what constitutes a "problem" or how one should be framed. For example, one plausible view is that a climate change problem is being caused by the way affluent people live and the solution is to radically restructure the lifestyles and energy systems of affluent people. One could memorize all of the great works in climatology without ever learning anything that bears very directly on whether climate change should be viewed in this way, or if it should, how this problem can be solved.

From the fact that climate change is not a scientific problem, it should not be inferred that it is a political problem. Political problems are basically about allocating goods and bads within a participatory system—they are about who gets what, when, where, and how. But climate change involves values and interests that run very far beyond the boundaries of any existing political system. Because science has global and universal pretensions, it can be a useful antidote to the shallow and narrow perspectives of many decision makers.

Ultimately, however, issues such as climate change are not "political," "scientific," or anything else in particular. Nor can they be usefully approached by conjunction—locking up a "political" expert with a "scientific" expert and expecting "solutions" to emerge. These problems should force us to think more holistically. They should return us to the insights of Greek philosophers such as Aristotle, who recognized that knowledge can be diverse, incoherent, and even apparently contradictory, but a good decision must be unified. Since there is obviously no algorithm for how to move from knowledge to action, what is needed, according to Aristotle and some of the other Greeks, is *phronesis*—usually translated as "practical wisdom" but perhaps better thought of as "sound judgment" or "good sense." Such a notion is difficult for us to grasp. As a deeply proceduralist society, we barely have a vocabulary for evaluating the quality of decisions as opposed to the process from which they emerged. Still, the fact that we may have little to say about such matters now should not lead us to think that there is nothing to say about them at all, or that we cannot develop insights into what constitutes good (and not just legitimate) decisions. Progress here will involve opening up the black box of both science and politics. These are difficult tasks, but there is encouraging work on both of these projects, sometimes under the name of "citizen science" (see, e.g., Irwin 1995) or "deliberative democracy" (see, e.g., Bohman and Rehg 1997).

The case studies in this volume also can help. Each pries open the box and peers at the science and politics inside. They show us what has worked and what hasn't. Perhaps most important, they show us that we must resist our lust for generalization. The cases are all different, and different skills and resources would be deployed by a wise decision maker in each case. Science is also diverse, and we need to develop a better vocabulary for discussing science-like activities carried out by people who have science degrees. While not all scientists are theoretical physicists in search of the fundamental laws of nature, there is still a distinction between those who are out to describe reality and those whose goal is to solve problems—it is the difference between a physicist and an engineer, or an ecologist and a conservation biologist. Another moral, which follows pretty directly from this one, is that texture is everything. In some cases some degree of prediction will be important; in other cases not. And the way prediction is important in cases in which it is will also be very different.

We are in a new world now. Our problems are different, our science is changing, and our politics have been transformed. Our decision making can be improved but not through grand generalizations and bold initiatives. Laplace is dead, and so are master narratives and unifying visions. We can do better in approaching the problems we face, but we will have to learn one step at a time.

References

Bohman, J., and W. Rehg, ed. 1997. *Deliberative democracy: Essays on reason and politics*. Cambridge, MA: M.I.T. Press.

Cartwright, N. 1989. *Nature's capacities and their measurement.* New York: Oxford University Press.

Casti, J.I. 1997. *Would-be worlds: How simulation is changing the frontier of science.* New York: John Wiley & Sons.

Gleick, J. 1988. *Chaos: Making a new science.* New York: Penguin Books.

Houghton, J.T., L.G. Meira Filho, B.A. Callander, N. Harris, A. Kattenburg, and K. Maksell, eds. 1996. *Climate change 1995: The science of climate change.* Cambridge, England: Cambridge University Press.

IPCC (Intergovernmental Panel on Climate Change). 1996. *Climate change 1995: The science of climate change,* J.T. Houghton et al., eds. Cambridge, UK: Cambridge University Press.

Irwin, A. 1995. *Citizen science: A study of people, expertise and sustainable development.* New York: Routledge.

Jamieson, D. 1992. Ethics, public policy, and global warming. *Science, Technology, and Human Values* 17(2): 138–153.

McNown, R. 1986. On the uses of econometric models: A guide for policy-makers. *Policy Sciences* 19: 359–380.

Rescher, N. 1997. *Predicting the future: An introduction to the theory of forecasting.* Albany: State University of New York Press.

Rorty, R. 1998. *Achieving our country: Leftist thought in twentieth century America.* Cambridge, MA: Harvard University Press.

Stewart, I. 1993. Comets and the world's end. In *Predicting the future,* L. How and A. Wain, eds. New York: Cambridge University Press.

Prediction and Characteristic Times

Radford Byerly, Jr.

Years ago in the mountains of West Virginia I began a backpacking trip in a storm. It was raining, snowing, and sleeting. The wind was blowing so hard I could not get my poncho snapped, it flew from my neck like a flag until I pulled it down and stuffed it. Walking away from my car, I questioned the wisdom of starting a trip in such weather. But I had driven far, it was still early November, I knew the area, and I would camp in a valley, out of the wind. So I kept going.

Sure enough, morning dawned cold but clear, and I had three beautiful days of crisp autumn hiking. Walking, I reviewed my decision not to abandon the hike at the very start. Basically, I had continued because the weather was almost certain to get better; there simply wasn't much room for it to get worse. Eventually, these thoughts led me to understand why the weather often worsens on a hike: most people don't start in a storm.

Living through weather teaches you some things.

*   *   *

If we consider the time over which a predicted event occurs, the characteristic time from its beginning to its end, we can better understand why some predictions are more useful than others for making decisions.[1] An event's characteristic time affects how we understand it both personally and scientifically, and thus how effectively we make decisions about it. As described in Sarewitz and Pielke (chapter 1), the predictions this book treats are scientific statements about the future made to inform policy or decisions. We further focus on predictions of events that fall within the earth sciences, such as those in the preceding case studies. Thus, we examine interfaces between natural events and society's response to them to improve policy and decision making (which hereafter is simply "decision making").

All of the events we consider are uncontrolled, that is, they occur in open systems subject to external forces. For example, groundwater contamination by mining often depends on distant political and economic decisions (chapter 9). This openness makes prediction difficult.[2]

Some events we consider do not have well-defined starting or ending points. The case of radioactive waste disposal provides an example (chapter 10). While it has a well-defined starting point, the production of atomic weapons in World War II, its end point is undetermined. Global greenhouse warming (chapter 13) started when modern society's burning of fossil fuels first began to overwhelm the "natural" sources and sinks of carbon dioxide in the atmosphere, thus causing the atmospheric concentration of carbon dioxide to increase sharply. Its end is undetermined but might occur when fossil fuels are exhausted, or their burning is regulated, or by fundamental change in global development. Living as we necessarily do on a human time scale, thinking of such phenomena as "events" subject to human planning may be difficult, for the times involved may be longer than any society has ever existed, longer even than recorded history. Important decisions regarding these and similar cases will be better informed if they recognize disparities of relevant time scales. Accordingly, we must also consider the characteristic time for making, implementing, maintaining, and evaluating decisions.

The following three sections consider our cases in terms of three interacting factors: First, "nature," which provides the events of interest (e.g., storms) and largely determines the characteristic times. Here nature includes the natural and human-impacted world, and humans themselves. Second, the "users" of predictions—decision makers (e.g., emergency managers and others who act on behalf of the lay citizenry, and who in their various ways deal with the events and sometimes with the science). "Users" include individuals and institutions. Third, earth scientists and their knowledge (e.g., seismologists, who provide the predictions of future events based on their analysis of information gathered from nature and seismology). We call scientists collectively with their knowledge, supporting institutions, resources, and data compilations "science."

Nature and Characteristic Times

Ignoring for now the difficulty of when some of our cases begin or end, let us define a "characteristic time" for a "complete" event: the time an event typically or characteristically takes to begin, occur, and end. Or, put another way, the characteristic time encompasses roughly the genesis, occurrence, and immediate sequellae of an event, e.g, the happenings and effects that concern an emergency manager. For example, the emergency manager of a city threatened by floods would be concerned first with genesis—upstream rain and snowmelt and flaws in dikes; second with occurrence—the river's rise and overtopping of the dikes, and rescue of stranded citizens; and finally with sequellae—cleanup of debris, recommissioning of critical infrastructure, and evaluation of decisions. This ensemble of genesis, occurrence, and sequellae is a complete event.

Physical happenings in the environment, like the rise and fall of a river's level, do not alone determine an event's characteristic time because that time represents the *complete* event and encompasses decisions and impacts directly related to the event. Thus, for a flood it encompasses the immediate sequellae including cleanup and recommissioning of critical infrastructure. Our purpose, to understand the use of prediction in making decisions, forces this broader consideration. The characteristic time for a flood reflects more than hydrology and hydrological times. It takes into account the time needed to make a prediction, to make a decision, to evacuate a threatened area, to raise dikes with sandbags, to clean up a flooded area, to restore electric power, to repair bridges, etc. It is not simply a sum of these sorts of times because things happen in parallel, and many things that happen quickly can occur while one slow thing is happening. For example, the lifetime of radioactive waste is so long that it dominates that case (chapter 10). Each event will have its own characteristic time, but similar events could be expected to have similar times. The characteristic time is useful because it forces thinking beyond the simple event, beyond the hydrology of a flood, to consider the many other aspects of the event relevant to making decisions.

Their characteristic times help explain why various events are more or less amenable to straightforward decision making. A complete thunderstorm event has a short characteristic time largely determined by the dynamics of the lower atmosphere. Thus, before, during, and after the storm, the context for decisions remains unchanged. Other things really are equal. Users of predictions deal with a storm in a familiar context.

Context refers to all the factors beyond the storm itself that might affect a decision, including economics, technology, politics, demographics, governance, and values. Context is important for decisions because, for example, a user needs to know how to value future outcomes of decisions. Although often taken for granted, a constant context represents a great deal of information relevant to decisions. Of course, a large storm can totally change the context for decisions, as

Hurricane Mitch did in Honduras in 1998. More typically, the flood of Grand Forks, North Dakota, affected the city in major ways, but did not affect the context on a national or even a broad regional scale.

At the other extreme Rayner (chapter 13) shows that all things will not be equal for greenhouse global warming, which is long term compared to human decision time scales and to the lifetimes of most human institutions.[3] Many things determine the characteristic time of global greenhouse warming—things as diverse as the rate at which people burn fossil fuels and the dynamics of deep ocean circulation. Its characteristic time is much longer than the lifetime of typical decision-making organizations, for example, a U.S. presidential administration. It could be centuries if ocean dynamics dominate. For comparison, the U.S. constitution is only two centuries old, but the Roman Catholic church has endured for almost two millennia. (Since greenhouse warming may involve decisions on new behavioral norms, comparison with the lifetime of a religious organization may be appropriate.) During the characteristic time of global greenhouse warming, human systems will change in substantial, unforeseeable ways. The future context in which today's decisions will be evaluated is unknown—a large and important information gap.

The characteristic time of the radioactive waste disposal issue, determined by how long the waste remains radioactive, is longer than recorded human experience. The relevant time could be hundreds of thousands of years. Metlay (chapter 10) shows the difficulties of making decisions to ensure safe disposal on such time scales—specifically, how a process often considered relatively slow, the migration of groundwater, can be relatively fast when such long times are in play.

The cases of beach erosion and of earthquakes each present us with two different characteristic times. For beaches, the short time is the time until the next big storm that washes away the added sand, say a decade, while the longer time arises from the slow rise in sea level, which has continued for decades already (chapter 8). For earthquakes, the shorter time is determined by the impacts of the quake itself (e.g., cleanup and repair), while the longer time arises from the slow movements of the earth's crust, which build strain and cause the next quake (chapter 7). Which time is relevant depends on the decision faced.

Finally, users also have a characteristic time. For officials, term of office strongly determines it—or perhaps the time to the next election. It also depends on current problems and opportunities, how effectively various constituencies are pushing them, and the options available for dealing with them.

In summary, our cases offer a range of characteristic times from short and amenable to decision making, to long and beyond the horizon of human experience. In cases like radioactive waste disposal, the characteristic time is not only longer than the time for *making* decisions, it is long compared with the lifetime of any conceivable system for *maintaining* and *evaluating* decisions. On the other hand, an unpredicted asteroid impact might have such severe consequences so quickly that users would transition virtually immediately from a well-known to a new, completely unknown context for making decisions.

Users and Experience

Users often have more than science-based information about events. They have information from personal experience—that is, the practical knowledge gained from what one perceives, meets with, or endures. Most people have experienced storms, perhaps major floods, possibly even destructive earthquakes. (None so far have experienced a catastrophic asteroid impact, but all have seen a "shooting star." That is, on a moment's reflection one knows that objects from space impinge on earth, usually harmlessly.) Sometimes users can literally look out the window for a timely personal assessment of an event. In such instances nature provides direct information, experience unmediated by science. A storm, earthquake, or flood can provide vivid and intense experience to both decision makers and the public; if a picture is worth a thousand words, such experience may be worth many megabytes of scientific data in terms of public awareness. People who see the rising waters, hear the winds begin to howl, and have endured previous catastrophes in this location may better appreciate and implement public safety decisions based on predictions. With experience they may also form an independent, perhaps skeptical, assessment of predictions.

A typical user, however, does not have salient experience with events such as greenhouse global warming that have long characteristic times. Of course, users have experienced components of climate, e.g., temperature. The distinction here is that they cannot perceive slow changes, e.g., in average global temperature. This lack of experience makes it difficult for the typical user to allocate attention to such problems, because she faces many other pressing problems and attention is her scarcest resource. Of course, users may have a vague sense of global warming—for example, a perception of summers becoming hotter or longer. Such a perception is likely to be unreliable and evanescent because normal year-to-year climate variability is larger than the expected annual rate of greenhouse warming. An unlikely but possible cool summer could erase such a perception, so its decision-making implications are not robust.

As Nigg shows (chapter 7), in some areas earthquakes occur often enough that experience accrues, while other areas have only historical evidence of earthquakes. In California, users have personal experience, but in the region where the three devastating New Madrid quakes of 1811–12 occurred, information comes to users virtually exclusively from science and history. Gautier (chapter 11) presents an interesting intermediate case dealing with the prediction of hydrocarbon reserves. Users have no experience of global exhaustion of hydrocarbons, but users in the oil industry have experienced the exhaustion of many individual oil fields. In the 1970s users saw the kind of impact that could occur when short-term geopolitics suddenly reduced oil supplies. The current situation of high global petroleum supply and low prices also is due to short-term political and economic factors that now drive consumption down and keep production up, not to any significant change in the size or fate of global hydrocarbon reserves. Again we see that characteristic times matter—the chatter of current events diverts attention from the relatively slow change in reserves, even as local oil fields reach the end of their production. Even when there is a constituency urging action, decision makers usually give slowly developing problems low priority.

Finally, a point raised in the previous section must be emphasized in this section on users and experience. When users have experienced many short-characteristic-time events in the same context, their earned familiarity with the event in context represents a large amount of nonpredictive information needed to make decisions. For events with long characteristic times, the context will change unpredictably before the event culminates, making information on the future context unavailable for present decisions.

Science and Prediction

As with users, characteristic times and the frequencies of occurrence of events determine the amounts and kinds of information they provide to science. When a potential problem with a long characteristic time first begins to emerge, science-based information, including predictions, may be the only information in play. Although important then, it can be problematic for making decisions. For example, science predicted global warming due to increasing atmospheric concentrations of carbon dioxide by using models to extend its current understanding into the future. Seeing a problem, science left the laboratory and mounted the policy stage to warn decision makers and thus can be said to have *created* greenhouse warming as both policy issue and event. This is important because scientists thus became advocates; they inevitably became politically interested in the outcomes of their models and may have compromised their objectivity.

In lieu of experience with climate change, science has proxy data from samples of ancient ice, showing earlier periods of warm and cool climate, which episodically come and go (we are now in a warm period). [4] Trying to compare the earlier warmings with the present situation raises questions, including, first, whether the warming was *caused by* increased carbon dioxide or *caused* the increase in carbon dioxide? Second, whether the ancient processes that increased carbon dioxide were comparable to our burning fossil fuels? Finally, why did the climate change back, i.e., become cool again? Proxy records cannot easily answer such questions. Compared to living through a change, when we can compare, analyze, repeat, and revise observations, proxy data offer thin, ambiguous, and scattered information. On the other hand, they offer information unavailable in the present, e.g., records of completed warming cycles. Scientists convincingly detect a change in our current climate through careful study of decades of temperature records, and ice cores credibly suggest relatively rapid climate changes in the past. Nevertheless, science has no experience of a complete global greenhouse warming event—its genesis, occurrence, and sequellae—and this makes useful, credible prediction difficult because science cannot compare predictions to outcomes, as it can for weather.

Radioactive waste is a case in which science, through laboratory measurements, unambiguously determined the characteristic time of the phenomenon, i.e., the time that the waste remains radioactive, to be hundreds of thousands of years. This time does not depend in any significant way on the context. Thus, these laboratory measurements are applicable to decision making, and science can definitively inform the decision process of an important fact: "This is an extremely long-term problem."[5] The issue then becomes finding an appropriate means of long-term disposal. If, as originally proposed, the waste is stored in relatively short-lived containers, then science must predict how it might migrate from the disposal site into the accessible environment, the essence of Metlay's case (chapter 10). When such migration is deemed possible or likely, science has to predict the life of containers used to keep the waste inside the site, which is where Metlay's case ends. In either case, science must deal with extremely long times based on little experience.

In summary, scientists can better understand and can create more information on frequent and short-term events such as thunderstorms than on long-term events such as climate change. When events are short-term and frequent, science can evaluate the accuracy of predictions and if they are not accurate, revise the procedures for testing against the next event. Like the individual user who evaluates weather predictions based on experience, science can learn from living through multiple events and improve its usefulness for decisions. Science's predictions of climate change are more speculative and are appropriately treated as such by users. For science to contribute effectively to practical decisions, its contributions must be timed to match the pace of decisions and to accommodate the characteristic time of the event. For example, a fast-paced decision process on a slowly developing event might suggest that science could best contribute by reviewing and synthesizing existing information rather than waiting to gather data on the event.

Making Decisions

Prediction aids decisions on some events more than others. Predictions are only one kind of information; so with events for which experience or alternatives to prediction are available, predictive information may be less important. Predictions will also be less important with events for which they are inaccurate or of unknown accuracy, in which case other information must be sought. Table 16.1 summarizes the two limiting situations, i.e., short versus long characteristic times.

The first limiting situation is that of short characteristic times. Here users live through many complete events and can get their own information about the event directly from nature without interpretation by science. In this limiting situation, the user's direct information usually reinforces or confirms what science says because the science is usually right—the science is mature because it has been repeatedly tested and corrected. Decision makers will more likely use predictive information well, in part because the many events allow users to advise scientists about what they need and in part because through experience users come to understand the uncertainty of predictions. Thus, over time science learns how to provide predictive information that is useful for decisions in both form and content, and decision makers learn how to use what science provides.

The paradigmatic example of this limit is weather. The science is relatively mature, though questions remain and research continues. Many weather-related decision processes function well and make effective use of predictive information. Indeed, so many different decision processes use weather predictions, and for the most part the predictions are so well understood, that they are broadcast to the general public. Put another way, weather predictions are so well and widely understood that the gain from general distribution apparently far outweighs any

TABLE 16.1

Some ways that characteristic time affects the creation and use of predictions.

	Short Characteristic Time	Long Characteristic Time
Role of nature	Obviously important.	May be partly subtle, invidious, unseen.
Decision process	Context known. Users and scientists develop processes for use of predictions.	Context unknown. Potential users may not understand problem or predictions; decision process may not exist.
Role of science in decision process	Can work interactively with decision makers.	Science may create and promote the issue as a policy problem. Little feedback from decision process.
Maturity of science	Experience leads to useful predictions.	Cannot develop *verifiable* predictions.
Role of predictions in decision process	Testable, used in light of experience, and less critical to decisions.	At first, predictions only information available. Predictions unreliable, so decision-making experience needed.

potential loss from inappropriate use. One reason decision makers use weather predictions appropriately is undoubtedly their experience with them—everyone has gotten wet on a day predicted to be fair. That is, the uncertainty in the prediction is intuitively acknowledged, and decisions allow for a margin of error. Another reason is that their accuracy is known and over the years has improved.

In some areas earthquakes occur often enough that users and scientists experience multiple complete events, although the frequency is low compared to weather events. Despite this experience and years of well-funded research and data gathering, earthquake science remains unable to make reliable predictions of individual events. Fortunately, experience teaches users that predictions are unreliable, and that therefore they need an alternative approach. Stated simply, their alternative is to use experience and laboratory science to make structures earthquake resistant. Note that this assumes that future earthquakes will be like past ones, which is a kind of prediction but not time specific, and so is easier for decision makers to use.

The second limiting situation includes those events with long characteristic times—which are often also infrequent, or slowly developing, or

ill-defined (which usually means having a long but poorly defined characteristic time), or some combination of all three. Examples include
greenhouse warming, radioactive waste disposal, and asteroid impact.
Users will have little or no personal experience of this kind of event and
are less likely, indeed less able, to make good use of a prediction of such
an event. For rare events no useful decision process is likely to exist (see
chapter 6); if one were developed, it would fall into disuse or obsoles-
cence, or simply disappear between events. The science is less likely to
be reliable because less frequent events mean less data.

Phenomena with poorly defined characteristic times raise the question Where's the event? This inhibits action; the lack of a clear deadline
means hard decisions can be postponed. We usually think of an event as
happening at a certain place and during a particular interval of time. But
global warming is everywhere by definition, so where is it? And when is
radioactive waste "disposed"? Such questions discourage simple
answers but need to be raised.

Greenhouse warming, a unique event with a long characteristic time,
receives large research funding to develop a predictive capability.
Because there is relatively little data on climate events comparable to
greenhouse warming, science is driven by models of the climate and
related systems. Lacking the ability to compare predictions with actual
events, scientists can never know the accuracy of their models. So
despite a decade of well-funded research on greenhouse warming, the
science is relatively immature (compared to weather, for example)
because of the lack of experience. Because the predictive science in this
limiting situation is essentially unverifiable, a prudent decision maker
does not rely on it and seeks alternatives to prediction.

How does one generate useful alternatives, given that atmospheric
carbon dioxide concentrations are virtually certain to increase for some
time, leading to significant global warming? We know we will have to
learn to adapt to a changed climate, and that such adaptation, because
it will be very pervasive—probably affecting virtually every aspect of our
lives—is not amenable to central direction and so must be local. Brunner (chapter 14) discusses how useful experience with alternatives can
be gained at the local level through implementing what he calls "distributed policies."

Although the United States has signed international agreements
promising to reduce emissions, it has no governmental decision process,
neither institutions nor programs, that could fulfill the promises made.
Brunner argues that given this situation, progress will come from making greenhouse warming another reason for doing what should already
be happening, such as conserving energy and improving our ability to

adapt to normal climate changes. For example, some companies are finding it in their interest to take actions that reduce the emission of greenhouse gases, i.e., to lower their costs of operation by conserving energy. Experience with the kinds of alternatives Brunner describes makes prediction less important.

Scientists behave differently with respect to decision making in the two limiting situations. In the case of weather, science stays within its expertise. It produces predictions, and otherwise largely stays out of the decision arena, because the users have shown that they will use prediction products well. At the other limit, because users have no experience of, for example, greenhouse warming (and much experience of the political dangers inherent in energy policy), users tend to avoid decisions. Scientists, therefore, ascended the policy stage to shout their alarm— "Global warming is coming!" To encourage reluctant decision makers to act, scientists feel they must assert the importance of global warming in human affairs from energy policy to foreign policy. In doing so, they risk going beyond their expertise. They take a similar risk in asserting that sound policy cannot be made without the predictive capability, which they alone can develop (chapter 14). This risk could be mitigated if users played a major role in the prediction process for greenhouse warming. Such involvement is unlikely because it might threaten current research programs, e.g., if users found them misdirected.

Three examples show how characteristic times affect decisions. The first deals with a long-characteristic-time event, greenhouse warming. In a warmed climate, earth systems would behave differently, and future climate patterns might neither occur nor behave as we have come to expect. Thus a new, warmed climate regime might call for different decisions than today's climate does. Specifically, that is, if weather conditions decrease global grain production, storing grain to get through one or two bad years might be an adequate policy response in our current climate regime, but it would be inadequate for a long-term change that depressed grain production for decades or centuries.

By contrast, "El Niño" is a different climate situation. Science has both experienced and collected data on complete climate events known as El Niños. El Niño's characteristic time is less than a decade and, often, two or three years. Science established causal links between abnormally warm Pacific water and apparently disparate weather phenomena occurring subsequently in widely separated places around the globe, and thus, much as for climate, *created* El Niño both as a climate event and as a policy issue. However, El Niño is not a shift in climate regime; rather, it is a variation in the current climate regime—one modality of our current climate. It occurs quickly enough that its context is fairly well known and its science is maturing, so in coming years decision makers will probably use El Niño predictions routinely. Decisions on adapting to El Niño conditions can be relatively short term, i.e., can assume that "normal" conditions will return in a few years. However, El Niño lessons can be applied to greenhouse warming decisions only with care because El Niño's characteristic time is so much shorter than that for greenhouse warming.

Second, consider another long characteristic time, that of radioactive waste disposal. U.S. legislation requires disposal that will be safe, i.e., prevent environmental contamination, for tens of thousands of years. However, no human decision framework has ever been in force for ten thousand years. The characteristic time for radioactive waste disposal is simply much longer than the characteristic times for human decisions, or for any human activity we know. In parallel with this uncertainty, we do not know how the repository or the containers will behave over thousands of years, so the physical disposal system is as uncertain as the decision system. This seems to argue that any present decision should provide for monitoring and, if necessary, future retrieval of the waste. Retrievable storage acknowledges the uncertainties by allowing for unforeseen changes in both the disposal and the decision systems. If we convince ourselves that we can dispose of the waste and safely forget it, we can be sure of one thing: We *will* forget it, thus putting future generations at risk of discovering a massive "new" problem if the repository ultimately leaks. This is not to argue for retrievable storage or any other particular approach, but rather to illustrate how a long characteristic time can be an important factor in a decision, bounding if not dominating it. Decision makers must realize that for events with long times, their decisions are only tentative because the long-term futures of the decision and the natural systems are equally unknown.

NAPAP and its support of acid rain regulation provide the third example. This case has several important characteristic times: First, the short times for precipitation of acid rain and development of its effects. Second, the ten-year lifetime of NAPAP. Other significant characteristic times include (1) the time to adopt and implement laws and regulations to limit emission of acid rain precursors, the "regulatory time," typically about a decade; and (2) the economic lifetime of a power plant, the "plant lifetime," several decades. The relatively short characteristic time of acid rain phenomena, a few days or weeks, means that the effects of emission reduction by regulatory implementation, perhaps installation and operation of sulfur dioxide scrubbers, can be seen quickly. On the other hand, the long plant lifetime, decades, means that industry tends to resist adoption and delay implementation of regulations that threaten its investment in generating plants.

Two times relevant to NAPAP's effectiveness are comparable; the statutory life of NAPAP, and the (approximate) regulatory time. When relevant characteristic times are comparable, consideration of the issue cannot be simplified by assuming that some things happen quickly in a more or less constant context, or that some other things happen so slowly that they only provide a constant context for faster events. That is, while it is true that reducing emissions does promptly reduce acid rain, and that replacement of existing, "dirty" power plants with new ones using inherently clean technology will happen only relatively slowly, these truths do not simplify the question of how NAPAP supported the mitigation of acid rain. One must look at what actually happened. In the actual case, the slowness of industry's cleanup and the quickness of nature's response to emission reductions led to pressure to accelerate mitigation. To accomplish this, Congress passed regulatory legislation before NAPAP released its final report.

So, was NAPAP effective? Herrick (chapter 12) sketches a complex situation, as one would expect from the characteristic times. While NAPAP released its final report after major regulatory decisions had been made, it had provided information to decision makers over its ten-year lifetime, and Herrick argues convincingly that such information significantly aided regulatory decisions to accelerate acid rain mitigation. The information NAPAP provided was not predictive; rather, it enabled legislators to understand acid rain well enough to develop an effective mitigation strategy *without* prediction. The final report was largely irrelevant, but the program succeeded because the pace of the ongoing decision processes was slow compared to the individual research projects studying the largely short-characteristic-time phenomena of acid rain. NAPAP could study these phenomena quickly enough to get useful information to decision makers on a timely basis. Put another way, although as a whole NAPAP was too slow to match the pace of legislation, its individual projects had characteristic times short enough to meet users' information needs.

In summary, the characteristic times of an event can help explain the varying effectiveness of scientific predictions in decision making. This is not prescriptive for decisions in any case, only a fruitful way to think about all cases.

Notes

1. This chapter is based on many years experience in science policy, after training in a predictive science.
2. As Sarewitz and Pielke, in chapter 1, also explain, our focus excludes the simpler predictions made in a controlled, e.g., laboratory, situation to verify or falsify a scientific hypothesis. These predictions usually have limited relevance to real-world decision making because they apply only in carefully controlled conditions and not in the real world with all its contingencies (chapter 2).
3. Rayner uses the phrase "global climate change," which can include natural climate variation. I use "global greenhouse warming" to focus on anthropogenic change.
4. Scientific "experience" refers to the same kind of practical knowledge that users gain. For example, science learns from perceiving that events don't match predictions and from enduring the consequent revision of cherished models.
5. Note that this is only a statement of how radioactivity works. Its futurity depends on the laws of physics being constant, not on the decision context.

A Decision Framework for Prediction in Environmental Policy

Charles Herrick and J. Michael Pendleton

In this chapter we attempt to synthesize common aspects or conclusions across the ten case histories presented in this book. Perhaps the most important consideration is the policy context, or *policy regime*, in which the prediction is being applied to an environmental problem. It is well established that mature policy regimes tend to exert a specific perspective on otherwise undifferentiated facts or characterizations of the physical world. Similar to the concept of *procedural rationality*, this phenomenon leads to situations in which facts and knowledge are relevant only if they can be cast in terms of particular legal or administrative constructs or frames of reference. For example, criminal court proceedings are governed by rules of evidence that stipulate what information is appropriate for consideration and how that information must be submitted. Policy regimes are subject to similar, if perhaps less overt, forms of procedural rationality.

Predictions addressing issues that fall within the purview of an established policy regime will likely be subsumed by the defining logic of that policy regime. Predictions that are not (or cannot be) cast in terms of constructs that are readily commensurate with those in use under the policy regime will likely be dismissed as irrelevant, sometimes in spite of a high level of technical rigor. In cases where the policy context itself is nascent or highly contentious, predictive information can play a role in shaping policy consensus and subsequent policy or regulatory implementation.

The value of predictive environmental information can be critically assessed in light of the policy context, or policy regime, in which it is to be placed. For example, we need to ask and answer questions such as the following: Is the prediction being used as a direct trigger for a prestipulated policy action or outcome; or, on the other hand, will the prediction be used to inform or edify public debate? In this chapter, we develop a framework for characterizing the application of predictive science to environmental decisions by considering the interaction of policy regime with three attributes, which we call *dimension, characteristic time,* and *science base and predictive capability.* We selected these attributes because they seem to have an especially strong bearing on how predictive information will be received, recognized, and utilized within the policy context. The three attributes are outlined below:

- *Dimension:* Environmental issues often involve complex, multifaceted phenomena, such as global warming, acid rain, or ecosystem management. Phenomena such as these are difficult to characterize and manage in terms of a single variable or metric of concern. Other issues considered in these case histories appear more discrete, or unidimensional—for instance, weather forecasts, earthquake predictions, or the projected path of an asteroid. The policy process tends to employ blunt instruments, capable of dealing more effectively with dichotomies than finely graded continuums. Complex, multivariate model outputs, or predictions that depend on multiple assumptions or scenarios, are thus likely to be broken up, or cannibalized, into simplified constituents in order to "fit" a specific policy regime or context.

- *Characteristic time:* In the previous chapter, Byerly introduced this term and defined it as "the time an event typically or characteristically takes to begin, occur, and end." It stands to reason that predictive information will be accorded a more serious hearing if the characteristic time of the predicted event somehow coincides with time frames that drive the policy process. The most obvious example is perhaps the time frame that falls within a typical term of office, say two to eight years. Characteristic times that exceed this time frame may become separated from political and/or bureaucratic accountability, or at least separated from accountability to individual decision makers. It may be that politicians and decision makers are more willing to either "game" or discount information derived from longer-term predictions because they can worry less about the political ramifications of a mistake.

- *Science base and predictive capability:* If the science base underlying an environmental issue is focused and mature and includes a refined and widely recognized body of modeling or other predictive activity, it stands to reason that predictions will carry more weight and be less subject to politically selective application or interpretation. The science of orbital mechanics provides a good example of a mature, predictive science base. If, on the other hand, the science base is developing and diffuse, with nascent or contentious predictive capabilities, one can expect that predictions will be less weighty and/or more subject to politically selective application. A diffuse science base, such as that for global warming, means that policy advocates can support their positions and contentions by picking and choosing among experts.

The following vignettes describe each case history in terms of the overall *issue* (the salient policy context or regime), the *modeling scenario* (the technical approach to prediction), and each of the three attributes described above. A final section will attempt to characterize patterns among the different cases, summarize comparative findings, and articulate implications for both science and policy practitioners.

Assessment of the Cases

Weather Forecasts

Issue.

In the United States National Weather Service offices produce and disseminate some twenty-four thousand hourly predictions, or over ten million predictions each year. Weather predictions are broadly utilized to help secure the following social benefits: (1) public safety, (2) reduced property loss, (3) job growth and economic prosperity, and (4) protection from enemies and support for military endeavors. Under its organic act, the National Weather Service is responsible for "forecasting of the weather, the issue of storm warnings, the display of weather and flood signals for the benefit of agriculture, commerce, and navigation, the gauging and reporting of rivers, . . . the reporting of temperature and rainfall conditions . . . and the distribution of meteorological information. . . ." Measurable improvments in forecast skill have not been matched by commensurate advances in the use of forecasts.

Modeling Scenario.

Weather forecasts are the result of monitoring data on the state of the atmosphere, computer models based on numerical representations of atmospheric processes, and human judgment. Because of the uncertainties associated with weather prediction, weather forecasts are inherently probabilistic (e.g., a 70 percent chance of rain). The frequency and widespread public dissemination of NWS operational forecasts results in a high degree of exposure and numerous opportunities to *evaluate* weather predictions. The effectiveness of weather forecasts is partly a function of predictive accuracy and rigor, but also a function of adequate dissemination and communication practices and "appropriate" behavior once the information is received.

Dimension.
Weather prediction involves multiple discrete, albeit probabilistic, events.

Characteristic Time.
Weather forecasts typically cover a temporal window of days to hours. Scientists estimate that the fundamental limits of weather prediction are about two weeks.

Science Base and Predictive Capability.
The past decade has been witness to significant improvements in short-term weather forecasting. New technologies such as Doppler weather radars and wind profilers have helped to enhance our understanding of the processes that drive weather. There is consensus within the meteorological community that continued improvements in short-term weather prediction are linearly related to research funding levels.

Flood Forecasts
Issue.
The United States has vastly improved flood forecasting and warning systems over the last 125 years and invested billions of dollars to build flood control structures, but flood losses have continued to grow. Flood damages in the United States totaled more than $30 billion during 1993–97, among the largest of any natural hazard.

The National Weather Service routinely issues flood forecasts for about four thousand locations in the United States. In the development of flood forecasts, the NWS relies on a number of other federal agencies, particularly the U.S. Geological Survey and the U.S. Army Corps of Engineers. In the United States, flood policy is a subset of water policy, which also involves navigation, irrigation, reclamation, and the supply of drinking water.

Modeling Scenario.
Developed on a routine basis by National Weather Service River Forecast Centers assigned to major river basins, flood forecasts typically involve three steps: (1) use of observations (precipitation, temperature, etc.) to estimate the net amount of water entering a river basin from rainfall and/or snowmelt; (2) conversion of the net amount of water into a volume that enters the stream (runoff); and (3) calculation of the volume rate of water (discharge) that flows from a point in the stream to points further downstream.

Dimension.
Flood predictions are typically issued for specific basins and include an estimate of how high flood waters will rise above more typical levels as the crest passes through affected communities.

Characteristic Time.

Flood watches and warnings are issued weeks, days, and hours in advance of the predicted event.

Science Base and Predictive Capability.

Much of the hydrologic knowledge necessary for useful predictions exists, but basin flood models need improvement to make more effective use of the science. Forecasters also lack near-real-time data on stream-flow and rainfall *during* floods.

Asteroid Impacts

Issue.

An impact by an asteroid or comet larger than 1 km in diameter could cause serious regional disasters (e.g., tsunami), and a collision between the earth and an object only slightly larger would have global environmental consequences. Although the probability that the earth will be struck by a 1-km or greater asteroid is minuscule—approximately one chance in a thousand per century—the potential consequences are enormous, threatening civilization as we know it. This so-called impact hazard was virtually unknown until the past two decades. It now seems likely that the discovery rate of near-earth objects (NEOs) will increase dramatically in the next ten years.

Although Congress has directed the National Aeronautics and Space Administration (NASA) to study our ability to detect and destroy or deflect Earth-orbit-crossing asteroids, the United States has nothing remotely resembling a policy for NEO response. The unusual nature of the NEO issue—potentially catastrophic consequences coupled with extremely low probability of occurrence—has presented difficulties in getting the issue considered alongside other natural hazards, especially by agencies with responsibility for disaster mitigation. It is not clear how the scientific community should advise officials and the public about the chances of an impact occurring shortly after discovery of a potentially threatening body but before it can be proven whether the object will actually strike the earth.

Modeling Scenario.

Different astronomers employ different computational approaches for plotting the orbits and impact probabilities of celestial objects. These diverse approaches tend to converge as more data become available. Put differently, the future trajectory of a newly discovered object becomes more certain the longer it is tracked.

Dimension.

This issue seems to have only two highly salient dimensions: (1) the trajectory of the object (either we can or cannot rule out a collision with earth), and (2) the size of the object and attendant impacts (moderate regional, serious, or global catastrophic).

Characteristic Time.

While it is remotely possible that a comet could suddenly emerge from the direction of the sun, most celestial objects are discovered long before they approach the earth. Modern telescopes can provide a window of as much as thirty years between discovery of an object and its potential collision with the earth.

Science Base and Predictive Capability.

With respect to asteroids and comets that have already been discovered, current understanding and technology can provide exceptionally precise predictions about when and where an impact may occur. Indeed, astronomers have been applauded for centuries for accurate predictions of various celestial phenomena. New telescopes combined with systematic survey efforts will enhance our ability to detect potentially threatening objects.

Earthquake Predictions
Issue.

Earthquakes represent a widespread and potentially catastrophic threat to human well-being and the economy for a number of regions in the United States. The ability to provide short-term predictions (hours to weeks in advance) for these low-probability, high-consequence events does not yet exist. The inability to identify a focused, near-term danger contrasts with and at times acts against the need to develop response plans with long-term programmatic components.

The National Earthquake Hazards Reduction Program was established by Congress in 1977. During the early years, efforts focused on short-term predictions, but by the 1980s it became clear that meaningful predictive capabilities were not forthcoming. It was at that time that the push toward long-term forecasts and emergency warning systems emerged, a thrust that continues today. The United States Geological Survey is the principal science agency in efforts to develop these approaches.

While scientific efforts to develop a short-term predictive capability have failed, early optimism regarding predictive capability resulted in politicians taking positive steps. In the absence of such a predictive capability, the current policy challenge is to determine how best to reduce social and structural vulnerabilities through construction codes, emergency preparedness, public education, etc.

Modeling Scenario.
From the early 1970s to the early 1980s, scientific efforts focused on advancing earthquake prediction science. Initial efforts focused on interpretation of "uplift" activity, magnetic field fluctuations, tilt data, and other geophysical anomalies. The goal was to develop the capability to accurately predict the location, timing, and magnitude of earthquakes. By the mid-1980s, optimism faded as scientists' efforts failed to bear fruit. Since the mid-1980s, scientific research has shifted its focus toward developing long-term forecasts to illustrate where earthquakes are likely to occur and early warning systems that provide announce- . ments to distant communities immediately following the initiation of an earthquake.

Dimension.
On one level, an earthquake is a unidimensional phenomenon: either it occurs, or it does not. On the other hand, policy relevant earthquake predictions should actually include at least six attributes: (1) a lead time, (2) a time window of occurrence, (3) an estimate of magnitude, (4) an assessment of location or epicenter, (5) a characterization of impact, and (6) an assessment of the uncertainty associated with items 1–5. An earthquake is therefore a multifaceted construct, subject to contention if only some of the attributes are accurately prescribed.

Characteristic Time.
Short-term predictions (hours to weeks) for earthquakes are not yet possible. At this time, long-term forecasts are currently being produced. Such predictions have no lead times and windows from one to several decades in duration. Currently under development in Southern California, early warning systems (not forecasts) could provide a warning of seven to twenty seconds.

Science Base and Predictive Capability.
The status of short-term earthquake prediction is described above under "Modeling Scenario." Long-term forecasts are primarily based on calculation of recurrence intervals from field analyses or historical records. There is consensus that the current science base does not support a predictive approach to mitigation of earthquake impacts.

Beach Erosion

Issue.

Beach nourishment has become the favored approach for dealing with erosion of U.S. shorelines. Most beach nourishment is conducted by the U.S. Army Corps of Engineers. In order for nourishment projects to be approved, favorable cost-benefit ratios and environmental impact assessments are needed, both of which require characterizations of future beach behavior. Accurate predictions are important in the societal process of weighing alternative responses to shoreline retreat.

Projects planned and conducted by U.S. government agencies, including the Corps of Engineers, are subject to the National Environmental Policy Act (NEPA) and its requirement to conduct public scoping, environmental assessments, or full-scale environmental impact statements. The NEPA regime is mature, litigation-defined and -tested, and subject to strong procedural rationalities in and among implementing agencies and stakeholder groups.

Modeling Scenario.

Mathematical models to predict beach behavior are used widely in coastal engineering. The most important use is to predict the lifespan, required sand volume, and cost of nourished beaches. One of the models on which the Corps of Engineers relies is GENESIS, the Generalized Model for Simulating Shoreline Change. GENESIS consists of two primary components: a longshore transport equation and a shoreline change equation. Physical data required to run the model include shoreline position, wave characteristics, beach and shoreface profiles, estimates of structural permeability, and engineered structures.

Dimensions.

Beach stability is actually a multifaceted construct, involving the volume and location of sand, the physical contour and appearance of the shoreline, the presence of geologic features that might impact the transport of sand and sediment, and the status of near-shore ecological communities. Contrary to popular understanding, "the beach" is by no means a simple, unidimensional construct.

Characteristic Time.

The common modeling application involves projections of beach behavior and status over a period of at least ten years.

Science Base and Predictive Capability.

The coastal zone is an immensely complex environment, linked to other earth environments in ways that we are only beginning to understand. Our ability to develop scientific models of beach behavior is very limited. This inherent scientific uncertainty is compounded by the fact that beach behavior models, while mathematical, are not designed to operate under statistical parameters, thus allowing no provision for probabilistic analysis. The stability of a nourished beach can be profoundly degraded by low-probability events such as a hurricane.

Mining Impacts

Issue.

Open-pit mines are often more than one thousand feet deep and as much as a mile wide. Construction of these huge structures inevitably involves moving and exposing massive volumes of residual material and mining well below the water table. Once mining ceases and water extraction pumps are shut off, lakes will form within the excavated holes, with depths that can approach one thousand feet. Water in the abandoned pits may leach trace metals and other contaminants from the exposed rock, migrate into the water table, and be transported into aquifers.

In the United States much mining is conducted on federally owned land, requiring an operating permit from the appropriate management agency, in this case the Bureau of Land Management (BLM). Federal management agencies are required to promote mining on government lands but to do so in a manner that minimizes environmental and socioeconomic impacts. The BLM usually mandates permittees to generate and present specific predictions of future water quality (and other impacts) in environmental impact studies prepared to meet the procedural requirements of the National Environmental Policy Act, discussed in the beach erosion case, above. In addition, the state of Nevada regulates the long-term status of water quality, regulating standards for twenty-three chemical constituents or attributes.

Modeling Scenario.

Mining companies use computer models to characterize future water quality in the abandoned pits. The models are used to derive estimates of the concentrations of minor and trace constituents of the pit lake for up to 350 years in the future. For the "Aguirre mine" in Nevada, modeling failed to include critical factors and a quantitative characterization of pertinent uncertainties. The model considered only conditions of chemical equilibrium and did not account for numerous other factors such as changing pH, evapoconcentration through time, reaction kinetics, changes in solubility due to water temperatures above standard conditions (the deep site water was geothermal in origin), the possible roles of microorganisms in the chemical reactions, and variations in chemistry with depth in the pit lake.

Dimensions.
As illustrated above, there are numerous variables that must be considered in assessing the long-term quality of water in an abandoned pit lake. However, a comprehensive assessment is by no means impossible, making it reasonable to expect fairly simple, if highly uncertain, characterizations of water over time.

Characteristic Time.
In the EIS, trace metal concentrations are predicted at intervals of 5, 50, 100, 150, 200, and 250 years from the anticipated date of mining cessation.

Science Base and Predictive Capability.
The basic chemistry underlying the nature, composition, and dynamics of water in the pit lake seems quite well understood. However, as described above ("Modeling Scenario"), the models employed by the mining company and its consultants simply failed to include many relevant variables, parameters, and processes. The predictive capability was therefore biased.

Nuclear Waste Disposal
Issue.
The United States lacks a means of isolating and "disposing" of nuclear waste. A wide variety of radioactive waste materials are currently dispersed across forty-three states, temporarily stored in metal tanks, dry storage casks, basins filled with water, and other holding mechanisms. A strong consensus has developed within the cognizant technical communities that the most effective method for long-term radioactive waste management is burial in a deep geologic repository. The policy objective is to demonstrate that a nuclear waste repository can be built in a way that safeguards human health, protects the environment, and ensures that risks are not "exported" to future generations. Based on model outputs, the U.S. Department of Energy (DOE) needs to project that there will be virtually zero release of any radionuclide to the environment for the first ten thousand years after closure of the repository.

The policy regime for this case study includes several established and mature mechanisms: (1) the environmental assessment (EA) process as practiced under the National Environmental Policy Act (as discussed in the previous synopsis), (2) U.S. Environmental Protection Agency and state of Nevada standards governing acceptable background radiation levels; and (3) background radiation standards promulgated by the Nuclear Regulatory Commission.

Modeling Scenarios.
The Department of Energy has developed models to project how a complex nuclear waste repository sited at Yucca Mountain, Nevada, might perform for as long as one million years. A key aspect of the modeling effort involves full characterization and prediction of "percolation flux," a measure of the amount of water that is likely to seep into the waste holding areas. The higher the percolation flux, the more water that will migrate into areas where the radioactive material is stored, potentially accelerating corrosion of the waste packages and transporting radionuclides from the repository to the accessible environment.

Dimension.
In comparison to other cases considered in this book, the issue of nuclear waste disposal is moderately complex, involving a wide range of assumptions and parameters dealing with the geology, hydrology, and geomorphology of the site; the long-term performance expectations of the engineered holding packages; and characterization of the transport and exposure dynamics of future releases of radioactive materials.

Characteristic Time.
The time frame for this modeling exercise is almost extrahistorical, on the order of tens of thousands of years.

Science Base and Predictive Capability.
Geologic and hydrologic characterizations of the Yucca Mountain site are evolving and have suffered from a paucity of relevant data. In some cases, key model parameters have been established through expert elicitation, with little or no observational data for the purpose of validation. This inability to characterize the site seriously weakens the authority of the predictive modeling exercise.

Oil and Gas Reserves
Issue.
Petroleum and petroleum-derived products remain near the heart of the world economic order. During most of the past one hundred years, the U.S. Geological Survey has had in its principal mission the characterization and evaluation of oil, gas, and other energy resources. Reduction

of uncertainties surrounding oil and gas reserves is critically important for national policy, with implications that drive other key policy arenas (e.g., defense, environment). As such, estimates of oil and gas reserves are watched closely by industry, the defense community, other economic and environmental stakeholders, and the public. Estimates of oil and gas reserves do not fall into a specific, time-honored regulatory regime; they are, however, utilized as justification for decisions on a wide range of issues pertaining to resource management, environmental policy, and energy policy. The case study makes it clear that strong policy predilections can buffet and act to discredit resource assessment approaches and findings.

Modeling Scenario.

Assessment of oil and gas reserves involves a mixture of science, economics, and prognostication of technology development and diffusion. It is difficult to make direct comparisons between estimates due to inclusion or exclusion of particular areas, such as the Arctic National Wildlife Refuge; differences in characterizing "accessible" offshore areas; estimates of quantities other than those deemed "recoverable"; and differing statistical approaches. On several occasions over the past two decades the U.S. Geological Survey has been embarrassed by the development and public release of estimates of oil and gas reserves that appeared to be radically divergent.

Dimensions.

Oil reserves are expressed in terms of estimates of billions of barrels of oil that remain to be found and developed. Likewise, gas reserves are expressed in terms of estimates of trillion cubic feet. Major policy decisions hinge on these unidimensional metrics, and there is little confusion or dispute regarding appropriate ways to frame the issue.

Characteristic Time.

Energy resource assessments seem to have a five- to twenty-year window of applicability. As mentioned above, these estimates are influenced by factors such as technology status, the cost of energy, environmental policy, and other economic issues.

Science Base and Predictive Capability.

A great deal is known about the basic geology of petroleum formation in the earth's crust. Scientists have long predicted the occurrence of petroleum-containing formations based on factors such as the number of remaining anticlines that could be drilled and the presence of other geological "traps." However, development of oil and gas reserve forecasts involves the integration of technological, economic, and geophysical factors. The "ensemble" nature of these forecasts makes them almost inherently contentious and subject to dispute among policy advocates.

Acid Rain
Issue.

In the early 1970s, concern arose that lakes, forests, and other natural and human-made resources were being damaged, even destroyed, by acidic rainfall. It was believed that acid rain was caused by sulfur dioxide emissions, mostly from coal-burning power plants located in the Midwest and the Ohio River Valley. Congressional leaders reacted to public concern by calling for emissions reductions. However, the cost of reducing sulfur dioxide emissions was predicted to be quite high, which in turn could lead to increased electricity costs and unemployment in the high-sulfur coal industry. Uncertainties surrounding acid rain causes, effects, and control costs prompted the Carter and Reagan administrations to sponsor a program of research, the National Acid Precipitation Assessment Program, or NAPAP, whose mission was to provide a scientific basis for national acid rain policy.

Modeling Scenario.

NAPAP either developed or adapted thirty-nine numerical models to enhance understanding of the acidification process and its effects. Models were developed in a wide variety of disciplinary categories to assist in (1) the characterization of future emissions based on projected energy and economic activity; (2) simulation of atmospheric processes leading to deposition of acidic substances; (3) characterization of the impacts of changes in deposition or air concentrations on aquatic and terrestrial systems; (4) characterization of biological effects on aquatic systems due to changes in water chemistry; (5) assessment of effects on materials, visibility, and human health due to changes in deposition and air concentrations; and (6) evaluation of the costs of control strategies and the economic benefits of effects mitigation. An integration framework was developed to link these disciplinary elements in a stepwise manner but was only partially utilized.

Dimension.

Acidic deposition is a multifaceted issue. Assessment of acid rain policy involves characterization and operationalization of many diverse factors including data and model-based analyses of atmospheric chemistry; the acid-based chemistry of lakes, streams, and other surface waters; and soil chemistry and watershed geology; as well as the long-term and/or short-term health of aquatic ecosystems; and terrestrial and economic impacts. Acid rain is by no means a discrete construct.

Characteristic Time.
The relevant time frame for model-based assessment of acid rain impacts ranges from several decades to episodic events of very short duration, possibly a matter of days. For instance, the acid-based chemistry of some watersheds is controlled by bedrock and/or soil chemistry, which may evolve over a period of many decades.

Science Base and Predictive Capability.
The multidisciplinary nature of acid deposition makes it very difficult to characterize the overall maturity or sophistication of relevant scientific understanding.

Global Climate Change
Issue.
Global climate change could alter the conditions under which humans live on earth. Although some of the outcomes are anticipated to be positive, analyses are primarily focused on the potentially negative impacts that temperature and precipitation changes could have on some human populations and ecological systems. The 1992 United Nations Framework Convention on Climate Change (FCCC) provides a policy framework for the global change issue. The objective of the FCCC is "to achieve . . . stabilization of greenhouse gas concentrations in the atmosphere that would prevent dangerous anthropogenic interference in the climate system." While 179 nations have ratified the FCCC, few have ratified implementation of the treaty through the Kyoto Protocol. There are many reasons for this, most driven by concern that a broad-based greenhouse gas control and abatement regime would have substantial and detrimental impacts on the economies of the United States and other developed nations. As also seen in the case of acid rain, the United States argues that it cannot develop sound policy without significantly reducing the scientific uncertainty about basic earth system processes. Further, developing countries, which are expected to contribute the most to emissions growth in the twenty-first century, are not parties to the Kyoto Protocol.

Modeling Scenario.
Scientific interest in the role of the atmosphere in regulating the earth's temperature can be traced back more than 150 years. Mathematical models of the atmospheric heat balance and its response to anthropogenic changes in carbon dioxide were developed and refined in the

1970s and 1980s. Present-day assessments rely on three-dimensional, general circulation models (GCMs) that represent the atmosphere, oceans, ice caps, and land surface characteristics. In spite of the sophistication of GCMs, numerous uncertainties remain hotly contested within the cognizant scientific communities. Moreover, these uncertainties "cascade" from the GCMs into model-based assessments of hydrogeology, ecology, and, of course, socioeconomic impacts.

Dimension.
The global climate change issue is nothing if not multidimensional. The problem can be operationalized in terms of numerous measures: carbon dioxide concentrations, temperature, precipitation, soil moisture content, ocean current dynamics, or the incidence of extreme weather events. Moreover, the impacts of climate change are expected to vary widely from region to region. The hyperdimensionality of this issue is exacerbated when the economic and social impacts of alternative control and mitigation regimes are considered.

Characteristic Time.
The relevant time frame for global change predictions falls in a range between several decades to over a century.

Science Base and Predictive Capability.
Some aspects of global climate change science and modeling are very well established; others are far less developed. The broad and diverse nature of the issue make it hard to characterize the overall status of the global climate change science base. However, the lack of understanding concerning climate change has been articulated in terms of a "cascade of uncertainty," wherein uncertainties about various aspects of the earth sciences are exacerbated due to uncertainties over future emissions of greenhouse gases. This, in turn, makes prognostication of impacts difficult, especially in light of global socioeconomic uncertainty and a distinct lack of clarity regarding how people, communities, and nations will respond to climate change impacts.

Assessment and Recommendations

As suggested at the beginning of this chapter, the case histories make a compelling argument that established policy or management regimes can exert enormous influence over the application of predictive information, especially in terms of the form and admissible content of model-based analyses. This can be seen through several examples including mining impacts, beach erosion, oil and gas reserve estimation, and nuclear waste disposal. Each of these case histories involves a situation in which weak or incomplete model outputs were somehow privileged due to their strong placement in the policy context. However, it also appears that the influence exerted by a policy regime can be countered if there is a strong possibility of near-term "mistake" recognition or widely recognized scientific conventions concerning appropriate predictive methodologies and applications. In other words, if stakeholders can quickly and unambiguously recognize and publicize a mistake or an implausible application of predictive information, then the influence of the policy regime will be checked or limited. This also seems true if the policy application of predictive information runs counter to well-established scientific consensus. This can be illustrated through the examples of weather prediction, flood forecasting, and to a lesser extent, earthquake predictions.

On the other hand, a contentious issue coupled with a lack of policy consensus or established management regime will frequently result in calls for research and, inevitably, predictive information. This means that predictive information can play an important role in helping to craft emerging policy and management regimes. This is illustrated through the examples of acid rain, global climate change, and asteroid impacts. However, if the science base is also developing and predictive conventions are nascent or comparatively underdeveloped, it is likely that predictions and other scientific assessments will be deconstructed by policy advocates seeking support for particular positions—a circumstance that applies to acid rain and global climate change, if not asteroid impacts.

Figure 17.1a–d shows how the ten cases in this book can be qualitatively viewed in terms of the four central attributes discussed in this chapter (policy regime, dimensions, characteristic time, and science base). This figure constitutes a decision framework, within which the case histories and relevent policy attributes appear to fall into three groups:

- Group 1—earthquakes, floods, weather, and asteroid impacts. These cases combine a relatively discrete event with a focused impact. All but asteroid impacts involve short characteristic times. In each case there is a strong possibility of immediate recognition of and repercussions from "mistakes" or inappropriate application of predictive information.

- Group 2—beach erosion, mining impacts, and oil and gas reserves. These cases combine compound phenomena with a somewhat diffuse impact. Each case involves a relatively to extremely long characteristic time, resulting in little possibility of immediate recognition

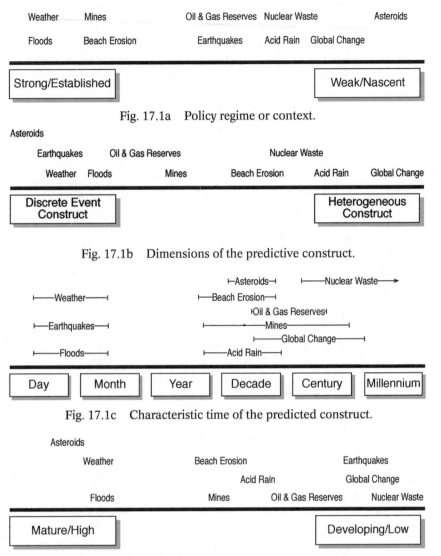

Fig. 17.1a   Policy regime or context.

Fig. 17.1b   Dimensions of the predictive construct.

Fig. 17.1c   Characteristic time of the predicted construct.

Fig. 17.1d   Science base and predictive capability.

of or repercussion from mistakes or inappropriate application of predictive information.

- Group 3—acid rain, global change, and nuclear waste disposal. These cases combine multifaceted events with broadly dispersed and vaguely defined impacts; there is almost no possibility of immediate recognition or repercussion from mistakes. In all three cases, we see predictive information being selectively extracted and "tailored" to the contours of competing policy regimes.

We recognize and acknowledge that other groupings can probably be identified and that the patterns we have identified are interpretive and arguable. However, we are convinced that this exercise provides a potent illustration of a fundamental heuristic: *Predictive information must be designed with policy application in mind*.

In almost all of the case histories discussed in this volume, the scientific modelers express disapproval, incredulity, and dismay that predictive information was "taken out of context" or otherwise used inappropriately. But there is nothing extraordinary about such a state of affairs, given the multidimensional character of many of the predicted phenomena. Does this mean that predictive science must "sell its soul" to attain policy relevance? Not at all. The following recommendations are intended to create greater compatibility between science products and policy needs.

1. *Emphasize communicability and rigor*. While the scientific community typically places a strong emphasis on technical rigor, it is perhaps equally important to assure that predictive information is transparent and broadly communicable. This helps to guard against inappropriate applications of predictive outputs because it increases the probability of mistake recognition. This might mean that science funding institutions and agencies should be encouraged or required to establish policies of technical peer review *as well as* formal review of process transparency and nonspecialist communicability.

2. *Assess policy relevant parameters in advance*. The likely policy or management context of predictive information should be assessed and characterized *before* the predictive model is designed or chosen. This will help to assure that model outputs fit as closely as possible with management metrics or key variables of an ongoing policy debate. Policy advocates will thus be less likely to deconstruct model outputs to fit their specific objective, and scientific inquiry will be easier to focus on the specific construct or metric of policy interest.

3. *Avoid a predictive bias*. To suggest that predictive modeling may not be especially useful in some situations (e.g., earthquakes) is not to argue that scientific information will not be useful. Not all "good science" is predictive. If a policy context is particularly contentious, nascent, or diffuse, it may make sense not to invest heavily in predictive models. It may be more efficient to invest in nonpredictive research such as mechanistic studies, historical reconstructions, monitoring networks and data development, and social vulnerability assessments.

PART SIX

Conclusion

Decision Making and the Future of Nature: Understanding and Using Predictions

Roger A. Pielke, Jr., Daniel Sarewitz, and Radford Byerly, Jr.

A Prediction Enterprise

The story, by now, is familiar. A danger or opportunity is lurking out there, perhaps ill defined, imminent or in the more distant future, and decision makers must take action. While our ten cases are diverse, each is rooted in an effort to mobilize predictive science to pursue desired outcomes on behalf of society.

We know what undesired outcomes look like: more than ten thousand deaths from Hurricane Mitch in Central America in 1998; losses of $20 billion as a consequence of the 1993 Midwest floods; serpentine lines of cars waiting for gasoline during times of shortage; acidified lakes in temperate forests. Other outcomes are yet left to the imagination: toxic effluent leaching from pit mines and nuclear waste repositories into groundwater supplies; huge conflagrations ignited by giant asteroid impacts; fragile ecosystems collapsing under the pressure of rapid climate change. Of course, the future of human interaction with nature does not offer only disaster. Changing weather patterns might allow for more efficient agricultural harvests; the discovery of new hydrocarbon reserves (or new energy technologies) might enhance economic well-being and lessen the incentive to drill for oil in ecologically sensitive areas. But underlying every such scenario, whether pessimistic or hopeful, is the assumption by those demanding action that knowledge of the future is necessary to prevent negative outcomes and to capitalize on opportunities for gain.

It is not surprising, then, that each year policy makers invest tens of billions of dollars of public funds into technologies ranging from satellite-based observational platforms in the sky, to stream gauges on the ground, to seismometers in the deep ocean in an effort to monitor the environment and provide an ever expanding database for scientific prediction of the future of nature. Prediction has been central to such organized efforts as the U.S. Global Change Research Program, the U.S. Weather Research Program, the National Earthquake Hazards Reduction Program, the Advanced Hydrological Prediction System, the National Acid Precipitation Assessment Program, the Yucca Mountain nuclear waste repository site assessment process, and the Near-Earth Asteroid Tracking Program. Each of these science programs has been justified in terms of the need to support decisions in the present through better scientific understanding of the future.

The quest for prediction of earth systems exists in a dynamic social and political milieu that we call the "prediction enterprise." The public demands action or useful information that can facilitate action. But because the public comprises a great diversity of interests and values, it rarely, if ever, speaks with one voice about what that action or information ought to be. Other participants in the prediction enterprise include policy makers looking to satisfy (or at least address) conflicting demands made by their constituents and a scientific community looking to help define and resolve problems while at the same time satisfying its own desire to expand the frontiers of knowledge.

The prediction enterprise also involves institutions. At the international level, the United Nations coordinates activities to address climate change and natural disasters. Within the United States, the Federal Emergency Management Agency and its state and local counterparts together help citizens prepare for and respond to disasters; the National Weather Service disseminates the latest meteorological information; the Bureau of Land Management seeks to manage public lands according to its legal mandate. Universities and federally funded laboratories are integral parts of the prediction infrastructure. Private-sector institutions are also involved: the insurance industry seeks profit from investments based on a balancing of risks, airlines depend on weather forecasts to maintain safety and schedules, and the construction industry implements building standards aimed at preventing damage from a variety of natural hazards.

How effectively does this prediction enterprise serve the common interest? Its sheer complexity—diverse participants, conflicting perspectives and values, numerous institutions representing different sectors of society, and significant resources at stake—makes evaluation a daunting task. In fact, the existence of a prediction enterprise has not been recognized as such, in part, perhaps, because prediction seems like such a "natural" part of science (chapter 2), society (chapter 15), and policy (chapter 1). Yet the prediction enterprise is as real and pervasive as "the economy" or "the medical system." As with the medical system, for example, one can look in many directions for accountability: to scientists, the media, government regulators, politicians, special interests, the nonexpert public. But unlike the economy or the medical system, little attention has been focused on the prediction enterprise. We therefore lack insight that can be applied to decision making at the intersection of predictive science and environmental policy. The cases in this volume begin the task of developing such insight.

Predictions are commonly viewed simply as pieces of information, as quantitative products of scientific research. From that perspective a prediction is understood as a "set of probabilities associated with a set of future events" (Fischoff 1994). To understand a prediction, one must understand the specific definition of the predicted event (or events), as well as the expected likelihood of the event's (or events') occurrence. When predictions are seen in this light, then the goal of the prediction enterprise is simply to develop *good predictions*, as evaluated by objective criteria such as accuracy and skill.

Yet once we have recognized the existence of a prediction enterprise, it becomes clear that prediction is more than just a *product* of science. Rather, it is a complex *process*. This process includes all of the interactions and feedbacks among participants, perspectives, institutions, values, interests, resources, decisions, and other factors that constitute the prediction enterprise. From this perspective, the goal of the prediction enterprise is *good decisions*, as evaluated by criteria of common interests.[1] The common interest is often invoked in areas such as social security and health care policy, but it should also be a rationale for the prediction enterprise.

Prediction as a Product

A central irony of this book is that the quest for prediction products can in some cases undermine the societal goals that originally motivate the quest. In the cases of earthquakes, global climate change, beach nourishment, nuclear waste, and mine impacts, for example, decision making might be improved through *less* reliance on predictions. Effective decisions are not necessarily promoted by "good" prediction products and not necessarily prevented by "bad" ones. Even so, there will be cases in which reliance on prediction is unavoidable. Knowing when to depend on predictions is itself a challenge of the prediction process and one taken up in greater detail below. (See box 18.1.)

BOX 18.1

Design of Critical Facilities without Time-Specific Predictions

Thomas L. Anderson, Construction Engineer

The engineering community is moving away from traditional prescriptive building codes toward performance-based design criteria. Experience shows that prescriptive codes do not ensure that a critical facility will continue to function in case of a natural hazard event, while performance-based codes are designed to ensure a specified level of performance in the face of specified hazards. Natural hazards of traditional concern include earthquakes and high winds.

Performance codes do not depend on prediction of specific events at specific times or places; rather, they use information based on past events and general understanding of hazard phenomena to determine the maximum expected level of stress placed on a building by a potential event.

Two examples illustrate the effective use of earthquake-related information in the design and construction of critical facilities. The examples involve the Fire Command and Control Facility (FCCF) in Los Angeles County, California, and the proposed trans-Alaska Natural Gas Pipeline (ANGP). The FCCF receives all fire and medical 911 calls for Los Angeles County and is responsible to over three million county residents. The proposed ANGP was to carry natural gas from the Prudhoe Bay fields to a terminal in southern Alaska along a route that generally paralleled the trans-Alaska oil pipeline.

The information needed for the FCCF design was the ground motions for the largest earthquake that could be expected from the several nearby faults in the region. The design requirement was for the FCCF to remain functional during and following such

Accuracy

Given that one has decided to rely on predictions for decision making, how does one know whether a particular prediction product is a good one? A critical assessment criterion is accuracy—a measure of how closely a specific prediction product conforms to the actual event (Ascher 1981). The value of accuracy may seem too obvious to merit discussion, but sometimes accuracy is impossible to evaluate; and other times, when evaluation is possible, decision makers fail to do it. The case studies of beach erosion and mining showed that once a forecast is an event. Design alternatives were subjected to detailed calculations of their response to expected earthquake motions. All uncertainty factors were provided to the design team so that they could compare building designs based on equal values of assumed parameters and other forms of uncertainty. There was a very close and collaborative relationship between the seismologists and the design engineers at every step in the process. The final decision was based on lowest life-cycle cost and lower first cost to achieve the performance level demanded by the county.

The information needed for the ANGP was the expected ground movement where the pipeline crossed active faults. The performance requirement was to lower the risk of pipeline rupture to less than 1/2,500 a year. The project geotechnical team provided the required fault motion descriptions in great detail based on extensive research and fieldwork, and uncertainties were fully disclosed and expressed in terms understandable to the design engineers. Design engineers have a "tool kit" of strategies that allow a buried pipeline to withstand a wide range of abrupt fault movements, i.e., without having to resort to placing the pipeline above ground on sliding supports, where it is exposed to many other hazards. But use of those tools requires the predicted fault motions to be fully defined, including the nature of uncertainties in those motions. Armed with those data, lowest life-cycle cost designs were readily developed for each of the fault crossings to keep the risk of rupture below the acceptable level.

These examples show how in many situations experiential information coupled with understanding can be more useful than uncertain predictions. The construction engineer wants to build safely, i.e., to an acceptable level of risk, no matter when an earthquake occurs.

produced and used in decision making, there may even be disincentives to looking back and assessing predictive accuracy.

Attempts to "retrodict" or "hindcast" past events can give a measure of the accuracy of predictive methods and has been central to assessment of global climate models (see chapter 13). Comparing different prediction methodologies can also give some indication of accuracy because if the results of independent predictions diverge, they cannot all be right (although they could all be wrong). However, the case of nuclear waste disposal showed that convergence of different predictions on a similar result is not necessarily a sign of accuracy, either. Shared scientific assumptions and political incentives may cause "independent" predictions to converge on a result that is palatable, even if incorrect (see chapter 10). Predictions of beach erosion and oil and gas reserves show similar evidence of such "convergence of convenience."

The ultimate test of a prediction, of course, is to evaluate its accuracy against actual events as they unfold, which is not as straightforward as it might seem. Consider the case of early tornado forecasts (Murphy 1996). In the 1880s a weather forecaster began issuing daily tornado forecasts in which he would predict "tornado" or "no tornado." After a period of issuing forecasts, the forecaster found his predictions to be 96.6 percent correct—a performance that would merit a solid A in school. But others discovered that simply issuing a standing forecast of no tornadoes would result in an accuracy of 98.2 percent. This finding suggested that in spite of the high accuracy, action based on the forecaster's predictions could result in costs rather than benefits. In other words, *simply comparing a prediction with actual events does not provide sufficient information to evaluate its performance.* A more sophisticated approach is needed.

Scientists use a range of techniques to assess the skill of a prediction—skill being defined as the improvement of a prediction over some standard (Murphy 1997). One way to evaluate skill is to compare the accuracy of a prediction with the accuracy of some naive baseline. For example, historical weather information provides such a baseline because it yields the best estimate of the future occurrence of weather events, absent any other information. Thus, a forecast is considered skillful if it improves upon a prediction based on such climatological data. For instance, the average high temperature over the past one hundred years in London on September 6 (the climatological mean for that date) might be, say, 10 degrees Celsius. Absent any other information, the best prediction of the temperature on the next September 6 is thus 10 degrees. Any forecast for that particular day would be considered skillful if it were closer than the climatological mean to the actual temperature recorded on that date.

Such considerations suggest that our capacity to evaluate prediction as a technical product depends strongly on what is predicted. The accuracy of some types of predictions is clearly amenable to evaluation. Weather is the best example, because of the huge number of forecasts, their wide use by decision makers, and the ease of comparing forecasts with actual events, which reflects what Byerly (chapter 16) has termed the short "characteristic time" of weather events. In contrast, if an event has a long characteristic time, predictive accuracy often cannot be evaluated. This situation applies to cases such as global climate change, long-term mining impacts, and nuclear waste disposal. Decisions on such issues will have to be made long before the skill of the prediction can be assessed. The case of floods (see chapter 5) represents an intermediate case, amenable to some evaluation of skill, yet considerably less than weather.

Uncertainty

The science fiction writer Isaac Asimov introduced, in his *Foundation* series, the notion of "psychohistorians," who could predict the future with scientific certainty based on complex mathematical models. We know that Asimov's characters lie squarely in the realm of science fiction—there can be no psychohistory. Yet the quest for a scientifically legitimated view of the future is no recent phenomenon; it dates back at least to the efforts of ancient Egyptian hydroengineers and astronomers to predict the stages of the Nile. Fifty centuries later, the future, as a weather predictor might say, still looks partly cloudy. Given today's circumstances, there are many possible ways that tomorrow might unfold (and even more possibilities for tomorrow's tomorrow). Prediction promises to narrow the range of possible futures so that decision making can be more successful. Occasional clearing can occur—we *can* predict some events with skill—but uncertainty can never be eliminated.

Stewart (chapter 3) distinguished between *aleatory* and *epistemic* uncertainty. Aleatory uncertainty is irreducible, because it is introduced by random processes in a closed system—for example, a deck of cards or a pair of dice. Epistemic uncertainty, on the other hand, derives from incomplete knowledge of a system—perhaps the dealer is a cheat, or the dice are loaded. Epistemic uncertainty can sometimes be reduced through more and better knowledge.

Even though epistemic uncertainty can be reduced, if one is dealing with open systems (as is generally the case for environmental predictions), the level of uncertainty itself can never be known with absolute certainty. Seismologists assigned a probability of 95 percent to their prediction of the Parkfield earthquake, but the earthquake never occurred (chapter 7). Were the scientists confounded by the unlikely but statistically explicable one-out-of-twenty chance of no earthquake? Or was their probability calculation simply wrong—i.e., was the uncertainty associated with the prediction in fact larger than initially thought? We would need many more Parkfield-like predictions to begin to answer such questions. Similarly, regardless of the sophistication of global climate models, many types of unpredictable events (changes in solar output, volcanic eruptions that cool the atmosphere, new energy technologies that reduce carbon emissions) can render climate predictions invalid and associated uncertainties meaningless (e.g., Keepin 1986). One way scientists deal with such "unknowable unknowns" is by introducing fudge factors into their predictions, as we saw with beach models, asteroid impact predictions, and global climate models.

Moreover, many of our cases show that efforts to reduce uncertainty reveal vast, previously unrecognized complexities. In such cases, decision-relevant uncertainties can actually *increase* with more knowledge. This dynamic of spiraling uncertainty can have the perverse effect of increasing political controversy rather than reducing it, leading to calls for even more research to reduce uncertainties, while the problem that motivated the research goes unaddressed. As Robinson (1982, p. 249) observes, "By basing present decisions on the apparent uncovering of future events, an appearance of inevitability is created that de-emphasizes the importance of present choice and further lessens the probability of developing creative policy in response to present problems." The counterintuitive lesson for decision makers is that *uncertainties about the future can often be reduced more successfully through decision making than through prediction.* (See box 18.2.)

Predictability

Asteroid orbits can be calculated from observations of the asteroid's positions combined with well-understood physical laws (chapter 6). But in the realm of earth science prediction, asteroid impacts are atypical. In most other cases, predictability is limited because knowledge of the future depends on knowing the present, which itself can never be completely or accurately characterized. For example, weather forecasts depend on knowing the present state of the atmosphere and then projecting the future behavior of the atmosphere, based on computer models. Because the future is dependent on initial conditions, small changes in these conditions can add up to large differences in outcomes. That is why maximum weather predictability is about two weeks: even though the system is well understood, measurement of initial conditions is invariably subject to error and omission.

The complexity of earth sciences phenomena of interest to policy makers increases when human and earth processes interact. Consider nuclear waste disposal. Predicting the performance of a waste facility ten thousand or more years into the future depends on knowing, among a multitude of other potentially relevant factors, how much precipitation might be expected at the site. Precipitation is a function of global climate patterns. And global climate patterns might be sensitive to human processes such as energy and land use. Energy and land use are functions of politics, policy, social changes, and so on. What at first seems a narrow scientific question rapidly spirals into unbounded complexity.

Finally, decision makers sometimes are led to believe that the sophistication of a prediction methodology contributes to greater predictive skill, i.e., that in a complex world, a complex methodology will enhance predictability. In reality, the situation is not so clear-cut. An evaluation of the performance of complex models in energy, economics, population, and other areas has shown that "methodological sophistication contributes very little to the accuracy of [predictions]" (Ascher 1981, p. 258; see also Keepin 1986). Yet energy, economics, and population are integral to any long-term, policy-relevant predictive capability in the areas of global climate change, nuclear waste disposal, and oil and gas reserve assessment. Overall, more sophistication can introduce more uncertainty and more sources of error into a prediction. Our case studies suggest that better prediction products arise more from the feedback between predictions and experience than from the introduction of more sophisticated predictive methodologies. The lesson for decision makers is that they should not be overly impressed by claims of sophistication, unless those claims are backed up by demonstrable increases in accuracy. (See box 18.3.)

The Interface of Product and Process: Understanding Predictions

Given the many factors that influence the generation of a scientific prediction, one can see why accurate, useful predictions are so hard to make. Those same factors also ensure that the predictions are hard to understand. How should the numerical or statistical output of a given predictive effort be interpreted? That is a problem that plagues scientists (with their "unknowable unknowns") as well as the decision makers who try to use predictions.

The challenge of understanding predictions was aptly illustrated in the case of the 1997 flooding of the Red River of the North (see chapter 4; also Pielke 1999). In February 1997, forecasters predicted that the river would see flooding larger than at any time in modern history. At Grand Forks, North Dakota, forecasters expected the spring flood to exceed the 1979 flood crest of 48.8 feet sometime in April. Forecasters issued a prediction that the flood would crest at a record 49 feet, hoping

BOX 18.2

Ranching and Prediction

Rob Ravenscroft, Rancher

As a rancher, I deal with physical and biological systems, as well as economic and social systems. Since these are dynamic and interconnected, attempts to predict their behavior are probably wrong more often than right. If honestly done, there are no "bad" predictions. It's my responsibility as a manager to use them properly.

"Proper" use can be measured only by progress toward a goal. Technology (including predictions), financial and biological capital, labor, and (most important) creativity pretty well sum up the tools a manager can use to devise and implement a plan of operation.

The dynamic nature of the systems involved, and the inherent possibility of errors in predictions and assumptions made, means that a plan must have two essential characteristics to be effective in achieving personal, family, and business goals. It must be monitored, and it must be flexible.

In ranching, as in most other businesses, the obvious monitoring areas are financially oriented. But cattle and beef production are just part of the entire biological system. Healthy plants, animals, and soils are critical to long-term sustainability of families, ranches, and communities. Biological alarms we watch for on our ranch are decreases in plant and animal diversity, which usually indicate some flaw in our plan that could hinder our ability to deal with future adversity.

Any plan that aims to achieve quality-of-life goals (which is the real need of the individuals and families implementing the

---

to convey the message that the flood would be the worst ever experienced. But the message sent by the forecasters was not the message received by decision makers in the community.

Decision makers in the community *interpreted the event* being predicted and the *probabilities* associated with the prediction within the context of their own experience. First, the prediction of 49 feet, rather than conveying serious concern to the public as the forecasters hoped, instead resulted in reduced concern. Local residents and officials interpreted the forecast in the context of the record 1979 flood, which caused plan) must be monitored for social impacts, too. If our ranching and business practices endanger our neighbors and community, our long-term goals can't be achieved. This is more difficult to measure but must be kept in mind.

Early warning is the most effective first step in reversing a planning mistake or reacting to a change in conditions. Flexibility built into the plan and the business is the next. For us, weather and prices are the major risk factors. Those are also factors that are regularly predicted. Experience shows that neither can be forecast with great reliability. This means that we can't afford to direct all our assets and efforts to best capitalize on any one set of predicted conditions. Here again, diversity enhances flexibility. Diverse plant communities support animals through a wider range of weather conditions. Diversity in the cattle enterprise can supply staying power as prices cycle from low to high. Monitoring lets us know when we're not achieving our goals; flexibility gives us the chance to replan and get back on track.

Science-based predictions can be enormously helpful. People are responsible for using such predictions appropriately. In most cases, that means recognizing that prediction is just one of the tools that can be used to help achieve goals. Quality-of-life –based, goal-driven plans that are economically, ecologically, and socially sound should be applied with flexibility and with a monitoring system that provides early warning when straying from the goal occurs. There are no bad predictions, only inappropriate uses of predictions.

damages but was not catastrophic. With the 1997 crest expected to be only a few inches higher than the record set in 1979, many expressed relief rather than concern, perhaps thinking: "We survived that one. How much worse can a few inches be?" Second, decision makers did not understand the uncertainty associated with the prediction. All flood forecasts are uncertain, but predictions of record floods, i.e., floods for which there is no experience, are especially uncertain. Yet forecasters issued a quantitative prediction with a simple qualitative warning about uncertainty. Hence, many decision makers could interpret the forecast

BOX 18.3

Perspective on Prediction Use in Funding Science

Jack Fellows, Government Executive

As someone who worked in the White House's Office of Management and Budget for many years and dealt with national public policy issues related to science, space, and the environment, the factors I would consider important for using or avoiding misuse of predictions include:

- *Problem dynamics.* Is using or improving a model's prediction even germane to the problem? In the situations I most faced, I was being asked whether (1) it was worth the cost of improving a predictive model, or (2) the output from a model would be relevant to a public policy decision. With respect to the first point, it was difficult to tell many times whether a model improvement would contribute to policy making or was only a challenging scientific topic. What could be a significant scientific advance might have little impact on those who might use the prediction in the real world. Depending on the nature of the situation, either outcome could have value, but if a model improvement was being proposed to address policy issues, then the value-added of the improvement to society needed to be demonstrated. Indeed, some problems

uncertainty in their own terms: Some viewed the forecast as a ceiling: "The flood will not exceed 49 feet." Others viewed the prediction as uncertain, with different individuals estimating uncertainty in the crest prediction to range from 1 to 6 feet. The historical record showed that average error for flood crest forecasts was about 10 percent.

On April 22, 1997, the Red River crested at 54 feet, inundating the communities of Grand Forks, North Dakota, and East Grand Forks, Minnesota, and causing $2 billion in damages. In the aftermath of the flood, local, state, and national officials pointed to inaccurate flood predictions as a cause of the disaster. In fact, the accuracy of the predictions was not out of line with historical performance by any objective measure. Instead, forecasters failed to express, and decision makers failed to understand, the *meaning* of the prediction, in terms of what was being forecast and the uncertainty associated with it. The failure was one of process, not of product. (See box 18.4.)

are so oriented toward mitigation or adaptation that improvement in prediction is of little consequence. For example, better warning and storm shelter improvements probably would yield significantly more return to society than small improvements in tornado model predictions.

- *Uncertainty and risk.* Can scientists adequately characterize the level of uncertainty associated with a prediction and how best to quantify the benefits or risks associated with those uncertainties?

- *Range of predictions.* This might be viewed as another form of uncertainty, but there are uncertainties associated with a specific model, and then there is uncertainty associated with a range of models addressing the same issue. If most models tend to have similar results (assuming the physics, etc., are believable), then I would be more likely to accept the results than if the models significantly differed.

- *Fidelity of the model.* How well does the model replicate the historical record? If not well, then I would discount the prediction or not use it at all. Also, does it fit the scale of the problem? Is the output global in nature when my problem is local or regional—can I scale down or up to my situation?

- *Affordability.* Can I afford the model, and do I have the tools and data to run it for my application?

Other cases presented in this volume further illustrate that decision makers' understanding of predictive products has a profound influence on how—and how well—the products are used. Consider the following three examples:

1. Debate has raged for more than a decade about the policy implications of possible future human-caused changes in climate. This debate has been about "global warming" expressed in terms of a single global average temperature. But no person and no ecosystem experiences global average temperature. Each policy advocate is thus free to interpret that prediction product in support of his or her particular interests, ranging from pending global catastrophe to benign (and perhaps beneficial) change. Uncertainty and the inability to compare predictions to experience allow even more interpretive freedom. Predictive science is thus used (and misused) to justify and advance the existing interests of contesting participants in the political process.

BOX 18.4
Accuracy of Flood Predictions

Dennis Walaker, Public Works

My area of emergency decision making relates to weather-related events regarding straight-line winds, tornadoes, blizzards, heavy rains, and river flooding. Accuracy is what I most expect of predictions. Timeliness, of course, is also an important element. More lead time provides the ability to react efficiently to reduce damage, save lives, and make our communities more disaster resistant.

A flood forecast is a guide, not an absolute. If you expect absolute accuracy, you will be disappointed because weather-related events have numerous variables that are all subject to change. The best flood forecast model can be wrong if conditions (e.g., rainfall, temperature, etc.) change dramatically.

In spring of 1997, when the flood of the Red River of the North occured, all flood forecasts were within acceptable ranges except in Grand Forks. We now focus on this one failure rather than the several other successes. In Fargo, we had severe weather, river gauge failures, and crest reversions, but we had time to adjust to those weather-related changes. Grand Forks had little time to react to a situation that overwhelmed them, as this was an unprecedented event.

We must rely on the National Weather Service and the Flood Forecast Center for reliable information. However, we must understand or at least better understand the variables, assumptions, and real value of their predictions. Society increasingly expects

2. In recent years scientists have increased their ability to observe asteroids and comets that potentially threaten the earth. In this case, the "event" is clear enough—possible extinction of life on earth if a large asteroid slams into the planet. But public reaction to the discovery of asteroid 1997 XF11, and the associated prediction that it could strike the earth on October 26, 2028, illustrates that scientists, as well as the public, can fall prey to misunderstanding. Blame for the misunderstanding can be apportioned among scientists who hastily issued an erroneous prediction; the media, which jumped on the prediction because it was spectacular; and the public, which responded to the magnitude of the potential event, rather than its uncertainties or probabilities.

science to solve all problems. But not all problems are easily solved, and answers are not always absolute. Today when serious events are not accurately predicted, society seeks someone to blame.

Could the city of Grand Forks have been saved from the disaster if the forecasts had been absolutely correct? In my opinion, contingency dikes, earlier evacuation, and loss control were options, but they required difficult, unpopular decisions. Previous victories over floods gave a false sense of security. It would have been difficult to construct emergency measures against a 56-foot crest when dikes that were supposed to handle 52 feet failed at 51 feet.

Responsibility for protecting against disaster is a local one. Predictions are but one tool, albeit one of the most important, used by local officials along with political influence, historical reflection, acceptable loss, and other considerations.

In summary, we must support the scientists in gathering the information to achieve the best predictions. We then must question the predictions when life and property in our communities are at risk. We can't simply assume that the prediction is completely accurate without our own review. Predictions are based in part on historical data. If an event significantly exceeds all previous levels, accurate predictions may be difficult. Blaming others when failure occurs isn't enough. If we have done everything possible, we must accept the consequences—some events are beyond our control. An elderly woman victimized by a flood summarized this by saying, "Even if we lose the flood fight, you must feel that we have done everything possible to be successful."

3. Weather forecasts afford decision makers the best opportunity to understand prediction products. It is well worth repeating that in the United States the National Weather Service issues more than 10 million predictions every year to hundreds of millions of users. (In contrast, we have seen less than a dozen scientifically legitimate earthquake predictions.) This activity provides a basis of experience from which users can learn, through trial and error, to understand the meaning of the prediction products they receive. Of course, in the case of weather prediction there is still room for confusion. People may fail to understand predictions even for routine events. Murphy (1980) documents that when forecasters call for a 70 percent chance of rain, decision makers understand the probabilistic

element of the forecast but do not know whether the rain has a 70 percent chance of occurring at each point in the forecast area, or whether 70 percent of the area is expected to receive rain with a 100 percent probability, and so on. Even so, one of the important lessons of weather prediction is that decision makers, including the public, are in general able to use probabilistic information, and such products can have significant value.

These examples illustrate how viewing prediction solely as a product is inherently problematic, because doing so conceals the context that gives the product meaning. Thus, if one wants to improve the use of prediction, one must do more than simply develop "better" prediction products, whether more precise (e.g., a forecast of a 49.1652-foot flood crest at East Grand Forks), more accurate (e.g., a forecast of a 51-foot crest), or more robust (e.g., a probabilistic distribution of various forecast crest levels). While better prediction products are in many cases more desirable, *better decisions* in the common interest require attention to the broader prediction process. From this standpoint, better prediction products may be neither necessary nor sufficient for improved decision making, and hence desired outcomes. To effect better decisions, it is necessary to understand prediction as a process.

Prediction as a Process

The prediction process can be thought of as three parallel decision activities:

- *Research.* Including science, observations, modeling, etc., as well as forecasters' judgments and the organizational structure—all of which go into the production of predictions for decision makers.

- *Communication.* Both the sending and receiving of information—e.g., who says what to whom, how is it said, and with what effect.

- *Use.* The incorporation of predictive information into decision making. Of course, decisions are typically contingent on many factors other than predictions.

These activities are not sequential. They are more accurately thought of as integrated components of a broader *prediction process*, with each activity proceeding in parallel, and with significant feedbacks and interrelations between them.

A robust conclusion of this book is that good decisions are more likely to occur when all three activities of the prediction process are functioning well—and research activity is often the least critical of the three. Open communication and consideration of alternative policy approaches can lead to successful decisions in the face of unsuccessful prediction products, but the opposite is unlikely to occur (see chapter 14). Consider the following examples:

- The case of the Red River flood illustrates how a technically skillful forecast that is miscommunicated or misused can result in costs rather than benefits. The overall prediction process broke down in several places. No one in the prediction process fully understood the uncertainty associated with the prediction, hence little attention was paid to communicating the uncertainty to decision makers, and poor assumptions were made about how decision makers would interpret and use the predictions. As a result poor decisions were made. Given that that region will to some degree always depend on flood forecasts, the situation can be improved in the future by including local decision makers in the research activity in order to develop more useful products (Pielke 1999).

- In the case of earthquake prediction, a focus on developing skillful predictions of earthquakes in the Parkfield region of California brought together seismologists, local officials, and emergency managers with the original goal of preparing for a predicted earthquake. A result was better communication among those groups and overall improved preparation for future earthquakes. In this case, even though the prediction product was a failure, the overall process adapted to that failure and made decisions that enhanced awareness, refocused attention on alternatives, and arguably reduced vulnerability to future earthquakes. (See box 18.5.)

- Global climate change seems to display attributes similar to the early stages of earthquake prediction. Policy making focused on prediction has run up against numerous political and technical obstacles, while alternatives to prediction are becoming increasingly visible. The prediction *process* will be said to work if it addresses the goals of climate policy—i.e., if it reduces the impacts of future climate changes on environment and society (Pielke 1998). More and better predictions are not a prerequisite for this desirable outcome.

- Nuclear waste disposal has also evolved from a situation in which the development of skillful predictions played a central role into

BOX 18.5

A User's Perspective on Earthquake Prediction and Public Policy

Shirley Mattingly, Emergency Management

As calamities go, earthquakes pose a special threat because the really disastrous events occur infrequently. Earthquakes challenge those who would predict them and those who would respond to predictions because uncertainty surrounds the science and the response and provides an excuse for no action.

Predictions, per se, can disrupt life in a city at risk, and they don't have to be valid or even scientifically based. The self-proclaimed clairvoyant Nostradamus predicted a devastating event in May 1988 in the "new city," assumed to be Los Angeles. Widespread publicity fed rumors, which led to near panic in one community. Hundreds of families took their children out of school and permanently left their homes, relocating to the relative seismic safety of Fresno or Oregon. I was disheartened that earthquake drills at school were regarded as proof that the coming catastrophe was inevitable.

Panic is not a good thing. But emergency managers took advantage of the public's heightened awareness of earthquakes to explain the science and promote simple safety measures. Nevertheless, the Nostradamus "prediction" and similar incidents negatively impact public policy makers' regard for the predictive science.

Even apparently legitimate scientific disagreements justify inaction as the preferred policy option. A so-called seismic deficit postulated in 1994 by an official science working group was believed to mean that destructive quakes were more likely to strike the region in coming decades. Subsequently, earthquake insurance rates quadrupled (Kerr 1988). Eventually, the deficit was debunked and suddenly disappeared. But for four years Los Angelenos thought that they faced either twice as many large earthquakes as normal or one huge quake many times more powerful than the last Big One. This incident didn't improve policymakers' regard for the predictive science, either.

Science is often foreign territory for politicians, and politics is often foreign territory for scientists. Scientists and public policy setters often don't even begin to speak the same language. They generally have very different backgrounds, motivations, and aims, and they work in different milieus. So communication does not come naturally.

There's art and science in both predictions and politics. In my experience, decision makers rely on input from people they trust, people with whom they have a history. They like their advisors to do their homework, define the problem, identify alternative approaches, evaluate potential solutions, and find solid answers. And they want advice that is clear and easy to understand. Then they act based on what they've heard and on other factors we don't know about. While they are pragmatic, they can be swayed by people with passionate beliefs.

Scientists and policy makers can help each other to understand the environment that will receive—perhaps eagerly, perhaps kicking and screaming—a prediction. They should collaboratively decide how to communicate information, how to frame it, and when to release it. Both should pursue good relationships with the local media, even when there is little news, because if the media are well informed, they'll do a better job of reporting accurately when there *is* a story.

For local decision makers to take predictions seriously, they must have faith in the prediction and the predictor. Scientists must have credibility with peers and with the people they hope to influence into action. That requires sustained dialogue and mutual trust. The responsibility lies with both.

I remember one highly respected seismologist coming downtown to Los Angeles' city hall, more than once, to discuss his research findings with any public official who would listen. He came on his own initiative. He moved us, not immediately, but over time. He changed public officials' perceptions and influenced public policy. I saw it happen, and ever since, I've been trying to make it happen again, anytime and anywhere that anyone will listen.

one in which decision making focuses on actions that can achieve desirable societal outcomes under various possible futures. Initially, the success of the repository seemed to be entirely dependent on predicting the hydrologic system at the disposal site over the next ten thousand years. Unanticipated complexities associated with this natural system led to decreased emphasis on prediction and increased emphasis on designing an engineered containment system. However, while the behavior of this engineered system is likely to be much more predictable than that of the hydrologic system, it will have its own problems, and uncertainty cannot be entirely eliminated. Additional options, such as monitored, retrievable storage, may be necessary to accommodate the remaining uncertainties.

A User's Guide to Prediction and Decision Making

When to Rely on Prediction Products

The case studies in this volume provide insight into when decision makers should look to prediction products and when they should look to alternative sources of information to help make decisions.[2] The conditions under which predictions should be relied on are easy to lay out in principle but may be difficult to apply in practice. In principle, predictions should be relied on when:

1. Predictive skill is known.
2. Decision makers have experience with understanding and using predictions.
3. The characteristic time of the predicted event is short.
4. There are limited alternatives.
5. The outcomes of various courses of action are understood in terms of well-constrained uncertainties (e.g., the likelihood and effects of false positives and false negatives).

Conversely, alternatives to prediction should be sought when:

1. Skill is low or unknown.
2. Little experience exists with using the predictions or with the phenomena in question.
3. The characteristic time is long.
4. Alternatives are available.
5. The outcomes of alternative decisions are highly uncertain.

Incorporating these principles into real-world decision processes may be difficult. Organizations often choose to gather more information (even if it is useless) rather than take action (e.g., Feldman and March 1981). Political incentives favor "the basing of policy on supposedly neutral forecasts [that allow] decision making institutions to assume a cloak of objectivity" (Robinson 1982, p. 240). Rejecting that cloak in favor of the hair shirt of realism requires decision makers to:

1. Be flexible.

2. Learn from experience.

3. Search for alternatives.

4. Hedge their bets.

5. Evaluate progress with respect to goals.

6. Evaluate predictive skill with respect to decisions.

7. Focus on good decisions, not just good predictions.

Each of these guidelines might seem obvious or common sensical—but as the cases here dramatically show, they are often neglected. Overcoming this neglect requires decision makers to change their focus—from predictions as a product to predictions as a process.

Creating a Successful Prediction Process
If society is to benefit from the predictive information products of the earth sciences, scientists and decision makers should together pay attention to the broad process in which predictions are made. In particular, participants in the prediction process must take action in six areas.

1. Above all, users of predictions, along with other stakeholders in the prediction process, must *question predictions*. For this questioning to be effective, predictions should be as transparent as possible to the user. In particular, assumptions, model limitations, and weaknesses in input data should be forthrightly discussed. Institutional motives must be questioned and revealed. Especially in cases where personal experience may be limited (such as asteroid impacts and global warming), both scientific rigor and public confidence in the validity of the prediction will benefit from this open questioning process. "Black boxes," i.e., closed processes, generate

public distrust, especially when a prediction can stimulate decisions that create winners and losers. They can also foster complacency among those doing the predicting. Even so, because of limited experience, many types of predictions will never be understood by decision makers in the way that weather predictions are understood. Table 18.1 lists seven general questions that can be asked about predictions and gives accompanying guidelines for seeking answers.

2. If users are to question predictions, then *the prediction process must be open to external scrutiny.* This means that policy makers must give procedural aspects of democratic openness, evaluation, and accountability the same priority as issues that may seem more directly connected to policy goals (e.g., funding predictive research or establishing environmental standards). Openness is important for many reasons, but perhaps the most interesting and least obvious is that the technical products of prediction are likely to be "better"—both more robust scientifically and more effectively integrated into the decision process— when predictive research is subjected to the tough love of democratic discourse. Scientists may reasonably fear that such attention could lead to politicization of research agendas, but many of our case histories show the opposite—that, in the absence of public openness, predictive science tends to converge on results that support the tacit assumptions of the administering organizations or policy regimes (see chapter 17). External scrutiny helps to reinvigorate the healthy skepticism that is supposed to be a part of the scientific process. Consider scientists working on the Yucca Mountain nuclear waste repository who converged on a predictive product that was consistent with their institutional interests. The presence of two oversight bodies provided the additional, outside scrutiny necessary to expose the technical flaws in those predictions. Similarly, Moran (chapter 9) shows that, in the effort to predict the environmental effects of mines, informed public scrutiny of environmental impact statements is necessary to ensure that significant uncertainties are brought to light. Pilkey (chapter 8) describes a failed process for making decisions about beach nourishment that is, perhaps predictably, neither open nor subject to evaluation. And as Gautier (chapter 11) explains, when the U.S. Geological Survey opened up its oil and gas assessment program to a range of interested customers, it improved both its own technical capability, and the utility of its prediction products.

3. In this same context of openness, *predictions must be generated primarily with the needs of the user in mind.* Television weather

TABLE 18.1 Questioning Predictions.

Questions to Ask	Guidelines to Follow in Seeking Answers
What are the policy goals (i.e., outcomes) that prediction is intended to achieve?	Specify the purposes of the prediction.
How does the process of developing predictions influence the policy process (and vice versa)?	Consider alternatives to prediction for achieving the purpose. Maintain flexibility of the system as work on predictions proceeds. Recognize that a choice to focus on prediction (as well as the choice of the specific predictive technique) will constrain future policy alternatives.
What are the direct societal impacts of the prediction?	Consider alternative societal impacts that might result from the prediction (including the different roles played by prediction). Evaluate past predictions in terms of impacts on society. Recognize that the prediction itself can be a significant event. If possible, assess the impacts of inadequate predictions relative to the impacts of successful ones.
What are the scientific limitations and uncertainties of the prediction?	Evaluate past predictions in terms of scientific validity. Recognize that different approaches can yield equally valid predictions. Recognize that prediction is not a substitute for data collection, analysis, experience, or reality. Recognize that predictions are always uncertain; assess the level of uncertainty acceptable in the particular context. Beware of precision without accuracy. Recognize that quantification and prediction are not (a) accuracy, (b) certainty, (c) relevancy, or (d) reality. Recognize that computers hide assumptions; computers don't kill predictions, assumptions do. Recognize that the science base may be inadequate for a given type of prediction.
What factors can influence how a prediction is used by society?	Recognize that prediction may be more effective at bringing problems to attention than forcing them to effective solution. Recognize that perceptions of predictions may differ from what predictors intend and may lead to unexpected responses. Recognize that the societal benefits of a prediction are not necessarily a function of its accuracy. Recognize that there are many types of prediction, and their potential uses in society are diverse.
What political and ethical considerations are raised by the generation and dissemination of a prediction?	Pay attention to conflicts of interest among those soliciting and making predictions. Understand who becomes empowered when the prediction is made. Who are the winners and losers? Pay attention to the ethical issues raised by the release of predictions.
How should predictions be communicated in society?	Make the prediction methodology as transparent as possible. Predictions should be communicated (a) in terms of their implications for societal response, and (b) in terms of their uncertainties.

predictions focus primarily on temperature, precipitation, and wind, rather than temperature gradients, behavior of aerosols, and barometric pressure. Scientists must understand the broader goals of the process, not the narrow goals of science; they must listen to stakeholders. Stakeholders must work closely and persistently with the scientists to communicate their needs and problems. To ensure useful prediction products, prediction research programs should be designed from their inception to include mechanisms of formal and informal, regular and frequent dialogue between prediction researchers and prediction users. More communication between producers and users of predictions always benefits the prediction process and the quest for good decisions, even if it introduces inefficiencies in the generation of prediction products.

4. *Uncertainties must be clearly understood and articulated* by the scientists, so that users understand their implications. If scientists do not understand the uncertainties—which is often the case—they must say so. Failure to understand and articulate uncertainties contributes to poor decisions that undermine relations among scientists and decision makers. *But merely understanding and articulating the uncertainties does not mean that the predictions will be useful decision tools.* For example, if policy makers truly understood the uncertainties associated with predictions of global climate change, they might decide that strategies for action should not depend only on predictions (e.g., Pielke 1998).

5. *Decision makers must realize that predictions themselves can be significant events.* Predictions can stimulate considerable action that can confer benefits or impose costs. False earthquake predictions have stimulated better earthquake preparedness, while false asteroid impact predictions have fueled needless alarm. More significantly, predictions can commit society to one course of action while foreclosing other options. The prediction of global warming, for example, has mobilized an international effort to reduce anthropogenic CO_2 emissions. Some would argue that such action is necessary to forestall disaster; others, that it is a fruitless and potentially dangerous distraction from more effective approaches to global environmental protection. In either case, the prediction itself has been a much greater catalyst for decision making than any unambiguous impact of global warming. A healthy prediction process depends on the recognition that predictions are themselves events.

6. Finally, *the quest for alternatives to prediction must be institu-tionalized in the prediction process*, especially when characteristic times are long, policy regimes are strong, and decision makers have limited (or no) experience with the predicted phenomenon. Alter-natives to prediction should be debated and evaluated (and perhaps tried on a pilot basis) at the earliest stages of the prediction process. As our case studies show, alternatives are in fact often available. Rather than trying to predict the impacts of hard-rock pit mines on water quality as a basis for environmental regulation, spreading risk through bonding or other types of insurance might be preferable. Rather than depending on predictions of acid rain mitigation to design a regulatory command-and-control system (see chapter 12), the U.S. Congress actually implemented a system of tradable emis-sions permits that did not depend on predictive earth science.

In Conclusion: Question Predictions!

The emergence of an environmental challenge appears to stimulate an almost automatic call for scientific prediction as the first step toward meeting the challenge. The possible sources of this reaction range from the desire to find an objective source of information that can dictate action while protecting against political backlash, to an unquestioning modern confidence in our technological ability to control the future (e.g., Heilbroner 1959). Whatever the cause, the scientific establishment of the United States focuses a not inconsiderable proportion of its intel-lectual energy and technological wherewithal on predicting the future of nature in order to promote a variety of desired societal outcomes.

Considered as a whole, the cases in this book portray a pervasive and energetic societal activity—a prediction enterprise supported by sub-stantial federal and private funds—that is unified by shared assumptions about the necessity and value of scientific foresight in environmental decision making, and rooted in a strong belief in predictability itself. The recognition that such an enterprise exists is a crucial first step toward fulfilling the goal set out at the beginning of this book: to improve environmental decision making. Only when the prediction enterprise is recognized can critical scrutiny begin.

Predictions—information products—lie at the heart of this enter-prise. They are its rationale, its currency, its legitimacy. For this reason, any effort to assess the prediction enterprise must openly and persis-tently *question predictions*. This questioning has to occur on two levels simultaneously. Of course it is important to question accuracy, uncertainty, and predictability. But we have seen that prediction products mean little by themselves. Predictions must also be questioned in the context—political, cultural, economic, environmental—of the larger enterprise. Given a particular environmental problem, we need to ask: How does the enterprise operate in this case? Who are the players, and how is power, legitimacy, and participation apportioned among them? What conflicting values are hidden by debate over technical matters? What criteria should be used for judging the output of the enterprise and the outcomes of that output?

Technically "good" predictions used in a healthy decision environment can of course facilitate better decisions, as illustrated by the case of weather predictions—our only candidate for the prediction hall of fame. But the "goodness" of weather predictions arises not just from their accuracy and skill, but also from the capacity of society to make effective use of them. A pretty good flood prediction did not forestall disaster in Grand Forks, North Dakota, and a pretty bad earthquake prediction did not prevent better earthquake preparedness in central and Southern California. These types of outcomes are paradoxical or confusing only if one persists in viewing predictions as simple information products. When the whole enterprise is seen, sense and order begin to emerge.

The central issue is an uncertain future. The cause of this uncertainty is a dynamic planet, an evolving society, and the interaction between the two. Scientific prediction is one tool for coping with this uncertainty— a tool with some promise, some problems, and much unacknowledged complexity, but only one tool among many. Given the uneven performance and our lack of understanding of the prediction enterprise, a good argument can be made for the following: First, our dependence on scientific prediction has become uncritical, and at times excessive and counterproductive. Second, we need to be more careful about how and when to make prediction a central activity in addressing environmental problems. Third, as soon as new environmental problems begin to command public attention, we need to resist the urge to immediately prescribe a predictive approach and should consider instead a range of possible actions. And finally, we should worry less about making good predictions and more about making good decisions.

Notes

1. As contrasted with narrow or parochial interests, which may conflict with common interests.

2. There are other such "user's guides" for understanding predictions. One notable example, which focuses on economic and technological forecasts, is Armstrong (1999). Also see Nicholls (1999).

References

Armstrong, J.S. 1999. Introduction. In *Principles of forecasting: A handbook for researchers and practitioners*, J.S. Armstrong, ed. Norwell, MA: Kluwer Academic Publishers.

Ascher, W. 1981. The forecasting potential of complex models. *Policy Sciences* 13:247–267.

Feldman, M.S., and J.G. March. 1981. Information in organizations as signal and sign. *Administrative Science Quarterly* 26:171–186.

Fischoff, B. 1994. What forecasts (seem to) mean. *International Journal of Forecasting* 10:387–403.

Heilbroner, R. L. 1959. *The future as history: The historic currents of our time and the direction in which they are taking America.* New York: Harper and Brothers.

Keepin, B. 1986. Review of global energy and carbon dioxide projections. *Annual Review of Energy* 11:357–392.

Murphy, A.H. 1977. The value of climatological, categorical, and probabilistic forecasts in the cost-loss ratio situation. *Monthly Weather Review* 105(7):803–816.

Murphy, A.H. 1996. The Finley affair. A signal event in the history of forecast verification. *Weather and Forecasting* 8:281–293.

Murphy, A.H. 1997. Forecast verification. In *Economic value of weather and climate forecasts*, R.W. Katz and A.H. Murphy, eds., pp. 19–74. Cambridge, UK: Cambridge University Press.

Murphy, A.H., S. Lichtenstein, B. Fischhoff, and R.L. Winkler. 1980. Misinterpretations of precipitation probability forecast. *Bulletin of the American Meteorological Society* 61: 695–701.

Nicholls, N. 1999. Cognitive illusions, heuristics, and climate prediction. *Bulletin of the American Meteorological Society* 80(7):1385–1397.

Pielke, R.A., Jr. 1998. Rethinking the role of adaptation in climate policy. *Global Environmental Change* 8(2):159–170.

Pielke, R.A., Jr. 1999. Who decides? Forecasts and responsibilities in the 1997 Red River flood. *American Behavioral Science Review*: 83–101.

Robinson, J.B. 1982. Backing into the future: On the methodological and institutional biases embedded in energy supply and demand forecasting. *Technological Forecasting & Social Change* 21:229–240.

Contributors

RONALD D. BRUNNER has been a professor of political science at the University of Colorado since 1981. He received a B.A. cum laude in 1964 and a Ph.D in Political Science with Distinction in 1971 from Yale University. He was a member of the faculty of the University of Michigan from 1968 to 1981, with tenure in political science and a joint appointment in the Institute of Public Policy Studies. Dr. Brunner has practiced the policy sciences through applications of central theory to specific problems in energy, welfare, space, geoscience, education policy, and, more recently, natural resources policy.

RADFORD BYERLY, JR. was trained at Rice University in experimental atomic and molecular physics (Ph.D., 1967). Subsequently, he moved to science management and policy at the National Institute for Standards and Technology, working on programs of environmental measurement and fire research. Byerly joined the staff of the U.S. House of Representatives Committee on Science and Technology in 1975, with initial responsibility for environmental research programs. In 1980 he took responsibility for space science and applications programs and was appointed staff director of the Space Subcommittee in 1985. In 1987, he became director of the University of Colorado's Center for Space and Geosciences Policy. The new chair of the House Science and Technology Committee, Representative George E. Brown, Jr. (D-CA) called Byerly back to Washington in 1991 to be committee chief of staff. Retired from that position in 1993, he remains active writing on science policy and serving on science policy committees (e.g., NASA Space Science Advisory Committee, NRC Board of Assessment of NIST Programs, NSF site visit committees, dissertation committees, etc.).

STANLEY A. CHANGNON has pursued atmospheric and hydrospheric research for forty-eight years. He directed the atmospheric research program of the Illinois State Water Survey for thirty years and served as survey chief for six years. Today, as chief emeritus, he also serves as a professor of geography and of atmospheric sciences at the University of Illinois and directs his own firm specializing in applied climate studies. His diverse research interests include investigations of floods and droughts; studies of severe weather and climate extremes; how climate impacts agriculture, water resources, and policy; and climate change.

CLARK R. CHAPMAN is a planetary scientist at the Boulder, Colorado, office of Southwest Research Institute. He edited the *Journal of Geophysical*

Research—Planets, is a past chairman of the Division for Planetary Sciences of the American Astronomical Society (DPS), and is a member of the science teams of the Galileo, NEAR, and MESSENGER deep-space missions. Chapman (M.S. meteorology, 1968, MIT; Ph.D. in planetary science, 1972, MIT) is a leading expert on asteroids, planetary cratering, and the impact hazard. He is the 1999 DPS Carl Sagan Medallist for excellence in public communication of planetary science. He is coauthor, with D. Morrison, of *Cosmic Catastrophes* (Plenum 1989).

DONALD L. GAUTIER is chief scientist for geologic mapping in the U.S. Geological Survey office in Menlo Park, California, where work focuses on the digital representation and analysis of three-dimensional spatial data. Born in Los Angeles, California, Gautier holds a Ph.D. in geology from the University of Colorado, Boulder, and has been employed by the USGS since leaving Mobil Oil Corporation in 1977. His principal research concerns properties and distribution of petroleum reservoir rocks and prediction of future discoveries of oil and gas resources. Gautier is the principal author of the most recent USGS assessment of the oil and gas resources of the United States and is currently involved in the evaluation of world oil and gas potential.

CHARLES HERRICK is vice president of PERI Environmental Associates, an international consultancy dealing with environmental and energy management issues. Working primarily for the U.S. Environmental Protection Agency, Herrick manages and conducts analyses of issues including environmental technology market assessment; environmental information management; local-scale environmental policy; and integration of behavioral, biophysical, and technological information. Prior to joining PERI, Herrick served as associate director of the White House Council on Environmental Quality (CEQ). He received his Ph.D. from the American University, School of Government and Public Administration.

WILLIAM H. HOOKE has worked for the National Oceanic and Atmospheric Administration (NOAA) and antecedent agencies since 1967. After six years of research in fundamental geophysical fluid dynamics and its application to the ionosphere, the boundary layer, air quality, aviation, and wind engineering, he moved into a series of management positions of increasing scope and responsibility. From 1973 to 1980, he was chief of the Wave Propagation Laboratory's Atmospheric Studies Branch. From 1980 to 1983, he rotated through a series of management development assignments. From 1984 to 1987, he directed NOAA's Environmental Science Group (now the Forecast Systems Lab) and was responsible for much of the systems R&D for the NWS modernization, as well as a range of other weather and climate research activities. For two decades he was an adjunct faculty member in the Department of Astrophysical, Planetary, and Atmospheric Sciences at the University of Colorado, teaching courses and supervising students. He was a Fellow of CIRES (Cooperative Institute for Research in Environmental Sciences) for six years and continues to serve as a CIRA (Cooperative Institute for Research in the Atmosphere) Fellow. He has worked on several NAS/NRC panels and committees. From 1987 to 1993, he served as the deputy chief scientist and acting chief scientist of NOAA. Dr. Hooke currently holds two national responsibilities: director of the U.S. Weather Research Program Office, and chair of the Interagency Subcommittee for Natural Disaster Reduction of the National Science and Technology Council's Committee on Environment and Natural Resources. He is a Fellow of the American Meteorological Society (AMS), has served on the AMS Council, and holds a special AMS Award. Dr. Hooke holds a B.S. (physics honors) from Swarthmore College (1964) and a Ph.D. (1967) from the University of Chicago.

DALE JAMIESON is Henry R. Luce Professor in Human Dimensions of Global Change at Carleton College, adjunct scientist in the Environmental and Societal Impacts Group at the National Center for Atmospheric Research, and adjunct professor at Sunshine Coast University College in Maroochydore, Australia. For nearly twenty years he taught at the University of Colorado, Boulder, where he was the only faculty member to win both the Dean's award for research in the social sciences and the Chancellor's award for research in the humanities. He regularly teaches courses in ethics, environmental philosophy, environmental justice, philosophy of biology and mind, and global change. He is the editor of *A Companion to Environmental Philosophy*, forthcoming from Basil Blackwell in 2000, and is working on the philosophical dimensions of global environmental change.

DANIEL METLAY is a member of the senior professional staff of the U.S. Nuclear Waste Technical Review Board, an independent federal agency set up by Congress to evaluate the technical validity of the Department of Energy's radioactive waste management program. Prior to joining the board, he served as a task force director on the Secretary of Energy Advisory Board. Dr. Metlay was a research scientist at Brookhaven National Laboratory. He also has taught public policy and organizational behavior at Indiana University and at the Massachusetts Institute of Technology. He received his Ph.D from the University of California at Berkeley.

ROBERT E. MORAN is a geochemical and hydrogeologic consultant with twenty-eight years of domestic and international experience (in the United States, Canada, Senegal, Guinea, Gambia, Oman, Pakistan, Kyrgyzstan, Mexico, Peru, and Chile) with the U.S. Geological Survey and several consulting firms. His activities involve mining, geothermal and nuclear issues, hazardous wastes, and water supply development. Dr. Moran has significant experience in the application of remote sensing to natural resources, development of resource policy, and litigation support. Clients include nongovernmental organizations, tribal groups, investors, industrial companies, law firms, and governmental agencies. He has a B.A. from San Francisco State College and a Ph.D. from the University of Texas.

JOANNE M. NIGG is professor of sociology and co-director of the Disaster Research Center at the University of Delaware. She has served on several National Research Council committees and was a member of the Board on Natural Disasters. She also served as president of the Earthquake Engineering Research Institute in 1997–98. She has been involved in research on societal response to natural hazards and disasters since 1975, with a special emphasis on earthquake threats. Her research has been published in five books and monographs, thirty journal articles or book chapters, and over fifty technical reports and conference papers.

NAOMI ORESKES is an associate professor in the Department of History and the Program in Science Studies at the University of California, San Diego. Having started her career as a field geologist in Australia, she now focuses her research on the historical and epistemic development of scientific methods and practices in the earth sciences. A 1994 recipient of the NSF Young Investigator Award, she has served as a consultant to the U.S. Environmental Protection Agency and the U.S. Nuclear Waste Technical Review Board on the validation of computer models. She is the author, most recently, of *The Rejection of Continental Drift: Theory and Method in American Earth Science* (Oxford University Press, 1999).

J. MICHAEL PENDLETON is a senior environmental research analyst at PERI Environmental Associates.. He has an M.S. in environmental science and policy from George Mason University.

ROGER PIELKE, JR., is a scientist with the Environmental and Societal Impacts Group at the National Center for Atmospheric Research in Boulder, Colorado. With a B.A. in mathematics and a Ph.D. in political science from the University of Colorado, he focuses his research on the relationship of scientific information and public- and private-sector decision making. His areas of interest are societal responses to extreme weather events, domestic and international policy responses to climate change, and United States science policy. He currently chairs the American Meteorological Society's Committee on Societal Impacts, and is a member of the Science Steering Committees of the U.S. Weather Research Program, and the World Meteorological Organization's World Weather Research Programme. He is a co-author (with Roger Pielke, Sr.) of *Hurricanes: Their Nature and Impact on Society* (John Wiley & Sons 1997).

ORRIN H. PILKEY is a James B. Duke Professor of Geology in the Division of Earth and Ocean Sciences, Nicholas School of the Environment at Duke University. He is a coastal geologist and director of the Program for the Study of Developed Shorelines, which focuses on beach nourishment, seawall impacts, and evaluation of mathematical models of beach behavior. Pilkey received the Francis Shepard Medal for Excellence in Marine Geology in 1991 and is an honorary member of the Society for Sedimentary Geology. He has co-edited, authored, and co-authored parts of the twenty-one-volume *Living with the Shore* series, as well as two 1996 volumes, *The Corps and the Shore* and *Living by the Rules of the Sea*.

STEVE RAYNER is professor of environment and public affairs at Columbia University, where he is also chief social scientist at the International Research Institute for Climate Prediction. Previously, he was a chief scientist and leader of the Global Change Research Group at Battelle, Pacific Northwest National Laboratory. He has received several awards for work on risk analysis and environmental policy. He has served on various national and international bodies concerned with the human dimensions of global change. He has testified before Congress on climate change research policy and prepared reports to Congress on climate policy and implementation. He is co-editor (with Elizabeth Malone) of the four-volume study *Human Choice and Climate Change*.

DANIEL SAREWITZ is senior research scholar for Columbia University's Center for Science, Policy, and Outcomes. His work focuses on understanding the connections between scientific research and social benefit, and on developing policies to strengthen such connections. He is the author of *Frontiers of Illusion: Science, Technology, and the Politics of Progress* (Temple University Press 1996), as well as many other articles, speeches, and reports about the relationship between science and social progress. Prior to taking up his current position, he was the first director of the Geological Society of America's Institute for Environmental Education. From 1989 to 1993 he worked on Capitol Hill, first as a Congressional Science Fellow and then as science consultant to the House of Representatives Committee on Science, Space, and Technology. His policy analysis responsibilities included federal research policy, international scientific cooperation, and science education. He was also principal speech writer for committee chairman George E. Brown, Jr. He received his Ph.D. in geological sciences from Cornell University in 1986.

THOMAS R. STEWART is director of the Center for Policy Research, Nelson A. Rockefeller College of Public Affairs and Policy, University State University of New York at Albany. He received his Ph.D. in psychology from the University of Illinois. He is a cognitive psychologist who specializes in theoretical, methodological and applied studies of judgment and decision making. His applied research interests focus on the application of judgment and decision research to problems involving scientific and technical expertise and public policy, including studies of visual air quality, medical decision making, weather forecasting judgment, and the use of weather forecasts in decision making.

Index